Sustainability in Plant and Crop Protection

Series editor
Aurelio Ciancio, Sezione di Bari, Consiglio Nazionale delle Ricerche Istituto per la Protezione delle Piante, Bari, Italy

More information about this series at http://www.springer.com/series/13031

Martin Lukac • Paola Grenni • Mauro Gamboni
Editors

Soil Biological Communities and Ecosystem Resilience

Springer

Editors
Martin Lukac
School of Agriculture, Policy and Development
University of Reading
Reading, UK

Faculty of Forestry and Wood Sciences
Czech University of Life Sciences
Prague, Czech Republic

Mauro Gamboni
Department of Biology, Agriculture and Food Sciences
CNR
Rome, Italy

Paola Grenni
National Research Council, Water Research Institute
CNR-IRSA
Rome, Italy

Sustainability in Plant and Crop Protection
ISBN 978-3-319-63335-0 ISBN 978-3-319-63336-7 (eBook)
DOI 10.1007/978-3-319-63336-7

Library of Congress Control Number: 2017951200

Printed on acid-free paper

This Springer imprint is published by Springer Nature
The registered company is Springer International Publishing AG
The registered company address is: Gewerbestrasse 11, 6330 Cham, Switzerland

Series Preface

This is the fourth volume of the series *Sustainability in Plant and Crop Protection*. It assembles the proceedings of the third annual meeting of the COST Action FP1305 "Biolink" (www.bio-link.eu), held in Rome on 17–19 November, 2015, and kindly funded by the EU Commission. The Action, titled *Linking Belowground Biodiversity and Ecosystem Function in European Forests*, aims at unraveling the complexity of the hidden side of forests and tree crops, focusing on how soil biodiversity affects ecosystem behavior and productivity. The contribution of biodiversity to aboveground resilience is indeed a topic at the leading edge in soil ecology, central to the Action, and attracting the interest of many scientists at the EU and international levels.

The volume considers various aspects of biodiversity management, with chapters including reviews or results of experimental assays, with description of related methods, technologies, and advanced approaches. It has been produced thanks to the valuable efforts of the editors, Martin Lukac, Mauro Gamboni, and, in particular, Paola Grenni. They all are acknowledged for this initiative, together with all the contributing authors.

Biodiversity conservation is a requisite for any sustainable management approach, both in natural and cultivated ecosystems. In this view, the richness of species present in a given space or environment represents a natural resource, whose irreversible loss may induce severe structural changes and also economically significant consequences. Many ecosystem services are in fact underpinned by the belowground species composition and diversity, as well as by the complex web of interactions that characterize their relationships. Challenging tasks include the conservation of services such as the management of pests, either native or invasive, the availability of nutrients and their recycling, or the exploitation of the numerous links that evolved among symbionts, endophytes, and their hosts.

The changes induced by man in the last century, first of all deforestation, modified the natural landscapes, with severe effects on the global scale. Only a few regions in the world may in fact be still considered as close to their pristine conditions. These issues are particularly important today due to the accelerating deterioration of the aboveground environments and to the losses of functions and services eventually provided.

Although belowground diversity is considered as less exposed to deterioration, thanks to the presence of redundant functional groups, mechanisms related to their persistence, loss, or conservation need to be elucidated at different scales. Studies dealing with the complexity of these issues are, therefore, welcome, to be added to the progress which has been already achieved. Many of the researchers involved in this volume have a long-term experience in field work, deploying innovative and advanced research approaches. It is hence with interest that the reader may look at the informations herein provided, which are flanked by rich bibliographies.

It is desirable that further studies could deepen in the future our knowledge about specific functions and services produced by soil microbial communities. These data will provide us a clearer insight on the huge complexity of underground food webs and on their role in sustaining the health of plants, their productivity, and, ultimately, our societies.

Bari, Italy

Aurelio Ciancio
(SUPP Series Editor)

Editors' Preface

Forests, and all other types of tree-dominated ecosystems, are of immense importance to both society and the environment, providing a range of products and ecosystem services. Over two-thirds of the world's terrestrial biodiversity occur in forests. Trees provide in fact habitats hosting numerous groups of organisms, each performing a function essential to the survival of the ecosystem. At the first Earth Summit, more than 25 years ago now, an overwhelming majority of nations agreed that humanity is dismantling earth's ecosystems and eliminating functional and genetic diversity contained within them. The intervening period has seen the rise of a great interest in answering the question of how the loss of biodiversity will affect ecosystems' functioning and ultimately their survival. Past research has shown that organisms can influence the physical formation of habitats, the fluxes of elements in biogeochemical cycles and the productivity of ecosystems. Such research resulted in the current awareness that loss of certain life forms can substantially alter the structure and functionality of whole ecosystems. Forests offer a stark example, as any simplification of tree species diversity or indeed a complete replacement of existing tree species by plantations alters productivity, resource utilisation efficiency, stability and the capacity to self-regulate. Many of these functions and services, including those currently unknown to humanity, are currently threatened by climate change.

Our understanding of biological diversity and of its functions, especially belowground, is currently limited. We understand that high biodiversity, whether genetic or functional, usually provides for high resource use efficiency and, by extension, for optimal ecosystem service delivery. We know that maintaining multiple ecosystem processes, at multiple places and times, requires higher levels of biodiversity than does needed by a single process, at a single place and time. We are also aware that functional diversity can enhance ecosystem stability and service delivery. There is an urgent need to connect existing knowledge of aboveground diversity and the rapidly developing understanding of the functionality of its belowground counterpart, to fully explain the connection between diversity, stability and function.

This volume provides an overview of some of the most recent developments in understanding the contribution of soil biodiversity to ecosystem function. The findings and methodologies presented in individual chapters present a window into the largely invisible world of soil-dwelling organisms. Their biodiversity and the range of functions they provide will be affected by environmental changes in the near future – likely agents including climatic change, nitrogen deposition or spread of invasive species. This volume aims to showcase to current knowledge of below-ground diversity and thus contribute to our effort to preserve biodiversity in all of its forms.

Reading, UK — Martin Lukac
Rome, Italy — Mauro Gamboni
Rome, Italy — Paola Grenni

Contents

Chapter 1
Introduction: The Role of Soil Biodiversity in Ecosystem Productivity and Resilience

Martin Lukac

Abstract Biodiversity, both above- and belowground, is positively linked with ecosystem productivity and stability. The relationship between biodiversity and ecosystem function is far better researched and understood for the aboveground part of terrestrial ecosystems. Nevertheless, recent developments of research methods and an increased effort aimed at understanding the contribution of belowground biodiversity indicate that it is at least as important as that of the visible part. Belowground ecosystem function is generated by an interaction between abiotic and biotic parts of the soil, both characterized by intricate and dynamic structure with specific spatial, temporal and organizational patterns. Reductionist experimentation and recent development of models which include soil biota and a representation of its diversity are starting to uncover the relationships and interactions between individual parts of the soil ecosystem, however understanding of the functionality of the whole system will remain a challenge for some time.

Keywords Soil biodiversity • Biodiversity loss • Ecosystem function

Aboveground, sustained research effort has shown that biodiversity – whether that of species, guilds, functional traits or even genes – has a strong and positive relationship with ecosystem productivity (Worm and Duffy 2003). Conversely, recent and accelerating loss of biodiversity may be linked to a reduction of ecosystem functions, stability, resilience and productivity (Cardinale et al. 2012). Past decades have witnessed an explosion of experimental and theoretical research effort aimed at understanding the role of biodiversity in the myriad of ecosystems on Earth. Due to the obvious advantage of being able to see it, the vast majority of studies and conceptual developments in terrestrial systems have focused on the aboveground part. Repeated and replicated manipulation experiments, sometimes surging ahead

M. Lukac (✉)
School of Agriculture, Policy and Development, University of Reading, Reading RG6 6AR, UK

Faculty of Forestry and Wood Sciences, Czech University of Life Sciences, Kamycka 129, 165 00 Prague, Czech Republic
e-mail: m.lukac@reading.ac.uk; lukacm@fld.czu.cz

M. Lukac et al. (eds.), *Soil Biological Communities and Ecosystem Resilience*, Sustainability in Plant and Crop Protection, DOI 10.1007/978-3-319-63336-7_1

of theory development, have shown that ecosystem function and biodiversity are intrinsically linked – at least aboveground (Grime 1997; Oliver et al. 2015; Schulze and Mooney 1994). Our understanding of the contribution of belowground biodiversity to overall ecosystem function is some way behind, despite the implied functional connection between the two parts. This lack of knowledge limits our ability to harness the ability of belowground biodiversity to deliver functions in the here-and-now and to predict its capacity to continue delivering in the future (Bardgett and van der Putten 2014).

Soil is the largest reservoir of biodiversity on Earth, one gram of soil has been shown to contain over 1 billion individual bacterial cells representing a large community of tens of thousands of species (Decaëns 2010), many of which are completely unknown at present and may fulfil a function we are not aware of yet. Soil biodiversity does not however include only bacteria, there are fungi, nematodes, earthworms, arthropods and even mammals who complete at least part of their life cycle in the soil and thus contribute to the overall biodiversity to be found in soils. One thing all of these life forms have in common is the fact that they all depend on aboveground primary production as their ultimate source of energy (Wardle et al. 2004). Organic matter rich in carbohydrates is either deposited on the soil surface or released into the soil profile from live and dead plant roots, to be utilised by a long chain of soil-dwelling organisms. This, together with the seasonal nature of organic matter input, leads to spatial, temporal and hierarchical arrangement of soil biodiversity. Correct understanding of the contribution of soil biodiversity to soil function thus implies knowledge of the input of each species, its variation in space and in time and the interactions between species and communities of organisms.

Recent developments and decreasing cost of molecular methods have led to the creation of reasonable evidence on the level of species diversity present in many soils on Earth, together with the identification and description of species performing key functions in soils. Long-standing assumptions guiding our approach to spatial distribution of soil biodiversity at global or biome scales have been challenged, it is now clear that soil organisms are spatially distributed and do not have the capacity to inhabit any soil (Callaway and Maron 2006). Global distribution of soil-dwelling species is restricted, probably by a combination of climatic, soil and plant composition factors (Öpik et al. 2006; Tedersoo et al. 2012). Spatial distribution of soil biodiversity is clearly linked to organism size, which ranges from single micrometres for bacteria, to tens of centimetres for earthworms, to several hectares for some soil fungi. At the smallest scale, the distribution of soil biota is governed by soil structure which determines resource availability (Elliott 1986; Rasche et al. 2011). Biotic interactions, such as competition and predation, root exudation by plants, soil turbation by larger organisms also play part, mainly by affecting chemical composition of soil solution and thus supporting or hindering growth of microbial populations (Langenheder and Prosser 2008; Stark and Firestone 1995). At centimetre to metre scale, diversity of soil biota is affected by local variation of soil conditions such as texture, organic matter content and water holding capacity, but also by the composition of plant community growing in the soil (Wardle et al. 2004). The latter determines the type and the timing of organic matter input to the soil, thus providing a clear indication of the close relationship between above- and belowground biodiversity. Finally, at ecosystem level, spatial patterns of soil biodiversity are governed by topography and related climatic factors, disturbance regime and landscape connectivity (Fierer and Jackson 2006; Reinhart and Callaway 2006). As a result, spatial distribution of soil biodiversity is fragmented and its description is very much dependent on the body size of the organism under consideration.

Similar to factors affecting spatial distribution, temporal changes in the diversity of soil organisms are very much dependent on the size of the organism in question. Population size of species with the smallest body size tends to be governed by minute availability of resources. An increase of water content in dry soil or a nutrient availability pulse can cause rapid microbial response (Austin et al. 2004), usually driven by fast multiplication and followed by an increase in population size. Nutrient pulses may be biotic or abiotic in their nature, root exudation by plants has been shown to trigger growth of specific groups of microorganisms at the expense of others (Bais et al. 2006; el Zahar Haichar et al. 2008) – a clear example of an interaction between above- and belowground diversity. Temporal changes in microbial community composition at the smallest scale can take minutes to hours and are usually accompanied by a rapid change of the dominant function performed by microbes at their location. Fluctuation of resource availability over seasons or years affects soil biodiversity at all levels (Lauber et al. 2013), mainly because all are dependent on aboveground productivity as their energy source – aboveground seasonality must be reflected belowground. At present, we have a very fragmented picture of this relationship, there is some evidence that microbial communities undergo a complete turnover from winter to summer (Hamel et al. 2006), are differentially affected by seasonal resource availability and that changes in plant community exert strong influence over the composition of soil biota. At the largest temporal scale – and the one least supported by evidence at present – there is indication that specific communities inhabit soils at different stages of natural succession or post-disturbance recovery (Neutel et al. 2002).

Vast numbers on individuals, a large span of body size of various species and often very tight specialisation in the food source or life strategy of soil biota gives rise to very complex hierarchical arrangement of communities within the soil. The simplest soil food-webs may consist of heterotrophic, nitrogen-fixing and autotrophic photosynthesising bacteria, with little reliance on plant productivity (Mikola and Setälä 1998). However, with any increase of aboveground productivity comes an influx of different soil-inhabiting life forms which self-organise into communities performing a variety of functions such as organic matter fragmentation, element cycling or water uptake. Over time, and in absence of disturbance, such communities become more and more complex by increasing food-chain length, accumulating larger levels of biodiversity and attaining higher level of stability (Neutel et al. 2007; Rooney et al. 2006). The mechanics of the relationship between complexity and stability of function of soil biological communities is not clear, however it is becoming evident that soil biodiversity might have direct implications for ecosystem stability, for example under environmental change (Isbell et al. 2015). Much is known about specific functions performed by specific groups of organisms, our knowledge of how widely are functions distributes among species within those groups is far from complete. Thus, it is difficult to assign an estimate of redundancy to any given species within a community – changes in species composition may or may not translate to changes in function (Strickland et al. 2009).

Humanity's need to preserve ecosystem productivity and resilience thus brings us to the issue of preserving soil biodiversity. Human land-use invariably introduces some level of soil disturbance, which usually interferes with soil biota and the processes it supports (Altieri 1999). One of the most striking effects of human land-use is simplification of soil biological communities and thus a decrease in soil biodiversity (Giller et al. 1997). The proportion of organisms with small body size expands at the expense of larger species as the latter tend to be severely affected by repeated soil disturbance (Wagg et al. 2014). It is also thought that modern soil working practices result in soils with fewer functional groups (Tsiafouli et al. 2015), presumably leading to a reduction in the capacity of soils to perform all of their functions. Processes such as carbon cycling, nitrogen fixation and cycling or phosphorus mining are impaired as soil biodiversity underpinning each and every function is reduced. It is, of course, possible to maintain acceptable level of aboveground ecosystem productivity by substituting some soil functions by addition of industrially produced nutrients (Vitousek et al. 1997), a practice that has been shown to lead to further erosion of soil biodiversity (Bai et al. 2010; Eisenhauer et al. 2012). The reverse side of this coin is the potential reduction in the resilience of soil biota in terms of supporting soil functions under changing environmental conditions (Folke et al. 2004). Less diverse communities will contain fewer redundant species which may have the capacity to sustain soil functions under a new set of conditions. Even when starting from a position where a single 3-dimentional soil profile may contain more biodiversity than the entire aboveground ecosystem (Myers et al. 2000), gradual and continuing reduction of soil biodiversity may lead to eventual loss of function.

This volume explores various strands of current research to illustrate the relationship between above- and belowground biodiversity, the role of the latter in ecosystem productivity and stability and the potential role of soil biodiversity in building resilience to environmental change. We explore how soil biota and its interactions contribute to ecosystem service delivery, including element and nutrient cycling within tree dominated ecosystems. We look at how soil management can be improved to benefit soil biota, and look at some of the latest methods used to uncover the contribution of soil biodiversity to ecosystem function.

Soils have been widely studied and classified in terms of their physical and chemical characteristics, however knowledge of soil biodiversity and its contribution to a wide range of **ecosystem services** derived from a variety of natural and managed ecosystems is still incomplete (Chap. 2). At the global scale, one of the key soil ecosystem services is carbon storage, there is some suggestion that soil biodiversity plays a key role in determining the amount of carbon in the soil. The evidence of this however is scant and one of the ways to address it is by the application of biodiversity to dynamic models (Chap. 16). Microbial communities are responsible for half of CO_2 emissions from terrestrial ecosystems, as well as many other functions essential for optimal functioning of terrestrial systems. Soil biota clearly has an effect on cycling and **bioavailability of soil nutrients**, including Ca. In an approach alternative to the accepted view, we introduce a new 'rhizo-centric' point of view where non-photosynthetic partners could be the dominant members (Chap. 4). Functional genetics can be used to explore the mechanistic link between tree species effects on soil microbes and the nitrogen cycling processes regulated by those microbes is discussed in Chap. 13, showing that the genetic potential for denitrification was strongly influenced by the abundance of ammonia oxidization potential within forest floors. Liming and wood ash effects on ectomycorrhizal fungal abundance are discussed in Chap. 14, with specific focus on diversity and community composition.

The understanding of soil biodiversity at various levels of organisation is critical to our ability to maintain the rage of ecosystem services delivered by soils. Lately, we have witnessed a rapid **development of methods** used to describe and to study soil biota. At the microscopic level, soil microbial community represents the biggest part of existing biodiversity, but it is still largely undiscovered. The use and integration of molecular methods makes it possible to improve knowledge about it. In this context, the application to soil samples of two different molecular phylogenetic techniques, such as FISH (Fluorescence In Situ Hybridization) and DGGE (Denaturing Gradient Gel Electrophoresis) is illustrated (Chap. 3). Next Generation Sequencing was used to show that samples from cultivated soils under olives and vegetables had higher frequencies (5-10%) of *Bacillales*, which were under-represented in the adjacent Mediterranean forest (Chap. 6). Chapter 11 illustrates the use of the intact core 'baiting' method to facilitate the study of natural fungal communities to experimentally uncover the functioning of ecologically-relevant fungal communities. β-glucosidase activity was used to assess the effects of Scots pine (*Pinus sylvestris*) and pedunculate oak (*Quercus robur*) on selected biochemical properties, physical and chemical characteristics of soil (Chap. 15). Finally, filamentous fungi are the producers of many secondary metabolites with wide spectra of biological effects. Among these, peptaibols represent a group of compounds produced mainly by members of genus *Trichoderma* and can thus be used to track their activity (Chap. 17).

Agricultural **soil management practices** clearly influence soil microbial community structure and function. A holistic approach taking into consideration all potential factors and drivers is thus necessary when examining the structure–function relationships of soil microbial communities to improve understandings of controls and functioning of below-ground processes (Chap. 3). Some soil-borne phytopathological agents (for instance fungus *Verticillium dahliae* Kleb.) are very difficult to control by methods alternative to those based on chemical products, biological control utilising soil biota and its diversity emerges as one of the most promising alternatives (Chap. 7). Aside from biological interactions, effects of biotic stress on soil fungal communities are largely unknown. In Chap. 8 we review existing information on the effects of drought episodes on mycorrhizal communities and associated ecosystem response. Due to the difficulty with remediation of sites characterized by multiple pollutants (e.g. organic and inorganic toxic compounds), the study of plant-microbial interactions is presented as a new and interesting challenge to discover more sustainability soil recovery strategies (Chap. 18). The occurrence of slowly desorbing fractions of hydrophobic pollutants, such as polycyclic aromatic hydrocarbons (PAHs), is well explained today by environmental organic chemistry. However, the microbial and plant interactions with such residues are less understood and are examined in Chap. 19.

Finally, the last thematic area covered in the volume concerns **biotic interactions** between various members of soil biological communities. Endophytes have an exciting potential to shape the biotic and abiotic stress tolerance in host plants. The abundance of ectomycorrhizal (EM) species in different pure beech and beech dominated stands across Central Europe is described in Chap. 8, while the response of AM fungal communities to a selection of permanent environment disturbances is dealt with in Chap. 9. To close, Chap. 20 provides an overview of various belowground community studies and their effect on the management of tree crops, identifying main drivers and services underpinning the implementation of sustainable strategies.

References

Altieri, M. A. (1999). The ecological role of biodiversity in agroecosystems. *Agriculture, Ecosystems and Environment, 74*, 19–31.

Austin, A. T., et al. (2004). Water pulses and biogeochemical cycles in arid and semiarid ecosystems. *Oecologia, 141*, 221–235.

Bai, Y., et al. (2010). Tradeoffs and thresholds in the effects of nitrogen addition on biodiversity and ecosystem functioning: evidence from inner Mongolia Grasslands. *Global Change Biology, 16*, 358–372.

Bais, H. P., et al. (2006). The role of root exudates in rhizosphere interactions with plants and other organisms'. *Annual Review on Plant Biology, 57*, 233–266.

Bardgett, R. D., & van der Putten, W. H. (2014). Belowground biodiversity and ecosystem functioning. *Nature, 515*, 505–511.

Callaway, R. M., & Maron, J. L. (2006). What have exotic plant invasions taught us over the past 20 years? *Trends in Ecology and Evolution, 21*, 369–374.

Cardinale, B. J., et al. (2012). Biodiversity loss and its impact on humanity. *Nature, 486*, 59–67.

Decaëns, T. (2010). Macroecological patterns in soil communities. *Global Ecology and Biogeography, 19*, 287–302.

Eisenhauer, N., et al. (2012). Global change belowground: impacts of elevated CO2, nitrogen, and summer drought on soil food webs and biodiversity. *Global Change Biology, 18*, 435–447.

el Zahar Haichar, F., et al. (2008). Plant host habitat and root exudates shape soil bacterial community structure. *The ISME Journal, 2*, 1221–1230.

Elliott, E. T. (1986). Aggregate structure and carbon, nitrogen, and phosphorus in native and cultivated soils. *Soil Science Society of America Journal, 50*, 627–633.

Fierer, N., & Jackson, R. B. (2006). The diversity and biogeography of soil bacterial communities. *Proceedings of the National Academy of Sciences, USA, 103*, 626–631.

Folke, C., et al. (2004). Regime shifts, resilience, and biodiversity in ecosystem management. *Annual Review of Ecology Evolution and. Systematics, 35*, 557–581.

Giller, K. E., et al. (1997). Agricultural intensification, soil biodiversity and agroecosystem function. *Applied Soil Ecology, 6*, 3–16.

Grime, J. P. (1997). Biodiversity and ecosystem function: the debate deepens. *Science, 277*, 1260–1261.

Hamel, C., et al. (2006). Seasonal and long-term resource-related variations in soil microbial communities in wheat-based rotations of the Canadian prairie. *Soil Biology and Biochemistry, 38*, 2104–2116.

Isbell, F., et al. (2015). Biodiversity increases the resistance of ecosystem productivity to climate extremes. *Nature, 526*, 574–577.

Langenheder, S., & Prosser, J. I. (2008). Resource availability influences the diversity of a functional group of heterotrophic soil bacteria. *Environmental Microbiology, 10*, 2245–2256.

Lauber, C. L., et al. (2013). Temporal variability in soil microbial communities across land-use types. *The ISME Journal, 7*, 1641–1650.

Mikola, J., & Setälä, H. (1998). Productivity and trophic-level biomasses in a microbial-based soil food web. *Oikos, 82*, 158–168.

Myers, N., et al. (2000). Biodiversity hotspots for conservation priorities. *Nature, 403*, 853–858.

Neutel, A. M., Heesterbeek, J. A., & de Ruiter, P. C. (2002). Stability in real food webs: weak links in long loops. *Science, 296*, 1120–1123.

Neutel, A.-M., et al. (2007). Reconciling complexity with stability in naturally assembling food webs. *Nature, 449*, 599–602.

Oliver, T. H., et al. (2015). Biodiversity and resilience of ecosystem functions. *Trends in Ecology and Evolution, 30*, 673–684.

Öpik, M., et al. (2006). Composition of root-colonizing arbuscular mycorrhizal fungal communities in different ecosystems around the globe. *Journal of Ecology, 94*, 778–790.

Rasche, F., et al. (2011). Seasonality and resource availability control bacterial and archaeal communities in soils of a temperate beech forest. *The ISME Journal, 5*, 389–402.

Reinhart, K. O., & Callaway, R. M. (2006). Soil biota and invasive plants. *New Phytologist, 170*, 445–457.

Rooney, N., et al. (2006). Structural asymmetry and the stability of diverse food webs. *Nature, 442*, 265–269.

Schulze, E. D., & Mooney, H. A. (1994). Ecosystem function of biodiversity: a summary. *Biodiversity and Ecosystem Function* (Springer). Series: Praktische Zahnmedizin Odonto-Stomatologie Pratique. *Practical Dental Medicine, 99*, 497–510.

Stark, J. M., & Firestone, M. K. (1995). Mechanisms for soil moisture effects on activity of nitrifying bacteria. *Applied and Environmental Microbiology, 61*, 218–221.

Strickland, M. S., et al. (2009). Testing the functional significance of microbial community composition. *Ecology, 90*, 441–451.

Tedersoo, L., et al. (2012). Towards global patterns in the diversity and community structure of ectomycorrhizal fungi. *Molecular Ecology, 21*, 4160–4170.

Tsiafouli, M. A., et al. (2015). Intensive agriculture reduces soil biodiversity across Europe. *Global Change Biology, 21*, 973–985.

Vitousek, P. M., et al. (1997). Human alteration of the global nitrogen cycle: sources and consequences. *Ecological Applications, 7*, 737–750.

Wagg, C., et al. (2014). Soil biodiversity and soil community composition determine ecosystem multifunctionality. *Proceedings of the National Academy of Sciences, USA, 111*, 5266–5270.

Wardle, D. A., et al. (2004). Ecological linkages between aboveground and belowground biota. *Science, 304*, 1629–1633.

Worm, B., & Duffy, J. E. (2003). Biodiversity, productivity and stability in real food webs. *Trends in Ecology and Evolution, 18*, 628–632.

Chapter 2
Ecosystem Services Provided By Soil Microorganisms

Maria Ludovica Saccá, Anna Barra Caracciolo, Martina Di Lenola, and Paola Grenni

Abstract Ecosystem services are the contributions that ecosystems provide to human well-being. They arise from the interaction of biotic and abiotic processes, and refer specifically to the 'final' outputs or products from ecological systems. Soil harbours a large proportion of Earth's biodiversity, and provides the physical substrate for most human activities. Although soils have been widely studied and classified in terms of physical and chemical characteristics, knowledge of soil biodiversity and functioning are still incomplete. Soil organisms are extremely diverse and contribute to a wide range of ecosystem services that are essential to the sustainable functioning of natural and managed ecosystems. Microbial communities (mainly composed by *Bacteria*, *Archaea* and microfungi) are vital to soil ecosystem functioning. This is because they exist in enormous numbers and have an immense cumulative mass and activity. Most of the phenomena observed in the visible aboveground world are steered directly or indirectly by species, interactions, or processes in the belowground soil. In particular, being microbial communities involved in nutrient cycling and organic matter degradation, they can affect biodiversity and productivity of aboveground ecosystems. Microorganisms can have stimulating or inhibiting effects on plants by the release of metabolites with a varying range of activities. Microbial communities are the main responsible of soil homeostatic capabilities removing contaminants and providing key ecosystem regulating and supporting services such as soil fertility, resilience and resistance to different stress. This chapter aims at describing the contributions provided by soil microbial communities to different ecosystem services and their potential use as indicators of ecosystem functioning. Understanding ecosystem functioning and predicting responses to global changes calls for much better knowledge than we have today about microbial processes and interactions, including those with plants in the rhizosphere.

M.L. Saccá
Council for Agricultural Research and Economics, Research Centre for Agriculture and Environment (CREA – AA), Via di Corticella 133, 40128 Bologna, Italy

A. Barra Caracciolo • M. Di Lenola • P. Grenni (✉)
National Research Council, Water Research Institute, CNR-IRSA, Monterotondo, Rome, Italy
e-mail: grenni@irsa.cnr.it

M. Lukac et al. (eds.), *Soil Biological Communities and Ecosystem Resilience*, Sustainability in Plant and Crop Protection, DOI 10.1007/978-3-319-63336-7_2

Keywords Microbial functional groups • Biodiversity • Microbial populations • Regulating and supporting services • Soil homeostasis

2.1 Introduction

Knowledge of soil complexity has enormously increased over the past 50 years. It is nowadays recognized that a series of drivers of global soil change, such as human population growth and urbanization, marketing land and climate change, have resulted in changes in soil health status, causing a diminished capacity of this ecosystem to provide goods and services for its beneficiaries (FAO and ITPS 2015). Soil preservation from multiple causes of degradation (erosion, salinization, compaction, sealing, landslides of soil and rock material, organic matter decline and local and diffuse contamination) is therefore receiving attention not only for an increased consideration of environmental issues, but also for an increased awareness of the economic value of soil functions. As defined by the Corine Land Cover classification, terrestrial ecosystems include urban systems, cropland, grassland, woodland and forest, heathland and shrub, sparsely vegetated land and wetlands (Maes 2013).

The extreme physico-chemical heterogeneity of soil can be translated in an extremely high heterogeneity of habitats. A single gram of soil can harbour up to 10^{10} bacterial cells and an estimated species diversity up to the order of 10^4 species (Roesch et al. 2007). High biodiversity reflects a wide spectrum of functional roles (Wagg et al. 2014) and soil organisms can be divided into three main broad functional groups (Turbé et al. 2010): chemical engineers, biological regulators and ecosystem engineers.

Chemical Engineers (mainly bacteria and fungi) are responsible for organic matter (OM) decomposition and cycling of readily available nutrients for plants, regulating 90% of energy flux. They play a key role in soil bioremediation, by degrading toxic chemicals through metabolic and co-metabolic pathways.

Biological regulators are responsible for population dynamic, regulation of plants, invertebrates and microorganisms through grazing, predation or parasitism. They comprise principally small invertebrates like nematodes, pot worms, springtails and mites.

Ecosystem engineers are responsible for physical modification of soil and regulating the availability of resources to other species, maintaining soil structure by the formation of pore networks and bio-structures, and particle aggregation or transport. They indirectly regulate resources availability for other soil organisms since soil structures become hotspots of microbial activities and affect water infiltration and distribution in soil. This group includes mainly earthworms, ants, termites and some small mammals.

Microbial communities have a key role as chemical engineers, and they can act as biological regulators and ecosystem engineers. The term microorganism generally denotes members of the *Bacteria* and *Archaea* kingdoms (Fig. 2.1), as well as microscopic *Eukarya* such as unicellular algae, fungi and *Protists*.

Fig. 2.1 Visualization under an epifluorescence microscope of soil microbial populations (1000×). Microbial cells are coloured in *blue* by DAPI stain, and soil particles are in *yellow*

A recent microbial census found that among bacterial and archaeal sequences, soil is the fourth most characterized category, after host-associated, aquatic, and built environment sources (Schloss et al. 2016). This surprising data indicate that today's technologies should be maximally exploited to reach a profound understanding of microbial soil biodiversity (Amann and Rosselló-Móra 2016). Bacterial biomass is particularly impressive in soil, with up to 1–2 t/ha in temperate grasslands (Orgiazzi et al. 2016). Soil is estimated to contain about 1000 Gbp of microbial genome sequences per gram (Vogel et al. 2009). Microbial abundance in the rhizosphere is at least three orders of magnitude higher than in bulk soil (Reinhold-Hurek et al. 2015; Thijs et al. 2016). This increase in microbial activity, known as 'rhizosphere effect', is due to the fact that bacteria, fungi, oomycetes, viruses and archaea feed on rhizodeposits that include nutrients, exudates, border cells and mucilage released by plant roots (Philippot et al. 2013). On the other side, it is recognized that plant nutrition is significantly influenced by rhizosphere microbial composition (Hartmann et al. 2008).

Even though the ability to explore complex environmental microbial communities has been rapidly increasing in the past few years, the relationship between soil biodiversity and functioning is not yet fully understood and many questions are still open (Graham et al. 2016; Widder et al. 2016).

Ecosystem services represent the benefits human populations derive, directly or indirectly, from ecosystem functions, including both goods and services. One of the first categorizations of ecosystem services brings us to the late 90's, when Costanza and colleagues estimated in economic terms their global value as twice the value of the gross national product of the whole world (Costanza et al. 1997). In 2000, the Millennium Ecosystem Assessment (MEA 2005) was the first large scale ecosystem assessment report to evaluate the consequences of ecosystem change for human well-being and the scientific basis for action to enhance the conservation and sustainable use of those systems. MEA report identified four main categories of ecosystem services: regulating, supporting, provisioning and cultural services. Other classifications have been proposed

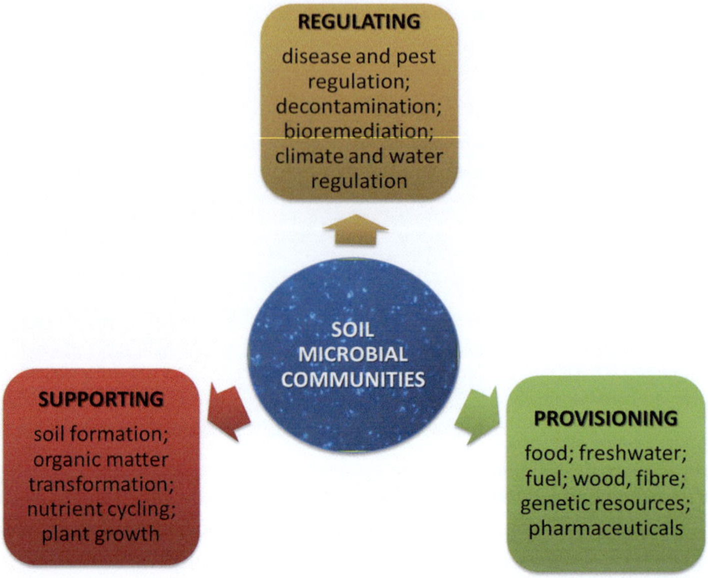

Fig. 2.2 Main regulating, supporting and provisioning ecosystem services provided by soil microbial communities

for global (Fisher et al. 2010; Maes 2013; Haines-Young 2016) and soil ecosystem services (Dominati et al. 2010; Aislabie and Deslippe 2013), all including provisioning and regulating services. Supporting services originally defined in the MEA report and in Turbé et al. (2010) have been omitted by other classifications (Fisher et al. 2010; Maes 2013), because considered as a subset of ecological processes.

Belowground microbial communities have a key role in the categories of regulating and supporting services of soil ecosystems, and they contribute to several provisioning services (Fig. 2.2).

2.2 Regulating Services

Regulating services provided by belowground microorganisms mainly include the ecosystem processes ensuring climate regulation, water regulation and purification, disease and pest regulation and bioremediation of organic pollutants.

Climate Regulation is one of the most important ecosystem services provided by soil microorganisms, since bacteria and fungi play a key role in land–atmosphere carbon (C) exchange. The climate regulating function of soils highly depends on land use and intensification, that are shown to have profound impacts on soil capacity to store C and on its emissions (Bailey et al. 2002; Fisher et al. 2010), implicating potential economic and ecological consequences (Lal 2004). On the other hand, a positive feedback on climate change is created by the temperature-induced acceleration of the rates of heterotrophic microbial activity (Bardgett et al. 2008). The latest Intergovernmental Panel on Climate Change (IPCC) Report states that continued high emissions would lead to mostly negative impacts for ecosystem services (IPCC 2014). Bacterial community structure would be affected by elevated CO_2 more than fungal and nematode communities, and responses may depend on plant type and soil nutrient availability (Drigo et al. 2007).

Water Regulation and Purification Soil acts as a water filter and reservoir, purifying water as it passes through the soil substrate and providing water storage also for plant uptake (Walthall et al. 2012). A recent study has calculated that the actual ecosystem service of water purification in Europe is worth €16 billion per year (La Notte et al. 2017). The role of microorganisms in the purification of water, as it passes through soil, is mainly due to their capacity to degrade various contaminants. As an example, an herbicide-degrading bacterium, *Rhodococcus wratislaviensis*, was detected both in soil and groundwater samples of a terbuthylazine-contaminated aquifer, suggesting its detoxification potential of contaminated soil-water systems (Grenni et al. 2009a). Furthermore, soil microorganisms can have indirect effects on water infiltration rates by modifying the quantity and quality of soil organic matter (Turbé et al. 2010).

Disease and Pest Regulation Microorganisms can control directly or indirectly pathogen diseases through a number of different mechanisms. Antagonism, competition, interference with pathogen signalling or stimulation of host plant defences have been proved in a wide diversity of soil bacterial (i.e. *Streptomyces, Pseudomonas*, *Bacillus*) and fungal (i.e. *Trichoderma*, *Fusarium*) species (Raaijmakers et al. 1997; Haas and Defago 2005; Cawoy et al. 2011; Palaniyandi et al. 2013; Ciancio et al. 2016).

Resistance to invasion by alien species is another major life support function provided by biologically diverse terrestrial microbial ecosystems, as compared to more simple ones (van Elsas et al. 2012). Under non-stress conditions, community evenness determines resistance to be invaded and preserves ecosystem functionality (De Roy et al. 2013).

Knowledge of the potentialities of beneficial organisms bearing ecosystem services can be exploited as a valid alternative to the use of agrochemicals, for a sustainable agriculture, but still many inconsistencies are observed when biological control agents are released under field conditions, especially when used for controlling woody plant diseases (Cazorla and Mercado-Blanco 2016).

Enhancement of ecosystem services associated both with organic matter decomposition and with the reduction of soil borne pathogens can be obtained using organic amendments. Pathogen suppressive composts have been used to control plant diseases through microbiota modification (Bonilla et al. 2012; Vida et al. 2016; Blaya et al. 2016), by enhancing soil suppressiveness, which is the natural capacity of a soil to suppress the development of a disease (Mazzola 2002). The most dynamic bacteria associated with disease suppression belong to γ-, β-*Proteobacteria* (*Pseudomonadaceae, Burkholderiaceae, Xanthomonadales*) and

Firmicutes (van Bruggen and Semenov 2000; Weller et al. 2002; Mendes et al. 2011), whilst Ascomycota phylum is considered the most active among fungi (Vida et al. 2016; Blaya et al. 2016).

It is interesting to note that a questionnaire on land user's awareness of ecosystem services in arable farms has shown that, among a number of selected services (i.e. water and nutrient retention and release, climate functions, natural attenuation), natural disease suppression was considered the most important ecosystem service (Rutgers et al. 2012).

Biodegradation of Organic Waste and Xenobiotics Soil contamination is one of the major threats to soil functions (CEC 2006). Main sources of soil contamination are inadequate and uncontrolled waste disposal, industrial and mining activities, spills, military and nuclear operations, agricultural practices. Natural attenuation of contaminants includes a variety of physico-chemical (e.g. dispersion, dilution, adsorption, and volatilization) and biological (transformation, mineralization, stabilization etc.) processes that under favourable conditions, act without human intervention, to reduce the mass, toxicity, mobility, volume or concentration of contaminants (US EPA 1999). The evaluation of microbial potential to degrade a wide range of pollutants in soil ecosystems can be tested in microcosm studies (Barra Caracciolo et al. 2013) as well as in field (Franchi et al. 2016). A chemical can be used in microbial metabolism as a source of energy, C, N or other nutrients or it can be transformed by constitutive microbial enzymes or cofactors (co-metabolism).

Acting as a selective force, exposure to a contaminant can promote the development of adapted microbial populations, that may exhibit increased tolerance and/or degrading capacity (Boivin et al. 2002). This latter possibility reflects the homeostatic capacity of natural microbial communities (Fig. 2.3).

Indeed, many laboratory studies have successfully identified specific bacterial strains able to transform contaminants (Grenni et al. 2009a; Franchi et al. 2016). However, the complete removal of a pollutant is often mediated by microbial consortia that conjunctly operate in degradation pathways (Barra Caracciolo et al. 2007; Pino and Peñuela 2011; Franchi et al. 2016). Moreover, soil contamination can be due to a mixture of contaminants as in the case of polychlorinated biphenyls (PCBs) and consequently their degradation is a very complex issue. It requires several microbial species performing different degradation pathways both in anaerobic and aerobic conditions (Ancona et al. 2016).

Soil microbial components provide regulating services of soil remediation both in bulk soil (bioremediation) and in the rhizosphere (plant-assisted bioremediation). Microbial degradation of contaminants in soil can be stimulated by the addition of nutrients/amendments (Luo et al. 2008; Grenni et al. 2009b, 2012), the augmentation of degrading microbes (Paul et al. 2006), or by promoting rhizosphere microbial degradation activity (Ancona et al. 2016; Franchi et al. 2016).

Environmental genomic approaches based on the identification of microbial species and catabolic genes involved in xenobiotic degradation can lead to set up useful bioindicators (species) and biomarkers (genes) that can be used as monitoring tools

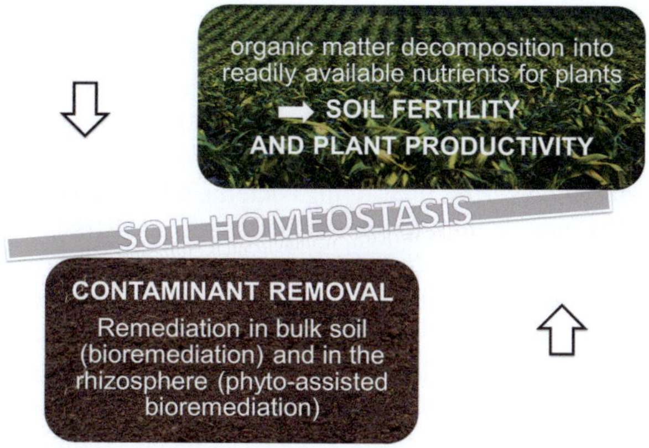

Fig. 2.3 Soil homeostasis is the capability to maintain an equilibrium between soil functions

(Eyers et al. 2004). Examples of the use of functional biomarkers of soil catabolic potential to degrade contaminants have been reported for triazines (Fajardo et al. 2012), phenoxyacetic acids (Dennis et al. 2003; Baelum et al. 2008), aromatic hydrocarbons (Gomes et al. 2005; Sipilä et al. 2008; Yagi et al. 2010; Franchi et al. 2016) and PCBs (Fava et al. 2003; Kjellerup et al. 2012; Hashmi et al. 2016).

The value of the ecosystem service of natural attenuation of soil pollutants provided by belowground microorganisms increases if we consider its consequences on other ecosystem services such as primary production and water purification.

2.3 Supporting Services

Supporting services are those that underpin ecosystem functions, which are not directly used by humans. Examples of supporting services in which soil microbial communities are involved are mainly soil formation, nutrient cycling, water cycling, primary production and habitat for biodiversity.

Soil fertility, defined as the ability to support and sustain plant growth by ensuring adequate storage of nutrients and recycling of organic matter (FAO and ITPS 2015), is mediated by chemical engineers. This functional group includes bacterial and fungal species involved in the catabolic reactions leading to the breakdown, transformation and mineralization of C and nutrient cycling. Key processes of element cycling are balanced by the property of microorganisms to maintain their biomass element ratio by the principle of stoichiometric homeostasis (Spohn 2016).

Soil microorganisms are recognized as drivers of plant diversity and productivity (van Der Heijden et al. 2008) and manipulation of soil communities through microbial inoculation can be used for the restoration of degraded terrestrial ecosystems (Wubs et al. 2016). Plant productivity is stimulated through a wide range of microbial mechanisms by a high portion of microbial diversity. A number of bacterial species are capable to produce organic compounds affecting the proliferation of plant root system. They include members of the *Actinobacteria* (i.e. *Arthrobacter* sp.*)*, *Proteobacteria* (i.e. *Pseudomonas* sp., *Agrobacterium* sp.) and *Firmicutes* (*Bacillus* spp.) (Haas and Defago 2005; Doornbos et al. 2011). The best-known symbiotic N-fixing bacteria are *Rhizobium* spp. (α-*Proteobacteria*) and *Frankia* sp. (*Actinobacteria*). Free-living bacteria, which significantly contribute to atmospheric N fixation, include ubiquitous species in terrestrial ecosystems, such as *Azotobacter*, *Azospirillum*, *Bacillus*, *Klebsiella* spp. and some *Cyanobacteria* species. Moreover, many bacterial species (*Pseudomonads*, *Frankia* and *Streptomyces* sp.) and fungi (*Aspergillus* spp.) have shown the ability to produce iron-chelating compounds increasing its availability for plants. Interesting potentialities in biofertilization and microbe-assisted phytoremediation of metal-contaminated soils have been found in siderophore producing *Streptomyces* species (Dimkpa et al. 2008). Microorganisms with phosphate-solubilizing capability have been described among bacteria (i.e. *Agrobacterium* sp., *Bacillus* sp. and *Paenibacillus* sp.) and fungi (i.e. *Aspergillus* sp. and *Penicillium* sp.).

Plant-rhizobacteria interactions can also alleviate abiotic stress such as drought, high soil salinity, extreme temperatures, nutrient deficiency and heavy metal toxicity (Dimkpa et al. 2009). The isolation and characterization of salt- and drought-tolerant strains may be of high practical interest to overcome yield losses due to water scarcity worldwide (Ali et al. 2014; Forni et al. 2017).

2.4 Provisioning Services

Provisioning services include products obtained from ecosystems, such as food, water, fibre, fuel, genetic resources, chemicals, medicines and pharmaceuticals.

Rhizosphere microorganisms are an important bioresource for bioactive substances such as antibiotics, biosurfactants, enzymes and osmoprotective substances. Antibiotics produced by soil microorganisms may find new uses as experimental pharmaceuticals, offering a resource for compounds to deal with the rapid development of multidrug-resistant human pathogenic bacteria (Compant et al. 2005). As an example, the METACONTROL project aimed at identifying the biotechnological potential of antibiotic producing soil microbiota by examining the microbial metagenome of disease suppressive soils, where the prevalence of antibiotic biosynthetic clusters is expected (van Elsas et al. 2008).

The multiple involvement of bacterial species in regulating, supporting and provisioning services shows that many of these services cannot be categorized strictly (Fig. 2.4).

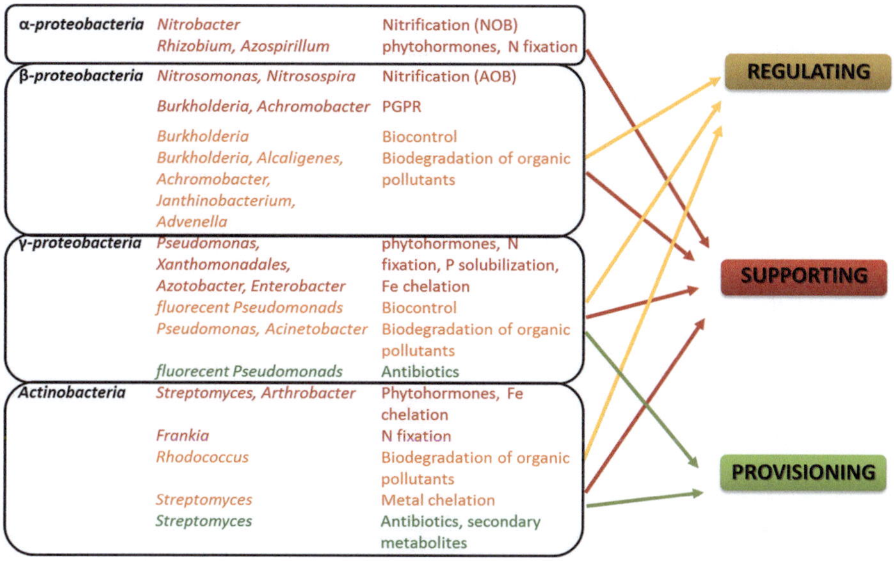

Fig. 2.4 Examples of overlapping of ecosystem services provided by some of the most active bacterial taxa present in soil

2.5 Soil Microorganisms as Indicators

Various biological indicators for soil health have been proposed in the past decades (van Bruggen and Semenov 2000; Griffiths et al. 2001; Lavelle et al. 2006; Ritz et al. 2009; Floch et al. 2011; Griffiths et al. 2016). Potential biological indicators should be measurable, cost-effective, policy-relevant, sensitive to changes and fit for use (Griffiths et al. 2016). In this context, microbial responses deserve particular attention (van Bruggen and Semenov 2000). Indeed, soil microorganisms meet most criteria to be valuable indicators for soil quality: they have sensitive responses to land management and climate, show correlation with soil and ecosystem functions, illustrate the cause and effect chain that links land management decisions to plant health and productivity, are comprehensible and useful to land managers and easy and inexpensive to measure (Doran and Zeiss 2000; Bennett et al. 2010).

Ecosystem functions such as food and fibre production, environmental interactions, supporting habitats and biodiversity have been used as indicators for monitoring soil quality. Using a multi-tier approach, a set of genotypic, phenotypic or functional level bioindicators was selected, including soil microbial community structure and biomass, soil respiration, C cycling and biochemical processes (Ritz et al. 2009). Furthermore, a recent study performed using this approach, stated that indicators for ecosystem functions related to the services of water regulation, C sequestration and nutrient provision would include microbial functional genes (Griffiths et al. 2016).

Currently soil microbial ecologists have availability of numerous biochemical and molecular methods for studying soil microbial diversity at different levels of detail (e.g. from a community fingerprint to species identification) and functioning (from overall enzymatic activities to specific functional gene identification). The importance of developing standard methods in soil microbiology to ensure comparable data across studies has been repeatedly underlined (Fierer et al. 2009; Philippot et al. 2012). A multiple polyphasic approach combining traditional cultivation methods with biochemical and molecular ones appears to be the best choice (Nardo et al. 2005; Pulleman et al. 2012). It has been recently recognized that although the -omics are unquestionably powerful approaches, a complete understanding of microbial processes cannot omit a combination with classic culture-dependent methods (Marx 2017).

2.6 Conclusions

The role of soil microbial communities in plant growth and health is widely acknowledged, but ecosystem services related to this biotic portion of soil are often undervalued because considered distant from the market and uncertain. Terrestrial services alone account for 60% of the value of total global ecosystem services, nevertheless a lack of understanding of the links between soil management, soil health and associated services and benefits, rarely leads to an establishment of soil management interventions for public benefits. Therefore, in order to reach an adequate protection goal of soil, a reliable environmental risk assessment should take into account also ecosystem services provided by microbial communities.

References

Aislabie, J., & Deslippe, J. R. (2013). *Soil microbes and their contribution to soil services* (pp. 143–161). Lincoln: Manaaki Whenua Press, Landcare Research.

Ali, S. Z., Sandhya, V., & Venkateswar Rao, L. (2014). Isolation and characterization of drought-tolerant ACC deaminase and exopolysaccharide-producing fluorescent Pseudomonas sp. *Annals of Microbiology, 64*, 493–502.

Amann, R., & Rosselló-Móra, R. (2016). After all, only millions? *mBio, 7*, e00999-16.

Ancona, V., Barra Caracciolo, A., Grenni, P., Di Lenola, M., Campanale, C., Calabrese, A., Uricchio, V.F., Mascolo, G., & Massacci, A. (2016). Plant-assisted bioremediation of a historically PCB and heavy metal-contaminated area in Southern Italy. *New Biotechnology*. doi:http://doi.org/10.1016/j.nbt.2016.09.006.

Baelum, J., Nicolaisen, M. H., Holben, W. E., Strobel, B. W., Sorensen, J., & Jacobsen, C. S. (2008). Direct analysis of *tfdA* gene expression by indigenous bacteria in phenoxy acid amended agricultural soil. *The ISME Journal, 2*, 677–687.

Bailey, V., Smith, J., & Bolton, H. (2002). Fungal-to-bacterial ratios in soils investigated for enhanced C sequestration. *Soil Biology and Biochemistry, 34*, 997–1007.

Bardgett, R. D., Freeman, C., & Ostle, N. J. (2008). Microbial contributions to climate change through carbon cycle feedbacks. *The ISME Journal, 2*, 805–814.

Barra Caracciolo, A., Grenni, P., Saccà, M. L., Falconi, F., Di Landa, G., & Ciccoli, R. (2007). The Role of the Bacterial Community of an Agroecosystem in Simazine Degradation. *Italian Journal of Agronomy, 2*, 163–169.

Barra Caracciolo, A., Bottoni, P., & Grenni, P. (2013). Microcosm studies to evaluate microbial potential to degrade pollutants in soil and water ecosystems. *Microchemical Journal, 107*, 126–130.

Bennett, L. T., Mele, P. M., Annett, S., & Kasel, S. (2010). Examining links between soil management, soil health, and public benefits in agricultural landscapes: An Australian perspective. *Agriculture, Ecosystems and Environment, 139*, 1–12.

Blaya, J., Marhuenda, F. C., Pascual, J. A., & Ros, M. (2016). Microbiota characterization of compost using omics approaches opens new perspectives for phytophthora root rot control. *PLOS ONE, 11*, e0158048.

Boivin, M. E., Breure, A., Posthuma, L., & Rutgers, M. (2002). Determination of field effects of contaminants—significance of pollution-induced community tolerance. *Human and Ecological Risk Assessment: An International Journal, 8*, 1035–1055.

Bonilla, N., Gutiérrez-Barranquero, J., Vicente, A., & Cazorla, F. (2012). Enhancing Soil Quality and Plant Health Through Suppressive Organic Amendments. *Diversity, 4*, 475–491.

van Bruggen, A. H. C., & Semenov, A. M. (2000). In search of biological indicators for soil health and disease suppression. *Applied Soil Ecology, 15*, 13–24.

Cawoy, H., Bettiol, W., Fickers, P., & Ongena, M. (2011). *Bacillus*-based biological control of plant diseases. In: M. Stoytcheva (Ed), *Pesticides in the modern world – Pesticides use and management* (pp. 273–302). InTech. Available at http://www.intechopen.com/books/pesticides-in-themodern-world-pesticides-use-and-management/bacillus-based-biological-control-of-plant-diseases

Cazorla, F. M., & Mercado-Blanco, J. (2016). Biological control of tree and woody plant diseases: an impossible task? *BioControl, 61*, 233–242.

CEC. (2006). Commission of the European Communities. COM(2006)231 final. Communication from the Commission to the Council, the European Parliament, the European Economic and Social Committee and the Committee of the Regions. In: *Thematic strategy for soil protection* (pp. 1–16). Brussels. Available at http://eur-lex.europa.eu/legal-content/EN/TXT/HTML/?uri=CELEX:52006DC0231&from=EN. Accessed April 2017.

Ciancio, A., Pieterse, C. M. J., & Mercado-Blanco, J. (2016). Editorial: Harnessing Useful Rhizosphere Microorganisms for Pathogen and Pest Biocontrol. *Frontiers in Microbiology, 7*, 1620.

Compant, S., Duffy, B., Nowak, J., Clement, C., & Barka, E. A. (2005). Use of plant growth-promoting bacteria for biocontrol of plant diseases: Principles, mechanisms of action, and future prospects. *Applied and Environmental Microbiology, 71*, 4951–4959.

Costanza, R., d'Arge, R., de Groot, R., Farber, S., Grasso, M., Hannon, B., Limburg, K., Naeem, S., O'Neill, R. V., Paruelo, J., Raskin, R. G., Sutton, P., & van den Belt, M. (1997). The value of the world's ecosystem services and natural capital. *Nature, 387*, 253–260.

De Roy, K., Marzorati, M., Negroni, A., Thas, O., Balloi, A., Fava, F., Verstraete, W., Daffonchio, D., & Boon, N. (2013). Environmental conditions and community evenness determine the outcome of biological invasion. *Nature Communications, 4*, 1383–1387.

Dennis, P., Edwards, E. A., Liss, S. N., & Fulthorpe, R. (2003). Monitoring gene expression in mixed microbial communities by using DNA microarrays. *Applied and Environmental Microbiology, 69*, 769–778.

van Elsas, J. D., Costa, R., Jansson, J., Sjoling, S., Bailey, M., Nalin, R., Vogel, T. M., & van Overbeek, L. (2008). The metagenomics of disease suppressive soils – Experiences from the METACONTROL project. *Trends in Biotechnology, 26*, 591–601.

van Elsas, J. D., Chiurazzi, M., Mallon, C. A., Elhottovā, D., Krištůfek, V., & Salles, J. F. (2012). Microbial diversity determines the invasion of soil by a bacterial pathogen. *Proceedings of the National Academy of Sciences, USA, 109*, 1159–1164.
Dimkpa, C. O., Svatos, A., Dabrowska, P., Schmidt, A., Boland, W., & Kothe, E. (2008). Involvement of siderophores in the reduction of metal-induced inhibition of auxin synthesis in *Streptomyces* spp. *Chemosphere, 74*, 19–25.
Dimkpa, C., Weinand, T., & Asch, F. (2009). Plant-rhizobacteria interactions alleviate abiotic stress conditions. *Plant, Cell and Environment, 32*, 1682–1694.
Dominati, E., Patterson, M., & Mackay, A. (2010). A framework for classifying and quantifying the natural capital and ecosystem services of soils. *Ecological Economics, 69*, 1858–1868.
Doornbos, R. F., Loon, L. C., & Bakker, P. A. H. M. (2011). Impact of root exudates and plant defense signaling on bacterial communities in the rhizosphere. A review. *Agronomy for Sustainable Development, 32*, 227–243.
Doran, J. W., & Zeiss, M. R. (2000). Soil health and sustainability: managing the biotic component of soil quality. *Applied Soil Ecology, 15*, 3–11.
Drigo, B., Kowalchuk, G. A., Yergeau, E., Bezemer, T. M., Boschker, H. T. S., & Van Veen, J. A. (2007). Impact of elevated carbon dioxide on the rhizosphere communities of *Carex arenaria* and *Festuca rubra*. *Global Change Biology, 13*, 2396–2410.
Eyers, L., George, I., Schuler, L., Stenuit, B., Agathos, S. N., & El Fantroussi, S. (2004). Environmental genomics: Exploring the unmined richness of microbes to degrade xenobiotics. *Applied Microbiology and Biotechnology, 66*, 123–130.
Fajardo, C., Saccà, M. L., Gibello, A., Martinez-Inigo, M. J., Nande, M., Lobo, C., & Martin, M. (2012). Assessment of s-triazine catabolic potential in soil bacterial isolates applying atz genes as functional biomarkers. *Water Air and Soil Pollution, 223*, 3385–3392.
FAO and ITPS. 2015). Status of the World's Soil Resources (SWSR). Main report. Food and Agriculture Organization of the United Nations and Intergovernmental Technical Panel on Soils, Rome, Italy, pp. 1–648. Available at http://www.fao.org/3/a-i5199e.pdf, Accessed April 2017.
Fava, F., Bertin, L., Fedi, S., & Zannoni, D. (2003). Methyl-b-cyclodextrin-enhanced solubilization and aerobic biodegradation of polychlorinated biphenyls in two aged-contaminated soils. *Biotechnology and Bioengineering, 81*, 381–390.
Fierer, N., Grandy, A. S., Six, J., & Paul, E. A. (2009). Searching for unifying principles in soil ecology. *Soil Biology and Biochemistry, 41*, 2249–2256.
Floch, C., Chevremont, A. C., Joanico, K., Capowiez, Y., & Criquet, S. (2011). Indicators of pesticide contamination: Soil enzyme compared to functional diversity of bacterial communities via Biolog® Ecoplates. *European Journal of Soil Biology, 47*, 256–263.
Forni, C., Duca, D., & Glick, B. R. (2017). Mechanisms of plant response to salt and drought stress and their alteration by rhizobacteria. *Plant and Soil, 410*, 335–356.
Franchi, E., Agazzi, G., Rolli, E., Borin, S., Marasco, R., Chiaberge, S., Conte, A., Filtri, P., Pedron, F., Rosellini, I., Barbafieri, M., & Petruzzelli, G. (2016). Exploiting hydrocarbon-degrading indigenous bacteria for bioremediation and phytoremediation of a multicontaminated soil. *Chemical Engineering and Technology, 39*, 1676–1684.
Gomes, N. C., Kosheleva, I. A., Abraham, W. R., & Smalla, K. (2005). Effects of the inoculant strain *Pseudomonas putida* KT2442 (pNF142) and of naphthalene contamination on the soil bacterial community. *FEMS Microbiology Ecology, 54*, 21–33.
Graham, E. B., Knelman, J. E., Schindlbacher, A., Siciliano, S., Breulmann, M., Yannarell, A., Beman, J. M., Abell, G., Philippot, L., Prosser, J., Foulquier, A., Yuste, J. C., Glanville, H. C., Jones, D. L., Angel, R., Salminen, J., Newton, R. J., Bürgmann, H., Ingram, L. J., Hamer, U., Siljanen, H. M. P., Peltoniemi, K., Potthast, K., Bañeras, L., Hartmann, M., Banerjee, S., Yu, R.-Q., Nogaro, G., Richter, A., Koranda, M., Castle, S. C., Goberna, M., Song, B., Chatterjee, A., Nunes, O. C., Lopes, A. R., Cao, Y., Kaisermann, A., Hallin, S., Strickland, M. S., Garcia-Pausas, J., Barba, J., Kang, H., Isobe, K., Papaspyrou, S., Pastorelli, R., Lagomarsino, A.,

Lindström, E. S., Basiliko, N., Nemergut, D., & R. (2016). Microbes as engines of ecosystem function: When does community structure enhance predictions of ecosystem processes? *Frontiers in Microbiology, 7*, 1–10.

Grenni, P., Gibello, A., Barra Caracciolo, A., Fajardo, C., Nande, M., Vargas, R., Saccà, M. L., Martinez-Iñigo, M. J., Ciccoli, R., & Martín, M. (2009a). A new fluorescent oligonucleotide probe for in situ detection of s-triazine-degrading Rhodococcus wratislaviensis in contaminated groundwater and soil samples. *Water Research, 43*, 2999–3008.

Grenni, P., Barra Caracciolo, A., Rodríguez-Cruz, M., & Sánchez-Martín, M. (2009b). Changes in the microbial activity in a soil amended with oak and pine residues and treated with linuron herbicide. *Applied Soil Ecology, 41*, 2–7.

Grenni, P., Rodríguez-Cruz, M. S., Herrero-Hernández, E., Marín-Benito, J. M., Sánchez-Martín, M. J., & Barra Caracciolo, A. (2012). Effects of wood amendments on the degradation of terbuthylazine and on soil microbial community activity in a clay loam soil. *Water, Air, and Soil Pollution, 223*, 5401–5412.

Griffiths, B. S., Bonkowski, M., Roy, J., & Ritz, K. (2001). Functional stability, substrate utilisation and biological indicators of soils following environmental impacts. *Applied Soil Ecology, 16*, 49–61.

Griffiths, B. S., Römbke, J., Schmelz, R. M., Scheffczyk, A., Faber, J. H., Bloem, J., Pérès, G., Cluzeau, D., Chabbi, A., Suhadolc, M., Sousa, J. P., Martins Da Silva, P., Carvalho, F., Mendes, S., Morais, P., Francisco, R., Pereira, C., Bonkowski, M., Geisen, S., Bardgett, R. D., De Vries, F. T., Bolger, T., Dirilgen, T., Schmidt, O., Winding, A., Hendriksen, N. B., Johansen, A., Philippot, L., Plassart, P., Bru, D., Thomson, B., Griffiths, R. I., Bailey, M. J., Keith, A., Rutgers, M., Mulder, C., Hannula, S. E., Creamer, R., & Stone, D. (2016). Selecting cost effective and policyrelevant biological indicators for European monitoring of soil biodiversity and ecosystem function. *Ecological Indicators, 69*, 213–223.

Haas, D., & Defago, G. (2005). Biological control of soil-borne pathogens by fluorescent pseudomonads. *Nature Reviews Microbiology, 3*, 307–319.

Haines-Young, R. (2016) Support to EEA tasks under the EU MAES process. Negotiated procedure No EEA/NSS/16/002. Report of Results of a Survey to Assess the Use of CICES, 2016 (Deliverable 2), pp. 1–22. Available at http://cices.eu/content/uploads/sites/8/2016/07/Report-on-Survey-Results_19072016_Upload.pdf, Accessed April 2017.

Hartmann, A., Rothballer, M., & Schmid, M. (2008). Lorenz Hiltner, a pioneer in rhizosphere microbial ecology and soil bacteriology research. *Plant and Soil, 312*, 7–14.

Hashmi, M. Z., Qin, Z., Yao, X., Ahmed, Z., Xiaomei, S., Shen, C., & Tang, X. (2016). PCBs attenuation and abundance of Dehalococcoides spp., bphC, CheA, and flic genes in typical polychlorinated biphenyl-polluted soil under floody and dry soil conditions. *Environmental Science and Pollution Research, 23*, 3907–3913.

IPCC. (2014). Climate change 2014 synthesis report. In: R. K. Pachauri, & L. A. Meyer (Eds.), *Contribution of Working Groups I, II and III to the Fifth Assessment Report of the Intergovernmental Panel on Climate Change*, Geneva, Switzerland, pp. 1–151.

Kjellerup, B. V., Paul, P., Ghosh, U., May, H. D., & Sowers, K. R. (2012). Spatial distribution of PCB dechlorinating bacteria and activities in contaminated soil. *Applied and Environmental Soil Science, 2012*, 1–11.

La Notte, A., Maes, J., Dalmazzone, S., Crossman, N. D., Grizzetti, B., & Bidoglio, G. (2017). Physical and monetary ecosystem service accounts for Europe: A case study for in-stream nitrogen retention. *Ecosystem Services, 23*, 18–29.

Lal, R. (2004). Soil carbon sequestration to mitigate climate change. *Geoderma, 123*, 1–22.

Lavelle, P., Decaëns, T., Aubert, M., Barot, S., Blouin, M., Bureau, F., Margerie, P., Mora, P., & Rossi, J.-P. (2006). Soil invertebrates and ecosystem services. *European Journal of Soil Biology, 42*, S3–S15.

Luo, W., D'Angelo, E. M., & Coyne, M. S. (2008). Organic carbon effects on aerobic polychlorinated biphenyl removal and bacterial community composition in soils and sediments. *Chemosphere, 70*, 364–373.

Maes. (2013). Mapping and assessment of ecosystems and their services. An analytical framework for ecosystem assessments under Action 5 of the EU Biodiversity Strategy to 2020, pp. 1–60.

Marx, V. (2017). Microbiology: the return of culture. *Nature Methods, 14*, 37–40.

Mazzola, M. (2002). Mechanisms of natural soil suppressiveness to soilborne diseases. *Antonie van Leeuwenhoek, 81*, 557–564.

MEA (2005). Millennium ecosystem assessment – Ecosystems and human well-being: Synthesis, pp. 1–155. Available at http://www.millenniumassessment.org/documents/document.356.aspx.pdf. Accessed April 2017.

Mendes, R., Kruijt, M., de Bruijn, I., Dekkers, E., van der Voort, M., Schneider, J. H., Piceno, Y. M., DeSantis, T. Z., Andersen, G. L., Bakker, P. A., & Raaijmakers, J. M. (2011). Deciphering the rhizosphere microbiome for disease-suppressive bacteria. *Science, 332*, 1097–1100.

Nardo, M., Saisana, M., Saltelli, A., Tarantola, S., Hoffman, A., Giovannini, E. (2005). Handbook on constructing composite indicators (pp. 1–162). Paris: OECD Publishing. Available at https://www.oecd.org/std/42495745.pdf. Accessed April 2017.

Orgiazzi, A., Bardgett, R. D., Barrios, E., Behan-Pelletier, V., Briones, M. J. I., Chotte, J.- L., De Deyn, G. B., Eggleton, P., Fierer, N., Fraser, T., Hedlund, K., Jeffery, S., Johnson, N. C., Jones, A., Kandeler, E., Kaneko, N., Lavelle, P., Lemanceau, P., Miko, L., Montanarella, L., Moreira, F. M. S., Ramirez, K. S., Scheu, S., Singh, B. K., Six, J., van der Putten, W. H., & Wall, D. H. (2016) *Global soil biodiversity atlas*. Luxembourg: Publications Office of the European Union. Available at http://esdac.jrc.ec.europa.eu/content/global-soil-biodiversity-atlas. Accessed Apr 2017.

Palaniyandi, S. A., Yang, S. H., Zhang, L., & Suh, J. W. (2013). Effects of actinobacteria on plant disease suppression and growth promotion. *Applied Microbiology and Biotechnology, 97*, 9621–9636.

Paul, D., Pandey, G., Meier, C., Roelof Van Der Meer, J., & Jain, R. K. (2006). Bacterial community structure of a pesticide-contaminated site and assessment of changes induced in community structure during bioremediation. *FEMS Microbiology Ecology, 57*, 116–127.

Philippot, L., Ritz, K., Pandard, P., Hallin, S., & Martin-Laurent, F. (2012). Standardisation of methods in soil microbiology: Progress and challenges. *FEMS Microbiology Ecology, 82*, 1–10.

Philippot, L., Raaijmakers, J. M., Lemanceau, P., & van der Putten, W. H. (2013). Going back to the roots: the microbial ecology of the rhizosphere. *Nature Reviews Microbiology, 11*, 789–799.

Pino, N., & Peñuela, G. (2011). Simultaneous degradation of the pesticides methyl parathion and chlorpyrifos by an isolated bacterial consortium from a contaminated site. *International Biodeterioration and Biodegradation, 65*, 827–831.

Pulleman, M. M., Creamer, R., Hamer, U., Helder, J., Pelosi, C., Pérès, G., & Rutgers, M. (2012). Soil biodiversity, biological indicators and soil ecosystem services—An overview of European approaches. *Current Opinion in Environmental Sustainability, 4*, 529–538.

Raaijmakers, J. M., Weller, D. M., & Thomashow, L. S. (1997). Frequency of antibiotic-producing *Pseudomonas* spp. in natural environments. *Applied and Environmental Microbiology, 63*, 881–887.

Reinhold-Hurek, B., Bünger, W., Burbano, C. S., Sabale, M., & Hurek, T. (2015). Roots shaping their microbiome: Global hotspots for microbial activity. *Annual Review of Phytopathology, 53*, 403–424.

Ritz, K., Black, H. I. J., Campbell, C. D., Harris, J. A., & Wood, C. (2009). Selecting biological indicators for monitoring soils: A framework for balancing scientific and technical opinion to assist policy development. *Ecological Indicators, 9*, 1212–1221.

Roesch, L. F., Fulthorpe, R. R., Riva, A., Casella, G., Hadwin, A. K., Kent, A. D., Daroub, S. H., Camargo, F. A., Farmerie, W. G., & Triplett, E. W. (2007). Pyrosequencing enumerates and contrasts soil microbial diversity. *The ISME Journal, 1*, 283–290. doi:10.1038/ismej.2007.53.

Rutgers, M., van Wijnen, H. J., Schouten, A. J., Mulder, C., Kuiten, A. M. P., Brussaard, L., & Breure, A. M. (2012). A method to assess ecosystem services developed from soil attributes with stakeholders and data of four arable farms. *Science of The Total Environment, 415*, 39–48.

Schloss, P. D., Girard, R. A., Martin, T., Edwards, J., & Thrash, J. C. (2016) Status of the archaeal and bacterial census: an update. *mBio*, 7, e00201-16.

Sipilä, T. P., Keskinen, A. K., Akerman, M. L., Fortelius, C., Haahtela, K., & Yrjälä, K. (2008). High aromatic ring-cleavage diversity in birch rhizosphere: PAH treatment-specific changes of I.E.3 group extradiol dioxygenases and 16S rRNA bacterial communities in soil. *The ISME Journal, 2*, 968–981.

Spohn, M. (2016). Element cycling as driven by stoichiometric homeostasis of soil microorganisms. *Basic and Applied Ecology, 17*, 471–478.

Thijs, S., Sillen, W., Rineau, F., Weyens, N., & Vangronsveld, J. (2016). Towards an enhanced understanding of plant-microbiome interactions to improve phytoremediation: Engineering the metaorganism. *Frontiers in Microbiology, 7*, 341.

Turbé, A., De Toni, A., Benito, P., Lavelle, P. A., Lavelle, P. E., Ruiz, N., Van der Putten, W. H., Labouze, E., & Mudgal, S. (2010). *Soil biodiversity: Functions, threaths and tools for policy makers*. Bio Intelligence Service, IRD, and NIOO, Report for European Commission (DG Environment), 254 pp.

US EPA. (1999). United States Environmental Protection Agency. Use of monitored natural attenuation at superfund, RCRA corrective action, and underground storage of tank sites. In: US Environmental Protection, Office of solid waste and emergency response. OSWER Directive 9200, 4–17. Available at https://www.epa.gov/sites/production/files/2014-02/documents/d9200.4-17.pdf. Accessed April 2017.

Vida, C., Bonilla, N., De Vicente, A., & Cazorla, F. M. (2016). Microbial profiling of a suppressiveness-induced agricultural soil amended with composted almond shells. *Frontiers in Microbiology, 7*, 1–14.

Vogel, T. M., Simonet, P., Jansson, J. K., Hirsch, P. R., Tiedje, J. M., van Elsas, J. D., Bailey, M. J., Nalin, R., & Philippot, L. (2009). TerraGenome: A consortium for the sequencing of a soil metagenome. *Nature Reviews Microbiology, 7*, 252.

Wagg, C., Bender, S. F., Widmer, F., & van der Heijden, M. G. A. (2014). Soil biodiversity and soil community composition determine ecosystem multifunctionality. *Proceedings of the National Academy of Sciences, USA, 111*, 5266–5270.

Walthall, C. L., Hatfield, J., Backlund, P., Lengnick, L., Marshall, E., Walsh, M., Adkins, S., Aillery, M., Ainsworth, E.A., Ammann, C., Anderson, C.J., Bartomeus, I., Baumgard, L.H., Booker, F., Bradley, B., Blumenthal, D.M., Bunce, J., Burkey, K., Dabney, S. M., Delgado, J. A., Dukes, J., Funk, A., Garrett, K., Glenn, M., Grantz, D. A., Goodrich, D., Hu, S., Izaurralde, R. C., Jones R. A. C., Kim, S- H., Leaky, A. D. B., Lewers, K., Mader, T. L., McClung, A., Morgan, J., Muth, D. J., Nearing, M., Oosterhuis, D. M., Ort, D., Parmesan, C., Pettigrew, W. T., Polley, W., Rader, R., Rice, C., Rivington, M., Rosskopf, E., Salas, W. A., Sollenberger, L. E., Srygley, R., Stöckle, C., Takle, E. S., Timlin, D., White, J. W., Winfree, R., Wright-Morton, L., & Ziska, L. H. (2012). *Climate change and agriculture in the United States: Effects and adaptation*. USDA Technical Bulletin 1935, Washington, DC, pp. 1–186. Available at https://www.usda.gov/oce/climate_change/effects_2012/CC%20and%20Agriculture%20Report%20(02-04-2013)b.pdf. Accessed Apr 2017.

Weller, D. M., Raaijmakers, J. M., Gardener, B. B. M., & Thomashow, L. S. (2002). Microbial populations responsible for specific soil suppressiveness to plant pathogens. *Annual Review of Phytopathology, 40*, 309–348.

Widder, S., Allen, R. J., Pfeiffer, T., Curtis, T. P., Wiuf, C., Sloan, W. T., Cordero, O. X., Brown, S. P., Momeni, B., Shou, W., Kettle, H., Flint, H. J., Haas, A. F., Laroche, B., Kreft, J. U., Rainey, P. B., Freilich, S., Schuster, S., Milferstedt, K., van der Meer, J. R., Groβ Kopf, T., Huisman, J., Free, A., Picioreanu, C., Quince, C., Klapper, I., Labarthe, S., Smets, B. F., Wang, H., Isaac Newton Institute Fellows, & Soyer, O. S. (2016). Challenges in microbial ecology:

Building predictive understanding of community function and dynamics. *The ISME Journal, 10*, 2557–2568.
Wubs, E. R. J., van der Putten, W. H., Bosch, M., & Bezemer, T. M. (2016). Soil inoculation steers restoration of terrestrial ecosystems. *Nature Plants, 2*, 16107.
Yagi, J. M., Suflita, J. M., Gieg, L. M., Derito, C. M., Jeon, C. O., & Madsen, E. L. (2010). Subsurface cycling of nitrogen and anaerobic aromatic hydrocarbon biodegradation revealed by nucleic acid and metabolic biomarkers. *Applied and Environmental Microbiology, 76*, 3124–3134.

Chapter 3
Comparison of Two Molecular Methods to Assess Soil Microbial Diversity

Martina Di Lenola, Paola Grenni, Diogo N. Proença, Paula V. Morais, and Anna Barra Caracciolo

Abstract Our knowledge of microbial soil biodiversity depends on the ability to combine different observation levels, ranging from the phenotypic to the molecular one, including direct visualization of microorganisms by using an epifluorescence microscope. Soil microbial communities, although still largely undiscovered, represent one of the biggest part of present biodiversity and play a key role in all soil processes. For example, microbial abundance, activity and composition largely determine the sustainable productivity of agricultural land or the degradation of organic pollutants. The diversity of microbial communities associated with plant roots is enormous, in the order of thousands of species. The simultaneous use of different molecular methods makes it possible to have a more holistic knowledge of the soil microbial community and to assess changes in it under different conditions. In this chapter we report the application to the same soil samples of two different phylogenetic molecular techniques, i.e. DGGE (Denaturing Gradient Gel Electrophoresis) and FISH (Fluorescence *In Situ* Hybridization). DGGE is a useful fingerprint technique of the overall microbial community based on amplification of 16S rRNA. The FISH technique identify, without extracting nucleic acids, active microbial cells at different phylogenetic levels (from domains to species) under an epifluorescence microscope. An example of the application of both methodologies to assessing the microbial community composition of soil samples from a phyto-assisted bioremediation experiment of a contaminated soil by polychlorinated biphenyls (PCBs) is reported.

M. Di Lenola • P. Grenni • A. Barra Caracciolo (✉)
National Research Council, Water Research Institute, CNR-IRSA,
Monterotondo, Rome, Italy
e-mail: barracaracciolo@irsa.cnr.it

D.N. Proença
CEMMPRE- University of Coimbra, Rua Luís Reis Santos, 3030-788 Coimbra, Portugal

P.V. Morais
CEMMPRE- University of Coimbra, Rua Luís Reis Santos, 3030-788 Coimbra, Portugal

Deparment of Life Sciences, FCTUC (Faculty of Sciences and Technology), University of Coimbra, Coimbra, Portugal

M. Lukac et al. (eds.), *Soil Biological Communities and Ecosystem Resilience*,
Sustainability in Plant and Crop Protection, DOI 10.1007/978-3-319-63336-7_3

Keywords Denaturing Gradient Gel Electrophoresis-DGGE • Fluorescence *In Situ* Hybridization-FISH • Bacterial 16S rRNA • Phyto-assisted bioremediation • Polychlorinated biphenyls (PCBs)

3.1 Introduction

Soil is a very complex system comprising a variety of microhabitats with different physicochemical gradients and discontinuous environmental conditions. Microorganisms adapt to microhabitats and live together in consortia with sharp boundaries, interacting with each other and with other parts of the soil biota (Wolińska et al. 2016). Torsvik et al. (1990a, b) estimated that in 1 g of soil there are 4,000 different bacterial 'genomic units' based on DNA–DNA re-association. Soil microorganisms are vital for the continued cycling of nutrients and for driving above-ground ecosystems (van der Heijden et al. 1998; Cairney 2000; Klironomos et al. 2000; Ovreas 2000). Soil biodiversity has been assumed to influence ecosystem stability, productivity and resilience towards stress and disturbance. Microbiota can directly regulate the structure and functioning of aboveground individuals and communities, by stimulating or inhibiting certain plant species more than others (Berendsen et al. 2012). Alternatively, plants can regulate aboveground communities indirectly by altering the dynamics of nutrients available to microorganisms (Wardle et al. 2004; Purahong et al. 2016). These indirect effects tend to involve less specific interactions and occur over longer durations than the direct regulations.

Research on microbial diversity should include multiple methods integrating holistic measurements at the total community level and partial approaches targeting structural or functional subsets (Kozdrój and van Elsas 2001; Johnsen et al. 2001).

Although there are several methods for studying microbial diversity, including *in situ* activity measurements and the most recent ones such as high-throughput sequencing, DNA microarrays and functional genomics, which provide huge amounts of new data, the association between diversity and functioning is nevertheless still not well known.

Soil degradation can be defined as a change in the soil health status resulting in a diminished capacity of the ecosystem to provide goods and services for its beneficiaries (Turbè et al. 2010; Tetteh 2015). Soil degradation is responsible for soil biodiversity loss and in turn for the disappearance of many ecosystem functions, thus altering plant productivity, soil resilience and resistance capacity. Anthropogenic pressures such as chemical pollution and agriculture practices can adversely affect microbial diversity and consequently above and belowground ecosystem functioning. Any loss in microbial diversity as a consequence of environmental changes is likely to alter the capacity of microbes to sustain multiple above- and belowground ecosystem functions (Delgado-Baquerizo et al. 2016).

Nucleic acid hybridization using specific oligonucleotide or polynucleotide probes, designed from known sequences ranging in specificity from domain to species and tagged with markers at the 5′-end, are an important qualitative and quantitative molecular microbial ecology tool for identifying microbial cells (Theron and Cloete 2000). The hybridization techniques can be performed either by using extracted DNA or RNA (PCR-based methods) or *in situ* (Fluorescence *In Situ* Hybridization). The latter provides valuable spatial distribution information on microorganisms in environmental samples.

The PCR method uses specific primers to amplify a DNA target sequence. Sequence variations in PCR fragments are detected either by a cloning/sequencing analysis, which produce a complete characterization of the fragments, or by an electrophoretic analysis, which provides a visual separation of the mixture of fragments. This separation is based on sequence polymorphism, in Denaturing Gradient Gel Electrophoresis (DGGE) or length polymorphism, Terminal-Restriction Fragment Length Polymorphism (T-RFLP) and Automated Ribosomal Intergenic Spacer Analysis (ARISA). DGGE is frequently used in environmental studies (Ibekwe et al. 2001; Guo et al. 2009). Although PCR is a very reliable technique, it has some limitations. In fact, although it is a highly sensitive technique, any sample contamination by even trace amounts of DNA can produce misleading results. Moreover, lysis efficiency of cells varies between and within microbial groups (Prosser 2002). The ability to separate bacteria from soil aggregates is crucial when studying biodiversity (Barra Caracciolo et al. 2005). For example, if the cell extraction method from particles used is too gentle, Gram-negative, but not Gram-positive bacterial cells are lysed. If the method is too harsh, both Gram-negative and Gram-positive cells are lysed but their DNA becomes sheared. The DNA extraction method used can also bias diversity studies. In fact, different methods of nucleic acid extraction result in different yields of nucleic acids product (von Wintzingerode et al. 1997). Moreover, with environmental samples such as soil, it is necessary to remove inhibitory substances such as humic acids, which can interfere with the PCR analysis. Subsequent purification steps can lead to loss of DNA, again potentially biasing molecular diversity analysis (Kirk et al. 2004). Differential target gene amplification can also bias PCR-based diversity studies. The DNA polymerase can be related to unspecific annealing of primers and to an incorrect nucleotide incorporation into the PCR sequence (Garibyan and Avashia 2013).

The use and combination of different molecular methods is necessary for understanding, in a complementary way, different aspects of the huge biodiversity in soil. In this chapter the analysis of PCB contaminated soil samples using two different phylogenetic molecular techniques, i.e. the PCR-based method, Denaturing Gradient Gel Electrophoresis (DGGE), and the epifluorescence microscope Fluorescence *In Situ* Hybridization (FISH) method, will be illustrated and discussed.

3.1.1 Denaturing Gradient Gel Electrophoreses

Denaturing Gradient Gel Electrophoresis (DGGE) is a well-known technique used in molecular microbial ecology for characterizing population structure and dynamics. It is a very common and easy fingerprinting technique for the fast evaluation of microbial genetic diversity (*Bacteria* and *Archaea*) in complex ecosystems such as soils, sediments, rivers, plants and biofilms (Ng et al. 2014; Proença 2014; Zhang et al. 2017).

PCR-DGGE analysis is a multi-step procedure that requires the total DNA extraction from a sample and PCR amplification with specific primers of particular DNA regions of taxonomic interest. The amplified product obtained is a mixture of amplicons from the species present in the initial sample which are separated on the basis of their sequence differences (DGGE) by gel electrophoresis; the patterns obtained are then analysed. It is a rapid and efficient separation technique of same length DNA sequences, which may vary as little as a single base pair modification. Separation is based on the decreased electrophoretic mobility of a partially melted double-stranded DNA molecule in polyacrylamide gels containing a linear gradient of DNA denaturants (a mixture of urea and formamide) or a linear temperature gradient. Identification of the species can be achieved by purifying and sequencing the bands in the DGGE profile (Salles et al. 2004). Before the sequencing of the distinct bands, it is possible to visualize the band profiles of each sample, compare the similarity in band composition between samples and thus analyse the changes occurring in the various microbial communities when undergoing different treatments or modifications (Fig. 3.1).

DGGE analyses separate double-stranded DNA fragments, produced via PCR amplification, which are identical in length, but differ in sequence. The technique mainly exploits the difference in stability of G-C pairing (3 hydrogen bonds per pairing) as opposed to A-T pairing (2 hydrogen bonds). A mixture of DNA fragments of different sequences is electrophoresed in an acrylamide gel containing an increasing gradient of DNA denaturants. By adjusting the primers used for amplification, both the major and minor constituents of microbial communities can be characterized. In general, DNA fragments richer in G-C content are more stable and remain double-stranded until reaching higher denaturant concentrations. Double-stranded DNA fragments migrate better in the acrylamide gel, while denatured DNA molecules in fact become larger (only attached to the GC clamp) and slow down and stop in the gel. In this way, DNA fragments of differing sequences can be separated in an acrylamide gel. The inclusion of DGGE markers on gel is necessary for comparing gels and for assessing gradient variations. Ideally, a marker should be chosen with a sufficient number of bands to span the entire gradient, since there is variation within a gradient.

It is possible to state qualitatively that two communities are different if the blurring of bands looks different; however, the contrary may not be true. The number of bands observed in a profile does not exactly correspond to the number of populations in a community. Several studies found that treatments that selectively enriched specific populations resulted in community structure differences that were visible in the PCR-DGGE profiles (Nakatsu et al. 2005; Saeki and Toyota 2004). The interpretation of profiles needs caution because the disappearance of bands does not always correspond to the complete disappearance of a species from the community. In fact, it may only reflect a change in the relative densities between populations within the community, where the increase in some populations puts other populations below the

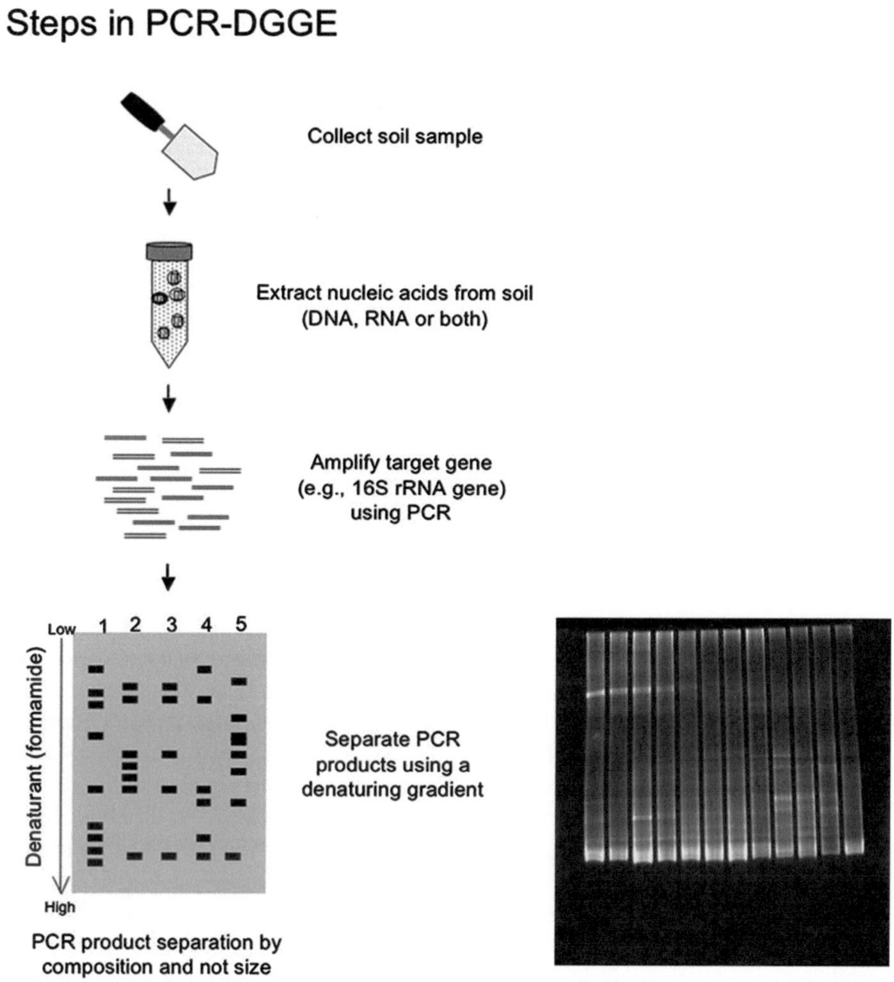

Fig. 3.1 Flow diagram of the steps for microbial community analysis using polymerase chain reaction (PCR)–denaturing gradient gel electrophoresis (DGGE), (Nakatsu 2007, modified). On the right an example of a DGGE gel showing band compositions (profiles) of various population samples, representative of complex microbial ecosystems. Each band in each lane represents a 16S rRNA gene amplified bacterial product migrating to a unique position in the gel, which melts in a sequence dependent manner

DGGE detection limit. In some cases, researchers have found that a single laboratory isolate can produce multiple bands with DGGE while, conversely, a single band may represent multiple populations (Yang and Crowley 2000; Nakatsu 2007).

Given that, the DNA preparation and amplification are crucial steps in a PCR-based method, as they represent the biggest source of bias. At the same time, the purity and quantity of the template DNA is crucial for efficient and reproducible PCR-DGGE analysis. DNA extraction efficiency greatly depends on the sample, as well as on the extraction procedure. Additionally, soil and environmental samples usually contain humic acids or other potential PCR inhibitors and the DNA obtained often needs further improvement in quality. On the other hand, efforts to remove PCR inhibitors for producing high-quality template decrease the quantity of DNA, which can be finally reflected in the variation of the PCR-DGGE pattern. Soil and other environmental samples thus represent challenging targets for PCR-based fingerprinting analysis (Balázs et al. 2013).

All traditional DNA-sequencing methods (e.g. DGGE, T-RFLP, ARISA) used to identify the organisms present in the profiles can only sequence specimens individually. They are therefore inadequate for processing complex environmental samples containing mixtures of DNA from hundreds or thousands of individuals and unable to read DNA from multiple templates at the same time. There are currently better tools for obtaining the information on the organisms that are in each profile, such as the next-generation sequencing (NGS) technologies which are costing ever less (Shokralla et al. 2012), although DGGE can be a good starting point.

3.1.2 Fluorescent In Situ *Hybridization*

The non-PCR-based methods (without DNA extraction) commonly used in environmental studies for quantifying microbial cell number and their phylogenetic characterization are epifluorescence microscope ones, such as total direct counts of microbial abundance (e.g. by DAPI stain), cell viability (Live/Dead assay) and Fluorescence *In Situ* Hybridization (FISH). They make it possible to characterize microbial populations *in situ* in their natural ecosystems (Grenni et al. 2009; Barra Caracciolo et al. 2010; Barra Caracciolo et al. 2015), allowing the detection of culturable and unculturable microorganisms and, can therefore help in understanding complex microbial communities. In particular, FISH combines the precision of molecular genetics with the direct visual information from microscopy, allowing simultaneous visualization, identification, enumeration and localization of individual microbial cells within their natural microhabitat. Because whole cells are hybridized, errors arising from biases in DNA extraction, PCR amplification and cloning are avoided (Amann and Fuchs 2008).

Fluorescent *In Situ* Hybridization is a well-established and reliable technique used to detect specific microorganisms at different hierarchical phylogenetic levels in both microbial cultures and environmental samples. FISH is used to phylogenetically detect microbial cells using fluorescent labelled oligonucleotides complementary to 16S or 23S ribosomal RNA (rRNA). Oligonucleotide probes are designed based on signature nucleotide positions in the bacterial 16S or 23S rRNA and may be used to target either a narrow or a broad group of organisms. Probes for kingdoms (*Bacteria*, *Archaea*), families, genera, species or strains can be differentially detected with different labelled oligonucleotide probes; the latter can be used in combination for simultaneously detecting the occurrence and spatial distribution of several taxonomic groups within a single soil sample (Grenni et al. 2009). Fluorescent dyes can be used or multiple probes can be designed to target different regions of the same 16S or 23S rRNA molecule, thus increasing the strength of the signal.

High levels of rRNA (equivalent to high ribosome numbers), which result in high detection signals with the FISH technique, are observed in active populations of bacteria in which protein synthesis is occurring either in non-dividing or dividing cells. The whole-cell identification of microorganisms together with information on the morphology of active cells is achieved quickly and with high specificity.

During the past few years different protocols have been developed in order to improve the sensitivity of FISH and to permit the detection of single active cells. Although there are some specific differences in protocols designed for water, soil and sediment (Barra Caracciolo et al. 2005; Barra Caracciolo and Grenni 2010), the methodology has the following basic steps: fixation and permeabilisation (with the aim of ensuring a proper penetration of oligonucleotide probes inside cells), hybridization of the probe with a rRNA complementary site, removal of nonspecifically bound probes and fluorescent detection of cells using for example an epifluorescence microscopy (Figure 3.2).

There are major steps in probe design: identifying short regions (usually 15–25 nucleotides in length) in a sequence alignment unique to the target group; keeping mismatches exclusively to non-target organisms (where possible), and modifying the sequence to meet probe design criteria such as a minimum melting temperature (Hugenholtz et al. 2002). The inaccessibility of probe target sites, mainly depending on the thermodynamic affinity of the probe to the target site, and the non-permeability of cell walls (especially Gram-positive bacteria) can be overcome by the application of pre-treatment protocols (Bottari et al. 2006). There is a constantly growing on-line data base available publicly (probeBase, http://probebase.csb.univie.ac.at) which contains the specification of more than 1200 probes and their hybridization conditions for use in a variety of environmental samples (Greuter et al. 2016).

The most commonly applied probes recommended for identification of higher taxa in soil, sediment and water microbial communities are presented in Table 3.1.

3.2 Example of Application of DGGE and Fish Techniques to Contaminated Soil Samples

3.2.1 *Experimental Set-Up*

The experiment aimed to restore a PCB-contaminated soil using a municipal compost and the leguminous plant *Medicago sativa*. The addition to soil of organic matter in the form of compost and the restoration of vegetation can be suitable strategies for improving soil quality and promoting recovery from contamination (Ancona et al. 2017; Turbè et al. 2010). Compost and a plant presence can act directly and indirectly to increase microbial diversity and functioning, thus stimulating contaminant biodegradation. DGGE and FISH were applied to soil samples in order to assess how the treatments affected the natural microbial community structure. The main results are illustrated and compared below.

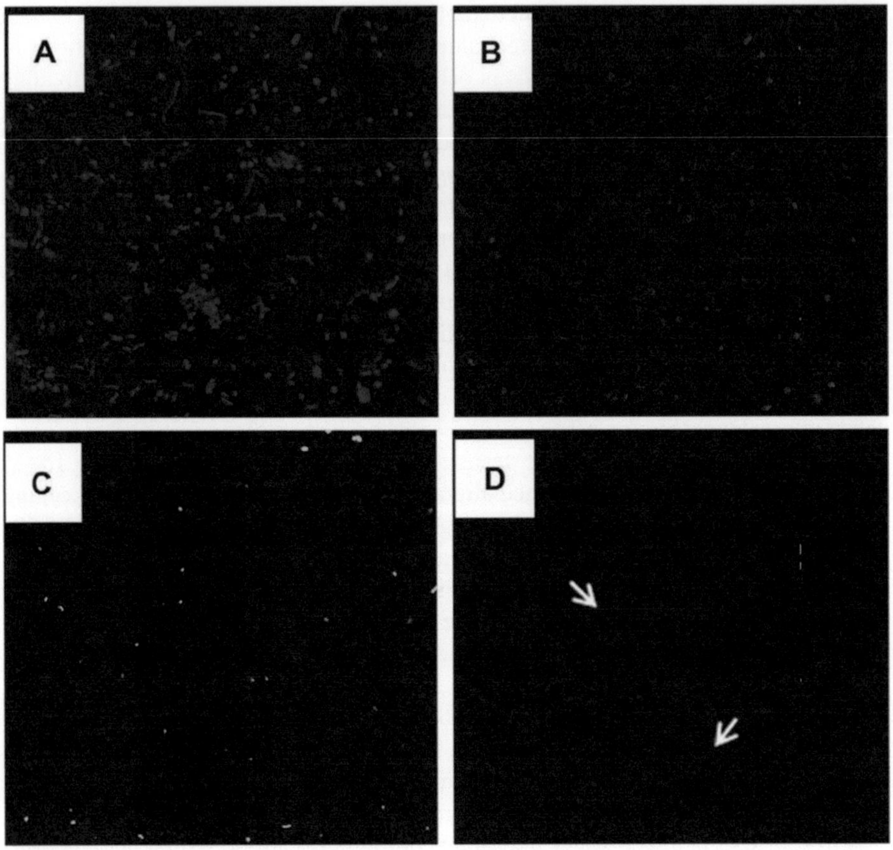

Fig. 3.2 Images of FISH assays under the laser confocal microscope. (**a**, **b**) pure culture of a bacterial strain (*Rhodococcus wratislaviensis* FPA1) capable of mineralizing the terbuthylazine pesticide: (**a**) DAPI-stained cells (*blue*); (**b**) *Rhodococcus wratislaviensis* FPA1 positive cells hybridized with a specific FISH probe (RhLu-probe) (*red*). (**c**, **d**) Detection of *Rhodococcus wratislaviensis* FPA1 in soil samples. (**c**) FAM-labelled EUB338 probe (*green*); (**d**) RhLu-positive bacteria. (From: Grenni et al. 2009)

The experiment consisted of soil laboratory microcosms containing samples collected from a PCB historically contaminated area. In order to simulate an accidental spill, soil samples were treated with Apirolio, a dielectric oil produced in Italy up to 1970 and mainly used in electric transformers. Each microcosm consisted of a polyethylene plant pot (10 × 11 cm, 1 L capacity) filled with an equal proportion (800 g) of PCB-treated soil (Apirolio, 100 mg/kg) or soil amended with Apirolio and compost (Apirolio+Compost) at 26 t/ha (corresponding to an agronomic dose). The 100 mg/kg concentration of Apirolio was obtained pouring it onto an extremely small solid particles of certified sand, which was subsequently dispersed in the appropriate amount of soil. The historically contaminated soil (HCS) was used as the control. Eight seeds of *Medicago sativa* were added to half the microcosms of the entire experimental set. In the end, the experimental

Table 3.1 Most common oligonucleotide probes used for microbial identification in water, sediment and soil samples

Probe and reference	Taxa	Sequence (5′-3′)	rRNA position
ARCH915 (1)	***Archaea***	GTGCTCCCCCGCCAATTCCT	16S(915–934)
EUB338 (2)	***Bacteria***	GCTGCCTCCCGTAGGAGT	16S(338–355)
EUB338II (3)	***Bacteria***	GCAGCCACCCGTAGGTGT	16S(338–355)
EUB338III (3)	***Bacteria***	GCTGCCACCCGTAGGTGT	16S(338–355)
ALF1b (4)	***α-Proteobacteria***	CGTTCGYTCTGAGCCAG	16S(19–35)
ALF968 (5)	***α-Proteobacteria***	GGTAAGGTTCTGCGCGTT	16S(968–985)
BET42a (4)	***β-Proteobacteria***	GCCTTCCCACTTCGTTT	23S(1027–1043)
GAM42a (4)	***γ-Proteobacteria***	GCCTTCCCACATCGTTT	23S(1027–1043)
DELTA495a (6)	***δ-Proteobacteria***	AGTTAGCCGGTGCTTCCT	16S(495–512)
DELTA495b (6)	***δ-Proteobacteria***	AGTTAGCCGGCGCTTCCT	16S(495–512)
DELTA495c (6)	***δ-Proteobacteria***	AATTAGCCGGTGCTTCCT	16S(495–512)
EPS710 (7)	***ε-Proteobacteria***	CAGTATCATCCCAGCAGA	16S(710–727)
SRB385 (2, 8)	***Desulfovibrionales* and other *Bacteria***	CGGCGTCGCTGCGTCAGG	16S(385–402)
EBAC1790 (9)	***Enterobacteriaceae***	CGTGTTTGCACAGTGCTG	23S(1790–1807
PLA46 (10)	***Planctomycetes***	GACTTGCATGCCTAATCC	16S(46–63)
PLA886 (10)	***Planctomycetes***	GCCTTGCGACCATACTCCC	16S(886–904)
CF319a (11)	***Bacteroidetes***	TGGTCCGTGTCTCAGTAC	16S(319–336)
LGC354a (12)	***Firmicutes***	TGGAAGATTCCCTACTGC	16S(354–371)
LGC354b (12)	***Firmicutes***	CGGAAGATTCCCTACTGC	16S(354–371)
LGC354c (12)	***Firmicutes***	CCGAAGATTCCCTACTGC	16S(354-371)
HGC69a (13)	***Actinobacteria***	TATAGTTACCACCGCCGT	23S(1901–1918)

(*1*) Stahl and Amann (1991), (*2*) Amman et al. (1990), (*3*) Daims et al. (1999), (*4*) Manz et al. (1992), (*5*) Neef (1997), (*6*) Loy et al. (2002), Lücker et al. (2007), (*7*) Watanabe et al. (2000), (*8*) Manz et al. (1998), (*9*) Bohnert et al. (2000), (*10*) Neef et al. (1998), (*11*) Manz et al. (1996), (*12*) Meier et al. (1999), (*13*) Roller et al. (1994)

set consisted of eight different conditions: HCS, Apirolio, HCS+Compost, Apirolio+Compost, HCS+Plant, Apirolio+Plant, HCS+Compost+Plant, Apirolio+Compost+Plant. Each condition was performed in three replicates (24 microcosms in total). All the microcosms were kept in a heated greenhouse under controlled temperature (20 ± 2 °C) for more than seven months (224 days), Figs. 3.3 and 3.4.

The pots were watered regularly and the soil maintained at about 60% of its field capacity throughout the experiment. The microcosms were sampled one day after the experimental set up (1 d) and at 224 days when the *Medicago sativa* roots filled the whole soil in each pot. Each microcosm was collected destructively, the soil was homogenized and immediately used for microbial analysis (DGGE; FISH). Other microbial and chemical analyses (total microbial abundance, cell viability, dehydrogenase activity, Ester-Linked Fatty Acids and PCBs) were performed but the results are not reported in this chapter. The data reported here are exclusively the results of DGGE and FISH applied to soil samples collected at days 1 and 224.

Fig. 3.3 Microcosms with *Medicago sativa* plant in the early stages of the experiment (*left*) and at the end of the experiment (*right*) during the destructive sampling

3.2.2 Microbial Community Structure Evaluated by DGGE

DNA was extracted from soil samples by using the E.Z.N.A.® Soil DNA kit D5626-01 (Omega Bio-Tek), in accordance with the manufacturer's instructions. DNA was used as the template for DGGE analysis of the 16S rRNA gene sequences from *Bacteria* and *Archaea*. The PCR reaction mix (50 μL) contained: reaction buffer (1.5 mM $MgCl_2$; 50 mM KCl and 10 mM TrisHCl; pH 8.3), 100 μM (each) deoxynucleoside triphosphates, 0.2 μM (each) primer, 1.5 U Taq polymerase (Bioline) and 6 μL of purified DNA solution as the template. The 16S rRNA amplicons, obtained by using primer pairs 27F and 1525R for the first PCR, were subsequently used as the DNA template for Nested-PCR by using the 341F-GC clamp and the 907R primers set (Muyzer et al. 1993; Santos et al. 2013; Proença 2014). Analysis of *Archaea* was performed by using the 21F and 958R primers for the first PCR, followed by Nested-PCR with an Arch915R-GC clamp and an Arch519F primers set (Vissers et al. 2009; Proença 2014). After the first PCR, the PCR products were analysed using 1% agarose gel electrophoresis to verify correct amplification of the desired fragment, using a 100 bp DNA ladder (NZYTECH DNA Ladder III) as a molecular weight marker. In samples where amplification was not obtained, PCR amplifications were repeated using half of the purified DNA solution used before (3 μL) as a template. In this way, we obtained bands with expected weight in all the samples and we could proceed with Nested-PCR.

DGGE was performed using a DCode™ Universal Mutation Detection System (Bio-Rad Laboratories, Hercules, California, USA). PCR samples were loaded into

Fig. 3.4 Picture of the experimental set. *HCS:* PCB historically contaminated soil, *HCS+Compost:* PCB historically contaminated soil + compost, *Apirolio:* PCB historically contaminated soil + Apirolio, *Apirolio+Compost:* PCB historically contaminated soil + compost + Apirolio. In the upper part of the picture, the same conditions with plant (*Medicago sativa*) are showed

8% polyacrylamide gels with a denaturing gradient ranging 30–70% (100% denaturant is defined as 7 M urea and 40% formamide). Gels were run at 70 V, for 17 h, at 60 °C and stained with ethidium bromide (Santos et al. 2013). The gels were photographed and then the similarity between the various microbial communities was compared based on the digitized DGGE profiles, using the cluster analysis technique in Quantity One 4.6.6 (Bio-Rad). The bands in each lane were manually detected and compared, allowing matching profiles to be generated. Matching profiles for each lane were used to produce a dendrogram by employing the unweighted pair group method using arithmetic averages (UPGAMA). The results are expressed as dendrograms. The two dendrograms obtained, one for *Archaea* and the other for *Bacteria*, provided information about the similarities between the microbial communities present in each sample.

3.2.3 Microbial Community Structure Evaluated by FISH

One gram (three replicates for each microcosm) of soil was placed in a sterile 10 ml test tube and fixed with 9 ml of filter-sterilized PBS (phosphate buffered saline; 130 mM NaCl; 7 mM Na_2HPO_4 * $12H_2O$; 3 mM NaH_2PO_4 * $2H_2O$) containing 0.5%

Tween 20, 2% formaldehyde, 1 g/L sodium pyrophosphate with a final pH of 7.4. Prior to FISH, an additional cell purification step was performed with a density gradient medium in order to detach cells from soil particles, as described in detail in Barra Caracciolo et al. (2005). Fluorescence *In Situ* Hybridization of the harvested cells, counter stained with DAPI, was performed using published protocols (Pernthaler et al. 2001; Barra Caracciolo et al. 2005).

Fluorescent Cy3-labelled oligonucleotide probes were used for the identification, under the epifluorescence microscope, of the *Archaea* and *Bacteria* domains. Within the *Bacteria* domain the major groups usually found in soil such as α-, β-, γ-, δ- *Proteobacteria* and the *Cytophaga-Flavobacterium* lineage of the *Bacteroidetes* (CFB group) were searched for. The following oligonucleotide probes were used: Arch915 (for *Archaea*), EUB338I-III (for *Bacteria*) and inside this domain ALF1b, BET42a, GAM42a, DELTA495a,b,c (for α-, β-, γ-, δ-, *Proteobacteria*, respectively), CF319 (for the *Cytophaga-Flavobacterium* lineage of the *Bacteroidetes*), HGC (Gram-positive Bacteria with high DNA G+C), LGC (Gram-positive Bacteria with low DNA G+C). All probes were synthesized by MWG AG Biotech, Germany. Further details of these probes are available at http://probebase.csb.univie.ac.at (Greuter et al. 2016). The slides were mounted with drops of Vectashield Mounting Medium (Vector Laboratories) and the preparations were examined and counted with a Leica DM LB 30 epifluorescence microscope at 1000 magnification, counting a minimum of 300 cells per section. The estimation of cells binding to the fluorescent probes is calculated as a percentage of the total DAPI positive cells.

3.2.4 Results and Comparison of the Two Microbiological Methods

In Fig. 3.5 the results of the two methods applied at the start (1 d) and at the end (224 days) of the experiment to identify *Bacteria* are shown.

In Fig. 3.6, the results (in the form of a dendrogram in the case of DGGE or expressed as percentage of the total DAPI positive cells in the case of FISH) of the two methods applied at the start (1 d) and at the end (224 days) of the experiment to identify *Archaea* are shown.

As is possible to see from the DGGE dendrogram for *Bacteria* (Fig. 3.5, left side) the lowest similarity (0.5) was found in the HCS 1d in line with the fact that it is the soil before the treatments. On the contrary, the highest similarity (>0.95) was observed between the Apirolio+Plant and HCS+Compost+Plant conditions; a lower similarity (>0.85) was found between the Apirolio+Compost and Apirolio conditions. Interestingly, all the conditions without plant at day 224 are linked in a branch with about 80% of similarities. The latter result is in line with the FISH results (Fig. 3.5, right side), reflecting the very similar community structure of the samples without plant.

Both DGGE and FISH results showed that the microbial community structure was influenced by the initial handling and experimental conditions (e.g. watering, constant temperature, etc.), no matter what the treatment, and that the combination of plant and Apirolio and/or compost marked the differences between the bacterial

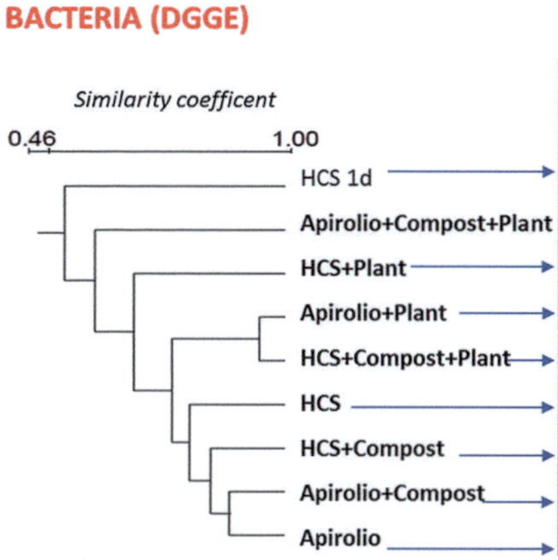

BACTERIA (FISH)

	EUB	Proteobacteria				CF	Others
		α	β	γ	δ		
HCS 1d	52%	3%	22%	2%	5%	5%	29%
Apirolio+Compost+Plant	62%	24%	7%	1%	0%	2%	45%
HCS+Plant	51%	25%	6%	3%	0%	3%	27%
Apirolio+Plant	58%	13%	3%	0%	0%	1%	71%
HCS+Compost+Plant	67%	20%	4%	3%	0%	2%	57%
HCS	39%	8%	0%	0%	1%	3%	69%
HCS+Compost	17%	3%	0%	0%	2%	0%	71%
Apirolio+Compost	23%	3%	0%	1%	3%	0%	70%
Apirolio	44%	0%	1%	0%	2%	1%	91%

Fig. 3.5 *Bacteria* detection in soil samples using the DGGE and FISH techniques at day 1 (HCS 1d) and at day 224 (in *bold*) of the experiment. DGGE results are represented as a dendrogram obtained from the 16S rRNA gene amplification by the 341F-GC clamp and 907R primers; FISH results are expressed as percentages of positives cells to specific oligonucleotide probes. *EUB: Bacteria*, *α:* α-*Proteobacteria*, *β*: β-*Proteobacteria*, *γ:* γ-*Proteobacteria*, *δ:* δ-*Proteobacteria*, *CF: Citophaga-Flavobacterium* of *Bacteroidetes*, *Others:* taxa not identified with the probes used. Soil treatments: *HCS:* PCB historically contaminated soil, *HCS+Compost:* PCB historically contaminated soil + compost, *Apirolio:* PCB historically contaminated soil + Apirolio, *Apirolio+Compost:* PCB historically contaminated soil + compost + Apirolio. The same conditions were analysed with and without a plant and at day 224 from the experimental set up

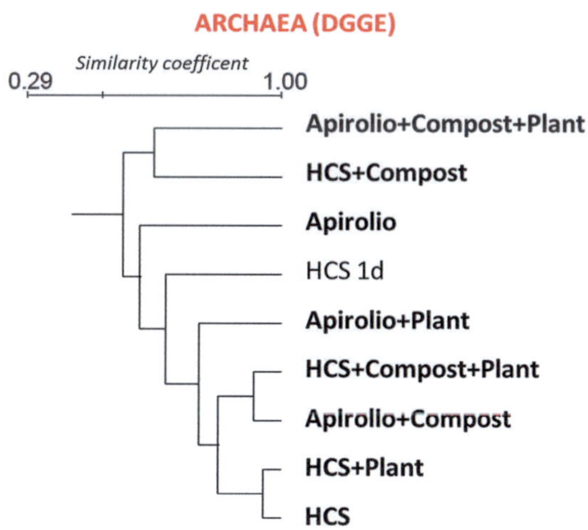

Fig. 3.6 *Archaea* detection by DGGE. Dendrogram obtained by cluster analysis for *Archaea* DGGE bands selected in DGGE profiles of soil archaeal community based on the 16S rRNA gene amplified by the 519F-GC clamp and 915R primers. Experimental conditions: *HCS:* PCB historically contaminated soil, *HCS+Compost:* PCB historically contaminated soil + compost, *Apirolio:* PCB historically contaminated soil + Apirolio, *Apirolio+Compost:* PCB historically contaminated soil + compost + Apirolio. The same conditions were analysed with (+Plant) and without plant and after 1 day (1 d) and 224 days (in *bold*) from the experiment set up

populations. Furthermore, the FISH analysis, which assessed the structure of the active microbial community, made it possible to investigate inside the bacterial taxon which was specifically affected by the treatments. In fact, at day 224 (Fig. 3.5, Table on the right side) the highest percentages of *Bacteria* were found when *Medicago sativa* was present and a shift in the dominance of several bacterial groups was observed. In particular, when comparing with the initial soil, α-*Proteobacteria* group significantly increased in the HCS+Plant, HCS+Compost+Plant and Apirolio+Compost+Plant conditions (Fig. 3.5, left side). The latter result is in line with the fact that the rhizosphere microbial populations had an availability of organic compounds, which stimulate their activity and diversity (Brimecombe et al. 2001; Nannipieri et al. 2008). *The Proteobacteria* phylum, which include most bacterial species involved in the main biogeochemical cycles, are typically the most abundant phylum found in a good quality state soil. As an example, *Rhizobiaceae*, which include both nitrogen-fixing and nitrifying bacteria such as *Nitrobacter*, belong to α-*Proteobacteria* (Brock et al. 2007). *Proteobacteria* dominance was observed in the rhizosphere in other experiments (Barra Caracciolo et al. 2015; Mocali et al. 2013; Buée et al. 2009; Singh et al. 2007).

The DGGE profiles for *Archaea* domain (Fig. 3.6) did not point to a clear cluster linking the different treatment effects, which were quite different inside each condition. While the DGGE method identified this taxon, FISH was not able to detect it, presumably because of its low activity.

3.3 Conclusions

There is a wide range of methods available for studying soil microbial diversity and each method has limitations and provides a partial picture of only one aspect of soil microbial diversity. On the contrary, the application of two or more techniques allows for a better understanding of the complex ecology of microbial communities. For example, the overall results of the experiment here reported show that the two techniques applied give information on soil microbial community at different levels.

DGGE is based on DNA extraction (viable and non-viable cells) and it is a screening technique to assess the genetic diversity of microbial communities or of particular populations without further characterization of the individual inhabitants. Nowadays, DGGE can be used as an initial technique for highlighting the main differences in a given microbial community and subsequently more specific ones to obtain a more detailed understanding of the microbiome composition.

With the FISH method (without DNA extraction and avoiding the bias due to it), the whole-cell identification of microorganisms is possible quickly and with high specificity (there are oligonucleotide probes from domain to species level). Although the information obtained are limited to the group investigated and there is a relatively low number of probes applicable, its great advantage is that it provides direct information on both the morphology of active cells and on their identity. This method is particularly useful for biomonitoring the population dynamics of microbial species of particular interest in their natural environment.

Acknowledgements This work was supported by COST Action FP1305 - Biolink: Linking belowground biodiversity and ecosystem function in European forests. Moreover, this study was partially supported by the "Fundação para a Ciência e a Tecnologia" postdoctoral fellowship SFRH/BPD/100721/2014.

References

Amann, R., & Fuchs, B. M. (2008). Single-cell identification in microbial communities by improved fluorescence in situ hybridization techniques. *Nature Reviews Microbiology, 6*, 339–348.

Amann, R. I., Binder, B. J., Olson, R. J., Chisholm, S. W., Devereux, R., & Stahl, D. A. (1990). Combination of 16S rRNA-targeted oligonucleotide probes with flow cytometry for analyzing mixed microbial populations. *Applied and Environmental Microbiology, 56*(6), 1919–1925.

Ancona, V., Barra Caracciolo, A., Grenni, P., Di Lenola, M., Campanale, C., Calabrese, A., Uricchio, V. F., Mascolo, G., & Massacci, A. (2017). Plant-assisted bioremediation of a historically PCB and heavy metal-contaminated area in Southern Italy. *New Biotechnology: Part B, 38*, 65–73.

Balázs, M., Rónavári, A., Németh, A., Bihari, Z., Rutkai, E., Bartos, P., Kiss, I., & Szvetnik, A. (2013). Effect of DNA polymerases on PCR-DGGE patterns. *International Biodeterioration & Biodegradation, 84*, 244–249.

Barra Caracciolo, A., Grenni, P. (2010). Microbial ecology methods for assessing the effects of xenobiotics in water and soil ecosystems. *Comparative Biochemistry and Physiology, Part A Molecular & Integrative Physiology, 157*(S1–S2), S56.

Barra Caracciolo, A., Grenni, P., Cupo, C., & Rossetti, S. (2005). In situ analysis of native microbial communities in complex samples with high particulate loads. *FEMS Microbiology Letters, 253*, 55–58.

Barra Caracciolo, A., Bottoni, P., & Grenni, P. (2010). Fluorescence In Situ Hybridization in soil and water ecosystems: an useful method for studying the effect of xenobiotics on bacterial community structure. *Toxicological & Environmental Chemistry, 92*(3), 567–579.

Barra Caracciolo, A., Bustamante, M. A., Nogues, I., Di Lenola, M., Luprano, M. L., & Grenni, P. (2015). Changes in microbial community structure and functioning of a semiarid soil due to the use of anaerobic digestate derived composts and rosemary plants. *Geoderma, 245–246*, 89–97.

Berendsen, R. L., Pieterse, C. M., & Bakker, P. A. (2012). The rhizosphere microbiome and plant health. *Trends in Plant Science, 17*(8), 478–486.

Bohnert, J., Hübner, B., & Botzenhart, K. (2000). Rapid identification of *Enterobacteriaceae* using a novel 23S rRNA-targeted oligonucleotide probe. *International Journal of Hygiene and Environmental Health, 203*, 77–82.

Bottari, B., Ercolini, D., Gatti, M., & Neviani, E. (2006). Application of FISH technology for microbiological analysis: current state and prospects. *Applied Microbiology Biotechnology, 73*, 485–494.

Brimecombe, M.J., De Lelj, F.A., & Lynch, J.M. (2001). The rhizosphere: The effect of root exudates on rhizosphere microbial populations. In: R. Pinton, Z. Varanini, P. Nannipieri (Eds.), *The rhizosphere. Biochemistry and organic substances at the soil-plant interface* (pp. 95–140). New York: Marcel Dekker.

Brock, T. D., Madigan, M. D., Martinko, J. M., & Parker, J. (2007). *Brock Biologia dei Microrganismi*. Milano, Italy: Casa Editrice Ambrosiana.

Buée, M., De Boer, W., Martin, F., van Overbeek, L., & Jurkevitch, E. (2009). The rhizosphere zoo: an overview of plant-associated communities of microorganisms, including phages, bacteria, archaea, and fungi, and of some of their structuring factors. *Plant Soil, 321*, 189–212.

Cairney, J. W. G. (2000). Evolution of micorrhiza systems. *Naturwissenschaften, 87*, 467–475.

Daims, H., Brühl, A., Amann, R., Schleifer, K. H., & Wagner, M. (1999). The domain-specific probe EUB338 is insufficient for the detection of all *Bacteria*: Development and evaluation of a more comprehensive probe set. *Systematic and Applied Microbiology, 22*, 434–444.

Delgado-Baquerizo, M., Maestre, F. T., Reich, P. B., Jeffries, T. C., Gaitan, J. J., Encinar, D., Berdugo, M., Campbell, C. D., & Singh, B. K. (2016). Microbial diversity drives multifunctionality in terrestrial ecosystems. *Nature Communications, 7*, 10541.

Garibyan, L., & Avashia, N. (2013). Research Techniques Made Simple: Polymerase Chain Reaction (PCR). *The Journal of Investigative Dermatology, 133*(3), e6.

Grenni, P., Gibello, A., Barra Caracciolo, A., Fajardo, C., Nande, M., Vargas, R., Saccà, M. L., Martinez-Iñigo, M. J., Ciccoli, R., & Martín, M. (2009). A new fluorescent oligonucleotide probe for *in situ* detection of s-triazine-degrading *Rhodococcus wratislaviensis* in contaminated groundwater and soil samples. *Water Research, 43*, 2999–3008.

Greuter, D., Loy, A., Horn, M., & Rattei, T. (2016). probeBase--an online resource for rRNA-targeted oligonucleotide probes and primers: new features 2016. *Nucleic Acids Research, 44*(D1), D586–D589.

Guo, Z., Megharaj, M., Beer, M., Ming, H., Mahmudur Rahman, M., Wu, W., & Naidu, R. (2009). Heavy metal impact on bacterial biomass based on DNA analyses and uptake by wild plants in the abandoned copper mine soils. *Bioresource Technology, 100*(17), 3831–3836.

Hugenholtz, P., Tyson, G. W., & Blackall, L. L. (2002). Design and evaluation of 16S rRNAtargeted oligonucleotide probes for fluorescence in situ hybridization. *Methods in Molecular Biology, 179*, 29–42.

Ibekwe, A. M., Papiernik, S. K., Gan, J., Yates, S. R., Yang, C. H., & Crowley, D. E. (2001). Impact of fumigants on soil microbial communities. *Applied and Environmental Microbiology, 67*, 3245–3257.

Johnsen, K., Jacobsen, C. S., Torsvik, V., & Sorensen, J. (2001). Pesticide effects on bacterial diversity in agricultural soils: a review. *Biology and Fertility of Soils, 33*, 443–453.

Kirk, J. L., Beaudette, L. A., Hart, M., Moutoglis, P., Klironomos, J. N., Lee, H., & Trevors, J. T. (2004). Methods of studying soil microbial diversity. *Journal of Microbiological Methods, 58*(2), 169–188.

Klironomos, J. N., McCune, J., Hart, M., & Neville, J. (2000). The influence of arbuscular mycorrhizae on the relationship between plant diversity and productivity. *Ecology Letters, 3*, 137–141.

Kozdrój, J., & van Elsas, J. D. (2001). Structural diversity of microorganisms in chemically perturbed soil assessed by molecular and cytochemical approaches. *Journal of Microbiological Methods, 43*, 197–212.

Loy, A., Lehner, A., Lee, N., Adamczyk, J., Meier, H., Ernst, J., Schleifer, K.-H., & Wagner, M. (2002). Oligonucleotide microarray for 16S rRNA gene-based detection of all recognized lineages of sulfate-reducing prokaryotes in the environment. *Applied and Environmental Microbiology, 68*, 5064–5081.

Lücker, S., Steger, D., Kjeldsen, K. U., MacGregor, B. J., Wagner, M., & Loy, A. (2007). Improved 16S rRNA-targeted probe set for analysis of sulfate-reducing bacteria by fluorescence in situ hybridization. *Journal of Microbiological Methods, 69*, 523–528.

Manz, W., Amann, R., Ludwig, W., Wagner, M., & Schleifer, K. H. (1992). Phylogenetic oligodeoxynucleotide probes for the major subclasses of Proteobacteria: problems and solutions. *Systematic and Applied Microbiology, 15*, 593–600.

Manz, W., Amann, R., Ludwig, W., Vancanneyt, M., & Schleifer, K. H. (1996). Application of a suite of 16S rRNA-specific oligonucleotide probes designed to investigate bacteria of the phylum cytophaga-flavobacter-bacteroides in the natural environment. *Microbiology, 142*, 1097–1106.

Manz, W., Eisenbrecher, M., Neu, T. R., & Szewzyk, U. (1998). Abundance and spatial organization of Gram-negative sulfate-reducing bacteria in activated sludge investigated by in situ probing with specific 16S rRNA targeted oligonucleotides. *FEMS Microbiology Ecology, 25*, 43–61.

Meier, H., Amann, R., Ludwig, W., & Schleifer, K. H. (1999). Specific oligonucleotide probes for in situ detection of a major group of gram-positive bacteria with low DNA G+C content. *Systematic and Applied Microbiology, 22*, 186–196.

Mocali, S., Galeffi, C., Perrin, E., Florio, A., Migliore, M., Canganella, A., Bianconi, G., Di Mattia, E., Dell'Abate, M. T., Fani, R., & Benedetti, A. (2013). Alteration of bacterial communities and organic matter in microbial fuel cells (MFCs) supplied with soil and organic fertilizer. *Applied Microbiology and Biotechnology, 97*, 1299–1315.

Muyzer, G., de Waal, E. C., & Uitterlinden, A. G. (1993). Profiling of complex microbial populations by denaturing gradient gel electrophoresis analysis of polymerase chain reaction-amplified genes coding for 16S rRNA. *Applied Environmental Microbiology, 59*, 695–700.

Nakatsu, C. H. (2007). Soil microbial community analysis using denaturing gradient gel electrophoresis. *Soil Science Society of America Journal, 71*, 562–571.

Nakatsu, C. H., Carmosini, N., Baldwin, B., Beasley, F., Kourtev, P., & Konopka, A. (2005). Soil microbial community responses to additions of organic carbon substrates and heavy metals (Pb and Cr). *Appl Environ Microbiol, 71*, 7679–7689.

Nannipieri, P., Ascher, J., Ceccherini, M. T., Landi, L., Pietramellara, G., Renella, G., & Valori, F. (2008). Effects of root exudates in microbial diversity and activity in rhizosphere soils. In *Molecular mechanisms of plant and microbe coexistence* (pp. 339–365). Berlin/Heidelberg: Springer.

Neef A. (1997). *Anwendung der in situ Einzelzell-Identifizierung von Bakterien zur Populationsanalyse in komplexen mikrobiellen Biozönosen*. Doctoral thesis, Technische Universität München, Germany.

Neef, A., Amann, R., Schlesner, H., & Schleifer, K. H. (1998). Monitoring a widespread bacterial group: in situ detection of planctomycetes with 16S rRNA-targeted probes. *Microbiology, 144*, 3257–3266.

Ng, E. L., Bandow, C., Proença, D. N., Santos, S., Guilherme, R., Morais, P. V., Römbke, J., & Sousa, J. P. (2014). Does altered rainfall regime change pesticide effects in soil? A terrestrial model ecosystem study from Mediterranean Portugal on the effects of pyrimethanil to soil microbial communities under extremes in rainfall. *Applied Soil Ecology, 84*, 245–253.

Ovreas, L. (2000). Population and community level approaches for analysing microbial diversity in natural environments. *Ecology Letters, 3*, 236–251.

Pernthaler, A., Pernthaler, J., & Amann, R. (2001). Fluorescence in situ hybridization (FISH) with rRNA-targeted oligonucleotide probes (review). *Methods in Microbiology, 30*, 207–226.

Proença, D. N. (2014). *Role of endophytic microbial community in pine wilt disease*. Coimbra: University of Coimbra, Portugal. PhD thesis. Available at http://hdl.handle.net/10316/25276.

Prosser, J. I. (2002). Molecular and functional diversity in soil microorganisms. *Plant Soil, 244*, 9e17.

Purahong, W., Wubet, T., Lentendu, G., Schloter, M., Pecyna, M. J., Kapturska, D., Hofrichter, M., Krüger, D., & Buscot, F. (2016). Life in leaf litter: novel insights into community dynamics of bacteria and fungi during litter decomposition. *Molecular Ecology, 25*, 4059–4074.

Roller, C., Wagner, M., Amann, R., Ludwig, W., & Schleifer, K. H. (1994). In situ probing of gram-positive bacteria with high DNA G + C content using 23S rRNA-targeted oligonucleotides. *Microbiology, 140*(10), 2849–2858.

Saeki, M., & Toyota, K. (2004). Effect of bensulfuron-methyl (a sulfonylurea herbicide) on the soil bacterial community of a paddy soil microcosm. *Biology and Fertility of Soils, 40*, 110–118.

Salles, J. F., van Veen, J. A., & van Elsas, J. D. (2004). Multivariate analyses of Burkholderia species in soil: Effect of crop and land use history. *Applied and Environmental Microbiology, 70*, 4012–4020.

Santos, B., Ribeiro, R., Domingues, I., Pereira, R., Soares, A. M. V. M., & Lopes, I. (2013). Salinity and copper interactive effects on perez's frog Pelophylax perezi. *Environmental Toxicology and Chemistry, 32*, 1864–1872.

Shokralla, S., Spall, J. L., Gibson, J. F., & Hajibabaei, M. (2012). Next-generation sequencing technologies for environmental DNA research. *Molecular Ecology, 21*(8), 1794–1805.

Singh, B. K., Munro, S., Potts, J. M., & Millard, P. (2007). Influence of grass species and soil type on rhizosphere microbial community structure in grassland soils. *Applied Soil Ecology, 36*, 147–155.

Stahl, D. A., & Amann, R. (1991). Development and application of nucleic acid probes. In E. Stackebrandt & M. Goodfellow (Eds.), *Nucleic acid techniques in bacterial systematics* (pp. 205–248). Chichester: Wiley.

Tetteh, R. N. (2015). Chemical soil degradation as a result of contamination: A review. *Journal of Soil Science and Environmental Management, 6*(11), 301–308.

Theron, J., & Cloete, T. E. (2000). Molecular techniques for determining microbial diversity and community structure in natural environments. *Critical Reviews in Microbiology, 26*, 37–57.

Torsvik, V., Goksoyr, J., & Daae, F. L. (1990a). High diversity in DNA of soil bacteria. *Applied and Environmental Microbiology, 56*, 782–787.

Torsvik, V., Salte, K., Soerheim, R., & Goksoeyr, J. (1990b). Comparison of phenotypic diversity and DNA heterogeneity in a population of soil bacteria. *Applied and Environmental Microbiology, 56*, 776–781.

Turbé, A., De Toni, A., Benito, P., Lavelle, P., Ruiz, N., Van der Putten, W. H., Labouze, E., & Mudgal, S. (2010). *Soil biodiversity: Functions, threats and tools for policy makers*. Bio Intelligence Service, IRD and NIOO, Report for European Commission (DG Environment).

Vissers, E. W., Bodelier, P. L. E., Muyzer, G., & Laanbroek, H. J. (2009). A nested PCR approach for improved recovery of archaeal 16S rRNA gene fragments from freshwater samples. *FEMS Microbiology letters, 298*, 193–198.

van der Heijden, M. G. A., Klironomos, J. N., Ursic, M., Moutoglis, P., Streitwolf-Engel, R., Boller, T., Wiemken, A., & Sanders, I. R. (1998). Mycorrhizal fungal diversity determines plant biodiversity, ecosystem variability and productivity. *Nature, 396*, 69–72.

von Wintzingerode, F., Gobel, U. B., & Stackebrandt, E. (1997). Determination of microbial diversity in environmental samples: Pitfalls of PCR-based rRNA analysis. *FEMS Microbiology Reviews, 21*, 213–229.

Wardle, D. A., Bardgett, R. D., Klironomos, J. N., Setälä, H., van der Putten, W. H., & Wall, D. H. (2004). Ecological linkages between aboveground and belowground biota. *Science, 304*(5677), 1629–1633.

Watanabe, K., Watanabe, K., Kodama, Y., Syutsubo, K., & Harayama, S. (2000). Molecular characterization of bacterial populations in petroleum-contaminated groundwater discharged from underground crude oil storage cavities. *Applied and Environmental Microbiology, 66*, 4803–4809.

Wolińska, A., Szafranek-Nakonieczna, A., Zielenkiewicz, U., Tomczyk-Żak, K., Banach, A., Błaszczyk, M., & Stępniewska, Z. (2016). Quantified characterization of soil biological activity under crop cultivation. *Journal of Advances in Biology, 8*(3), 1655–1665.

Yang, C. H., & Crowley, D. E. (2000). Rhizosphere microbial community structure in relation to root location and plant iron nutritional status. *Applied and Environmental Microbiology, 66*, 345–351.

Zhang, C., Wang, B., Dai, X., Li, S., Lu, G., & Zhou, Y. (2017). Structure and function of the bacterial communities during rhizoremediation of hexachlorobenzene in constructed wetlands. *Environmental Science and Pollution Research, 24*(12), 11483–11492.

Chapter 4
Towards Integrated Understanding of the Rhizosphere Phenomenon as Ecological Driver: Can Rhizoculture Improve Agricultural and Forestry Systems?

Luis G. García-Montero, Pablo Manzano, Deaa Alwanney, Inmaculada Valverde-Asenjo, Amaya Álvarez-Lafuente, Luis F. Benito-Matías, Xavier Parladé, Sigfredo Ortuño, Marcos Morcillo, Antonio Gascó, Carlos Calderón-Guerrero, Francisco Mauro, Mercedes Méndez, Alvaro Sánchez-Medina, María P. Andrés, José R. Quintana, Cristina Menta, Stefania Pinto, Liliana Pinto, Pilar Pita, Cafer Turkmen, Cristina Pascual, Esperanza Ayuga, Fernando Torrent, José C. Robredo, Pablo Martín-Ortega, Joan Pera, Luis Gómez, Gonzalo Almendros, Carlos Colinas, and Eric P. Verrecchia

Abstract Agriculture and forestry traditionally focus on improving plant growth traits based on an anthropocentric point of view. This paradigm has led to global problems associated to soil overexploitation such as soil losses, reductions of the C stock in soils, and the generalized use of fertilizers, which particularly increases the costs of production and pollution treatment. This view may also have limited our understanding of mutualistic symbioses of plants and microorganisms assuming

---

L.G. García-Montero (✉) • D. Alwanney • S. Ortuño • A. Gascó • C. Calderón-Guerrero
F. Mauro • M. Méndez • A. Sánchez-Medina • M.P. Andrés • L. Pinto • P. Pita • C. Pascual
E. Ayuga • F. Torrent • J.C. Robredo • P. Martín-Ortega • L. Gómez
Technical University of Madrid, ETSI Montes, Madrid, Spain

University of Lleida, Lleida, Spain

University of Lausanne, Lausanne, Switzerland
e-mail: luisgonzaga.garcia@upm.es

P. Manzano
IUCN Commission for Ecosystem Management, Gland, Switzerland

I. Valverde-Asenjo • J.R. Quintana
Complutense University of Madrid, Madrid, Spain

A. Álvarez-Lafuente • L.F. Benito-Matías
National Forest Genetic Resources Centre El Serranillo, Madrid, Spain

X. Parladé • J. Pera
IRTA, Sustainable Plant Protection Centre de Cabrils, Cabrils, Spain

M. Lukac et al. (eds.), *Soil Biological Communities and Ecosystem Resilience*, Sustainability in Plant and Crop Protection, DOI 10.1007/978-3-319-63336-7_4

that the main role of non-photosynthetic symbionts is to mobilize the nutrients that are necessary for plant growth and development, and being plants the dominant agents of the symbiotic relationship. In response to these issues, this chapter offers an alternative approach taking advantage of the "rhizo-centric" point of view, where non-photosynthetic partners are the main protagonists in play; and secondly, it builds a multidisciplinary body of knowledge that could be called "rhizoculture", which includes techniques focussing on the intensification of the development and activity of roots, mycorrhizae, and other symbiotic and free living rhizosphere organisms. In short, rhizoculture may lead to decrease plant production dependence on fertilization and provides other benefits to agriculture, forestry, and the environment. Within this conceptual framework, the first objective of this book chapter is to explore whether there is a "paradox of calcium salts" (i.e., Ca^{2+} and its salts are simultaneously nutrients, promoters, and stressors for the host plants) that would explain a dominance of mycorrhizal fungi over plants based on inducing a Ca(pH)–mediated chlorosis to the host plants. If this paradigm shifting hypothesis were finally fully verified, it would provide conceptual bases to reconsider our current technologies in agriculture and forestry by introducing the "rhizocultural" approach, based on the management of roots (introducing alternative cultural practices), Ca^{2+} salts (using liming and other techniques), rock-eating mycorrhizae, organic matter, and the soil microbiome (increasing the presence of symbiotic microorganisms against saprophytes), N and P contents (by aquaculture and smart recycling of organic waste), and the physical properties of the soil (by the activity of soil symbiotic microorganisms and soil fauna, such as ants, termites and earthworms). The development of such new technological approaches in rhizoculture would significantly decrease the high cost and associated pollution of the application of fertilizers and phytochemicals; as well as it would increase soil C stocks, improve the resilience of agricultural and forest systems to environmental disturbances, such as climate change, and enhance food production and security.

Keywords Rhizoculture • Rhizosphere • Mycorrhizae • N-fixing bacteria • Soil fauna • Agriculture • Forestry

---

M. Morcillo
Micología Forestal Aplicada – Micofora, Sant Antoni De Vilamajor, Barcelona, Spain

C. Menta • S. Pinto
Evolutionary and Functional Biology Department, University of Parma, Parma, Italy

C. Turkmen
Canakkale Onsekiz Mart University, Canakkale 17020, Turkey

G. Almendros
Spanish National Research Council (CSIC), Madrid, Spain

C. Colinas
University of Lleida, Lleida, Spain

E.P. Verrecchia
University of Lausanne, Lausanne, Switzerland

4.1 Introduction

Since the 1930s, the Green revolution has replaced traditional agriculture with conventional agriculture (CA) by the application of high inputs of chemical fertilizers and power derived from fossil fuels. These technologies saved over a billion people from starvation, but nowadays entail serious sustainability issues due to the limitation of fertilizers supply from current sources, greenhouse gas (GHG) emissions, and diffuse pollution (Farmer 1986; Te Pas and Rees 2014). Moreover, worldwide farmers only collect 50% of the yield they would obtain under optimal conditions, with 60–70% of the losses attributable to abiotic factors such as nutritional deficiencies or drought. The remaining 30–40% losses are attributed to biotic factors connected to the rhizosphere (Peleg et al. 2011). While organic agriculture (OA) promotes organic fertilization and outperforms CA in environmental and social aspects, critical questions arise on whether OA benefits compensate for its lower crop yield or if it will preserve soil fertility over time (Andersen et al. 2015; Hansen et al. 2001; Meier et al. 2015).

New solutions and policies are urgently needed to improve availability of fertilizers, OA and CA yields, and meet the "greening" requirements set by, for example, the Common Agricultural Policy (CAP) (Eurostat 2015). In this regard, European societal challenge analyses indicate that new nature-based solutions considering the multi-functional use of ecosystems are required to perform the transition towards a circular economic eco-innovative management system changing the traditional model based on production and consumption towards a green society.

4.1.1 Rhizoculture: Strategies for the Intensification of Roots and Rhizosphere Activity to Reduce Fertilization and Improve Resilience in Agriculture and Forestry

To include the multiple functions of ecosystems in both OA and CA, the rhizosphere is essential to soil, vegetation and ecosystem functioning (Akeem 2012). The term rhizosphere refers to the portion of soil surrounding roots where soil organisms are influenced by their presence (Killham 1994). Rhizosphere is the main bottleneck for closing loops on N, P and C-cycling in ecosystems because roots allow the integration of plant photosynthesis (leading to C inputs) and the activity of symbiotic organisms, such as mycorrhizae and N-fixing bacteria (leading to P and N inputs, respectively). Therefore, mutualistic plant-microbe interactions offer a novel approach to enhance agricultural productivity while reducing dependence on fertilization and environmental costs. In concert with other novel agronomic technologies and management practices, plant-microbial mutualisms can help increasing crop production from 115% and up to 300%, and reduce yield losses by improving resistance and/or resilience to edaphic, biologic, and climatic variability from both bottom-up and top-down perspectives (Adesemoye and Kloepper 2009; Baum et al. 2015; Hamilton et al. 2016).

In this regard, García-Montero et al. (2015a, b) proposed the development of a multidisciplinary body of knowledge called "rhizoculture" (i.e. *farming the rhizosphere*), which could be defined as a set of techniques focused on intensifying the abundance, biomass, depth and activity of roots, symbiotic and free-living organisms associated to the rhizosphere. Therefore, the so-defined rhizoculture can lead to decrease fertilization and diffuse pollution, and increase soil C storage, plant resilience, and agricultural and forest production by: (i) intensifying multiple symbioses among soil organisms and the host plants; (ii) managing roots and rhizosphere in connection with biogeochemical processes; (iii) rising soil organic matter content to increase the ratios of symbiotic to saprophytic communities in the soil; (iv) managing roots and rhizosphere impacts on soil diversity and C pools; (v) enhancing root and rhizosphere development to improve the ability of plants to cope with environmental disturbances such as climate change; and (vi) orienting policies aiming to incorporate the socioeconomic value of the positive externalities associated to rhizoculture.

Assuming the challenge of shifting production and consumption management patterns in agriculture and forestry towards a sustainable green economy and society, the value-added provided by this type of rhizoculture might include a reduction of the high costs and pollution associated to current fertilizers and phytochemicals inputs, which should result into improving both "environmental health" and the performance of agricultural and forestry systems.

In this way, developing new rhizoculture techniques connecting root and fertilizing management practices would be beneficial to agricultural and forestry sustainability, both supported by four R D areas: (1) agrotechnological support to root development, (2) "soft" fertilization application, (3) the management of soil organisms, and (4) and active management of soil water availability. The first RD area would incorporate a variety of small-medium industries as beneficiaries, and the other three areas would strengthen green agricultural and forestry policies, such as CAP.

4.2 Bases and Hypotheses for Developing Rhizoculture in Agricultural and Forestry Systems

Since the development of agriculture in the early Neolithic, the models of agriculture and forestry performance have primarily been focused on improving plant growth (especially focused on the development of parts of plants that were economically valuable) following anthropocentric and economical points of view. This approach may have neglected the understanding of plants´ interactions with their mutualistic symbionts, such as *ectomycorrhizal fungi* (ECMFs), arbuscular *mycorrhizal fungi* (AMFs) and N-fixing *bacteria* (NfBs), and commensal organisms, such as earthworms (García-Montero et al. 2015a, b).

Therefore, the problem under study is assessing whether *mycorrhizae* and other symbiotic organisms affect agriculture and forestry in a different way to the one assumed from a "phyto-centric" point of view, which is based on the traditional approach *viz.*: "Symbiotic organisms are related to plants following ecosystem models, but assuming that the role of non-photosynthetic symbionts is to mobilize the nutrients necessary for the primary production of plants, which appear to be the dominant actors of the symbiotic relationship. This conceptualization may result from phyto-centric and anthropocentric points of view which have led to the current agricultural and forestry models" (García-Montero et al. 2015a, b).

This traditional approach has led to high-intensity agricultural and forestry overproduction models, which are leading to soil degradation and loss, declining biodiversity and of soil C to the atmosphere. In response to these issues, this chapter addresses an alternative hypothesis, improving on the phyto-centric point of view, where plants behave as dominant agents in the symbiosis, with a "rhizo-centric" (Phillips 2007) point of view, where non-photosynthetic partners could play a predominant role.

Within this conceptual framework, the first objective of this chapter is revising in the literature whether the "paradox of Ca^{2+} salts" (Ca^{2+} and its salts simultaneously act as nutrients, promoters and stressors of plants, in the framework of the symbiotic organism-host plant relationship) could contribute to explain the postulated dominance of some mycorrhizal fungi over host plants (García-Montero et al. 2009; 2015a, b). As far as this hypothesis would be corroborated, the second objective of this chapter would portray a suitable scenario to improve our technologies in agriculture and forestry by introducing "rhizoculture".

4.2.1 A Calcium (pH)-Induced Chlorosis Hypothesis in Some Ecmf-Host Plant Symbiosis Based on the Ability of ECMF to Mobilize Calcium Salts

There is still insufficient knowledge of ECMF functional structure, on its precise role in ecosystem processes and on its biogeochemical cycling. Numerous *in situ* studies on the ecological function of ECMFs have been hindered by the difficulty of monitoring ECMFs in nature, and limited research has already been conducted on the successional processes of ECMF communities (Courty et al. 2010; García-Montero et al. 2012; Nara 2008; Smith and Read 2008). In order to minimize the research difficulty of performing ECMF field studies, a potential valid approach would be to study disturbed sites with simple ECMF/host-plant communities developed on relatively homogeneous substrates (Nara 2008), such as truffle brûlés.

Plants that have been infected by some ECMF species exhibit a zone around their trunk where the growth of other plants is inhibited, either permanently or periodically. Some examples of fungi producing these vegetation clearings on the soil (brûlés) are *Tuber melanosporum* Vittad. (black truffle), *T. aestivum* Vittad.

(summer truffle) and several species of *Scleroderma* Pers. In particular, *Tuber melanosporum* brûlé occurs because its mycelium and fruit bodies release multiple compounds that adversely affect seed germination and young plants development; as well as producing necrosis in grass roots (Plattner and Hall 1995).

García-Montero et al. (2009, 2012, 2013) have monitored natural *Tuber melanosporum* brûlé*s* in Mediterranean calcimorphic soils for over 15 years. A principal component analysis of 40 soil samples relates the soils inside the brûlés with high active carbonate content and lower total organic carbon (TOC) and total carbonate contents than in soils outside the brûlés. The disappearance of grasses in the brûlé causes modifications in soil surface layers affecting soil organic matter content (Castrignano et al. 2000). Callot et al. (1999) and Ricard et al. (2003) propose that the succession over time of different mycorrhizal fungi and microorganisms in *T. melanosporum* brûlés may be the result of the evolution of both the quantity and quality of the organic matter. Those authors report that the soils bearing *T. melanosporum* contain a small amount of coarse-fraction organic matter, and a high proportion of fine-fraction stable organic matter, in comparison with non-productive soils.

Statistical comparisons between carbonated fractions inside and outside these brûlé*s* have indicated that the concentration of total $CaCO_3$ inside the mature brûlé is significantly lower than in the outside soil (an average 0.64 times lower), which provides evidence of the "rock-eating" activity of *T. melanosporum*. However, the concentration of active $CaCO_3$ (smaller than 50 μm in size) inside the brûlé*s* is significantly higher (2.17 times greater on average) due to the soil physical, chemical and ecophysiological conditions that are associated to the existence of a mature brûlé, thus favouring secondary carbonate precipitation that produces a net increase of active $CaCO_3$ and exchangeable Ca^{2+} contents in the soil. The active $CaCO_3$ concentration inside the brûlés shows a positive correlation with the concentration of exchangeable Ca^{2+} (accounting for up to 42% of its variance). All these observations have led Prof. S. Rivas-Martínez to propose the term "Calcideserta" (calcic wilderness) for describing the brûlé of *T. melanosporum* from a geobotanical point of view (personal communication 2014).

Menta et al. (2014) have confirmed the ability of *T. aestivum* to modify the soil biogeochemical status in the brûlé. The TOC and water contents of soil tended to be significantly lower inside the brûlé in comparison to its outside. Mello et al. (2013) indicated that the lack of vegetation that appears in the brûlés could cause significant changes in light exposure and soil moisture level, leading, in turn, to water stress in the brûlé soil, especially in the summer. Finally, inside *T. aestivum* brûlés, the pH of the soil was significantly higher (0.22 to 0.28 units on average) than outside it.

On the other hand, some authors have reported comparable ECMF soil activity patterns resulting into significant accumulation of Ca, as calcium oxalate (CaC_2O_4), associated to ECMF mycelium presence on different soils. Oxalate and siderophore secretion are considered as the main mineral weathering means of ECMFs (Rineau and Garbaye 2010; Rineau et al. 2010). The ability to exude oxalate by some ECMFs has been widely documented, for instance as the means to acquire P by weathering mineral substrates (Kluber et al. 2010).

Based on the above findings, we upgrade a soil-plant nutrition hypothesis proposed by García-Montero et al. (2009) associated to some ECMF-host plant mutualisms: "*some ECMF communities could use different metabolic strategies to promote two biogeochemical changes, (i) increase soil Ca, and its salts, availability and/or (ii) increase soil pH in the rhizosphere of their host plants. Both strategies have been observed in carbonated soils as well as in acid soils. This ECMF activity could be a case of "ecosystem engineering" because increased pH and Ca^{2+} concentration in the rhizosphere may lead to host plant (i) chlorosis (via the paradox of Ca^{2+} salts) and (ii) modification of root-growth patterns (i.e., a significant root tips generation increase). Both, the stimulation to produce root tips and the induced nutrient deficiency, could lead to a greater ectomycorrhizal colonization activity and promote further growth of ECMF mycelia, which in turn induces the formation of further content of Ca^{2+} or increase soil pH in the rhizosphere on the basis of a feedback model*".

The impact of liming and soil pH change on tree growth patterns reinforce this soil-plant nutrition hypothesis. Monfort-Salvador et al. (2015) reviewed the impact of calcareous amendments on the soil-ECMF-host plant system. They assessed: (i) the influence of Ca^{2+} salts of biological origin on ECMF communities; (ii) the impact of Ca^{2+} on the growth patterns of the host tree roots; (iii) the importance of rock-eating processes associated to the presence of ECMFs for the host plants; and (iv) the impact of the soil Ca^{2+} cycle (associated to litterfall) on ECMFs activity. Tree root systems react to liming improving their ability to explore the soil through an enhanced absorption-efficient soil-root interface with extensive fine root development and a greater number of mycorrhizal tips.

In summary, ECMF populations and communities could be acting as "ecosystem engineering" agents at a global scale, changing soil properties in the rhizosphere of the host trees in order to cause them stress (via the paradox of Ca^{2+} salts); therefore, modifying tree growth patterns and driving mycorrhization processes and nutrient exchange mechanisms. This, in turn, would make the host plants more dependent on the fungi to obtain nutrients. However, simultaneously, host plants will be more resistant to biological and environmental stresses (including climate change), and in the case of agriculture, host plants will be less dependent on fertilizers and irrigation.

4.2.2 The Oxalate-Calcium Carbonate Pathway in the Interactions Between Fungi and Oxalotrophic Bacteria

Verrecchia et al. (2006) analysed the interactions between fungi and oxalotrophic bacteria, remarking the biogeochemical impact of fungi on many elemental cycles. Fungi are not only biologically important as saprophytes for the recycling of organic matter, but also play a geological role by excreting notable amounts of organic acid, among which oxalic acid is particularly important, and can be a source of secondary calcium carbonate in the soil. In contrast, calcium oxalate crystals precipitate and may constitute almost the 25% of soil hyphae and rhizomorphs dry weight in some ecosystems (Cromack et al. 1977). Verrecchia et al. (2006) demonstrated that oxalotrophic *bacteria* could effectively use these abundant oxalate crystals as C, electron, and energy sources by oxidizing them into calcium carbonate. This secondary $CaCO_3$ may then accumulate modifying soil conditions and increasing pH in the soil solution. Therefore, when associated with plant biomineralisation, fungi and bacteria contribute to long-term C storage by transforming half of the organic C from oxalate (or low-molecular-weight organic acids) into carbonate, with longer residence time in the soil than organic substrates. The other half is released into the atmosphere as CO_2. In conclusion, because oxalate salts are of organic origin, the oxalate-carbonate pathway represents a potentially major C sink, and probably acts as a regulator of atmospheric pCO_2.

Moreover, the oxalate-calcium carbonate pathway associated to these interactions probably also acts as a regulator of soil Ca^{2+} content and pH level in the rhizosphere environment of mycorrhizal-plant systems. This could have a major impact at the global scale. Therefore, it may be necessary to further develop another new hypothesis on mutualistic symbiotic relationships between *mycorrhizae* and oxalotrophic *bacteria* associated to the "paradox of Ca^{2+} salts" because the interaction of both soil microorganisms could increase their capacity to control host plants (based on the described Ca(pH)-induced-chlorosis).

On the other hand, calcium oxalate content showed a significant negative relationship with agricultural yield and crop quality attributes, emphasizing that increasing the levels of oxalate in the soil had a detrimental effect on soil microbial variables (Laxminarayana 2016). In summary, oxalotrophic *bacteria* may have positive impacts on agricultural systems, because they reduce the level of soil calcium oxalate, which benefits agriculture; and simultaneously increase the content of calcium carbonate, which stimulates the mycorrhization of plants. Therefore, this particular deserves conducting further research.

4.2.3 Interactions Between Soil Fauna and Mycorrhizae-Host Plant Systems Associated to the "paradox of Ca^{2+} salts"

Other relevant patterns of atmospheric CO_2 sequestration as soil carbonate (SIC) could be led by earthworms´ activity (i) producing calcite granules and (ii) the soil aeration generated by their tunnels. Spherical soil concretions up to the millimetric diametric size and composed of calcite crystals are found in soils containing earthworms. These calcite granules ranging from single crystals to aggregations up to 2.5 mm in diameter are produced by the calciferous glands present in all the species of the *Lumbricidae* family (Briones et al. 2008a; Canti and Piearce 2003; Lee et al. 2008; Versteegh et al. 2014; Wiecek and Messenger 1972). Jongmans et al. (2001)

found earthworm calcite granules down to 50 cm deep in the soil profile. Briones et al. (2008a, b), Gago-Duport et al. (2008), and Lee et al. (2008) described the microstructural transformations involved in $CaCO_3$ precipitation from the calciferous glands of earthworms. Furthermore, these biomineral granules are stable. In particular, Canti (2009) reported that earthworm calcite granules have been found in archaeological soils and ancient sediments.

Coleman et al. (2004)), Briones et al. (2008b) and Versteegh et al. (2014) indicated that the calciferous glands of earthworms could provide a mechanism for regulating CO_2 and Ca^{2+} in their blood and tissues. Canti (2009) showed that the C present in calcite granules mainly came from dietary inputs (litter), old soil C (soil OM), and from atmospheric CO_2 (although not in quantities larger than about one third of the total C in the granule), but very little quantity seemed to come from soil $CaCO_3$. Therefore, earthworms could lead to the promotion of significant amounts of secondary precipitation of $CaCO_3$ in the soil, whose C comes from the atmospheric CO_2 that is fixed by plant photosynthesis (in the same way as the fungal-bacteria oxalate-carbonate pathway). In any case, the use of fresh OM, soil OM, and atmospheric CO_2 to fix C in the earthworms' calciferous glands as granules formed from calcite crystals would result in a net soil C fixing (as SIC) and the recarbonation of soil. In addition, Fraser et al. (2011) conclude that earthworms truly synthesize $CaCO_3$ and they do not merely recycle ingested material.

García-Montero et al. (2008, 2013) confirmed Canti's (2009) laboratory results in truffle orchards. They found a significant increase of carbonate contents (and a higher pH value) in earthworm casts associated to calcite granules production that cannot be just explained by the existing levels of original carbonate in weakly calcareous soils. Chan (2003) showed that the pH of casts is significantly higher (pH >6) than that of the bulk acid soil material (pH = 4.0) that was used in laboratory experiments. Wiecek and Messenger (1972) also indicated that *L. terrestris* casts are less acid than the ingested food, and proposed that the weathering of earthworm calcite granules is partly responsible for pH values greater than 7.0 in the A_1 horizons of acid soils under forests covering some American woodlands. Elsewhere, Lambkin et al. (2011) reported from a laboratory experiment that an increased soil pH could promote a higher production of calcite granules by *L. terrestris*. In consequence, García-Montero et al. (2013) pointed to a feedback loop of varying intensity with the different types of soils and earthworms.

Versteegh et al. (2014) addressed that, the observed increase of $CaCO_3$ production rate with temperature can be explained by an increase in their metabolic activity. Lee et al. (2008) highlighted that earthworms produce calcite granules in sufficient volume to have a measurable impact on soil C cycling (as SIC). Lambkin et al. (2011) estimated that annual granule production by *L. terrestris* lies in the range of 18-3,139 mol-$CaCO_3$ ha^{-1} $year^{-1}$, with an average value of 438 mol-$CaCO_3$ ha^{-1} $year^{-1}$. These authors pointed out that their values are higher than those reported by Wiecek and Messenger (1972), who calculated that excreted $CaCO_3$ could contribute up to 11 mol-calcite ha^{-1} $yearr^{-1}$ in forests on acid soils; which are also higher than the results addressed in Canti (2007), who estimated *L. terrestris* production rates of more than 0.02 mol-calcite day^{-1}. However, Lambkin et al. (2011) indicated that there is significant variation in the mass of granules produced by different earthworms exposed to the same soil, and explain that a large proportion of this variation should reflect biological variations between individuals. Recently, Versteegh et al. (2014) reported that *L. terrestris* is a major $CaCO_3^-$ producing species in temperate soils. Production rates range from 0.8 to 2.9 mg earthworm^{-1} day^{-1}. Considering 1.9-61.8 individuals m^{-2}, this equates to precipitating 2-261 kg C ha^{-1} year^{-1}, which represents a potentially significant contribution to carbon sequestration. These authors also highlighted the on-going debate on whether earthworms increase soil greenhouse-gas emissions or contribute to carbon sequestration, as well as on the timescale and nature of the experiments that would be required to determine this particular (Zhang et al. 2013). Particularly, Versteegh et al. (2014) showed that it is likely that, at higher temperatures and atmospheric [CO_2], $CaCO_3$ earthworm synthesis will increase. As granules can last in soil for >300,000 years, the potential sequestration of C in the form of $CaCO_3$ is on a longer-time scale than roots and soil organic matter, for example. On the other hand, the precipitation of soil $CaCO_3$ increases when the partial pressure of CO_2 decreases in soil due to the aeration that is generated by the activity of earthworms themselves. In this regard, Callot et al. (1999) concluded that besides the contribution made by the respiration of roots and microorganisms, pCO_2 in the soil atmosphere depends on soil aeration, which is strongly conditioned by earthworms.

On the other hand, Verrecchia et al. (2006) showed that the accumulation of oxalate crystals by fungi may also produce unexpected consequences associated to soil fauna, because calcium oxalate can be disseminated inside soil by the action of oribatid mites. Such mites feed on calcium oxalate crystals produced by fungi, and then re-precipitate the mineral in their hardened cuticles (Norton and Behan-Pelletier 1991). In fact, they are thought to process a significant portion of the Ca^{2+} pool in some ecosystems (Gist and Crossley 1975). Therefore, it would be necessary to draw additional models on a mutualistic symbiosis existing between *mycorrhizae* and soil fauna (such as earthworms and mites) associated to the "paradox of Ca^{2+} salts" because the interaction of these organisms could increase the control over host plants based on the described Ca(pH)–induced-chlorosis. Menta et al. (2014) furthermore indicated that soil fauna could have an impact on fungal growth, dispersion and fruit body production, altering fungal fitness, and therefore, their combativeness in interaction with other soil microorganisms (Rotheray et al. 2011). Several authors have described some mechanisms explaining the impact of soil microarthropods on fungal communities. For instance, Hanlon and Anderson (1979) reported that microarthropod's feeding activity can exert a strong differential effect on fungal and bacterial populations; as well as some reviews on AMFs and soil fauna interactions suggest that *Collembola* have the potential to restrict mycorrhizal functioning in the field (Fitter and Garbaye 1994; Fitter and Sanders 1992). On the other hand, Menta et al. (2014) addressed that some ECMFs, such as *Tuber aestivum*, may modify soil fauna communities; that is, *Folsomia* sp. showed higher abundance inside the soil of the *T. aestivum* brûlé than in the outside.

4.2.4 The Liming Effect on AMFs in Agriculture

AMFs are among the most ubiquitous and abundant symbiotic organisms, and colonise roots of almost all plant species spreading over upland grasslands (Sparling and Tinker 1978; Yang et al. 2016). AMFs enhance nutrient translocation and uptake by developing an extensive network of external mycelium acting as an extension of the root absorption system (Mosse 1981). Moreover, Guo et al. (2012) have hypothesized that the ability to form AMF associations may be important to crop species adaptation to low-input systems. The AMFs can also alleviate stresses, improve soil structure, and provide plants with tolerance to heavy metals (Augé 2001; Clark and Zeto 1996; Smith and Read 2008).

We have reviewed the impact of liming on AMFs populations, thus opening the field to future studies to (*i*) assess possible cause-effect relationships between liming and the composition of AMF communities, (*ii*) addressing the suitable lime dose, and (*iii*) establishing their interactions with other soil organisms, such as diazotrophs (i.e. NfBs), and the whole nitrogen cycle. AMFs seem to show different responses to lime amendments. They indicated that all three possible effects –positive, negative and neutral– were observed:

(i) *Negative impact of liming on AMFs*: associated to the reduction of AMF spores, depressing root colonization by some AMF species and negative impacts on AMF community richness and diversity. In short, apparently, long-term application of lime might influence the colonization of AMF and, thus, the sustainability of farming systems on acid soils (Guo et al. 2012).
(ii) *Non-significant impact of liming on AMFs*: very few authors have reported the absence impacts on fungal communities, i.e., Shah et al. (1990).
(iii) *Positive impact of liming on AMFs*: in general, liming increases the intensity and abundance of mycorrhizal colonisation (but with significant differences among plant species), which may be crucial for conservation management systems, such as no-till farming (Guo et al. 2012; Johnson et al. 2005; Schneider et al. 2011). Liming increase the AMF colonisation of barley, corn, *Brachiaria decumbes,* and *Stylosanthes guianensis* and, moreover, increased the ability of AMFs to enhance the efficiency of P use by soybean and *Stylosanthes guianensis*. These authors also explained that liming increases the number of propagules, germination, and colonisation of several AMFs.

In summary, when calcareous amendment has been applied to crops, negative and positive effects on AMFs have been reported. However, our review allows concluding that, in the majority of cases, liming improves AMFs spore germination, the colonisation of AMFs, and their overall performance, although their diversity may be reduced.

In addition, liming can promote nodulation in N_2–fixing *bacteria* (NfBs), and N_2 fixation itself as well (Redecker et al. 1997). Legumes experiment ineffective nodulation in low pH soils (Munns 1986); however, liming and the inoculation of plants with AMFs and *Rhizobium* may solve this problem (Guo et al. 2010). For these reasons, we also reviewed the impact of liming on N-fixing *bacteria* (NfBs) in agricultural systems. Liming affects the interactions between legumes and *rhizobia*, resulting in an increased nitrogen fixation, which can duplicate N_2 fixation in some legume fields; i.e., liming can increase nodulation of *Trifolium repens* and *Alnus rubra* by *Rhizobium* spp. (Newbould and Rangeley 1984) and *Frankia* spp. (Crannell et al. 1994), respectively. Therefore, further research should help understanding how liming practices could impact on the whole system of symbioses of plants with soil organisms; as well as to establish analogies and differences with the model based on Ca(pH)–inducing chlorosis to the host plants that has been hypothesized above for ECMFs-plant systems.

4.3 Basis and Applications of Rhizoculture in Some Strategic Approaches

4.3.1 Food Production and Forest Resources at the Global Level

The Earth sums up 1,600 M ha of cropland, 3900 M ha of pastures and 3,900 M ha of forests, holding more than 1.400 M of farmers. These figures highlight the actual importance of the rural sector to the world economy. It is predicted that 50% more food will be needed in 2050 to feed a population of 9,000 M inhabitants, that is, 900 M tons more of cereals, 200 M tons more of meat, and 70 M ha more of cropland. In addition, it cannot be forgotten that there are currently 800 M people poorly nourished in the world. Even though they were almost 1,000 M in 2000, there is a lot to improve in this matter of concern (FAO 2017). Moreover, climate change may reduce up to a 25% of the yield of current crops in many areas of the planet. Therefore, there is still a big challenge for sustainability to overcome by investing and innovating in agricultural development. Within the agricultural sector, cereals production is remarkable, covering 60% of the cropland worldwide, along with oleaginous crops and fruit trees; whereas bovine, including cow milk production, is the most important within the livestock sector together with porcine. Considering the forest sector, its industrial activity is remarkable taking into account cellulose and paper, wood boards, round wood, and bioenergy production. In fact, forest production has experienced an important concentration within forestry plantations and crops, so they nowadays account for the 35% of the total forest outcome, although it is predicted to grow up to 75% in 2050. As a result, FAO (2017) estimates that wood demand will be increased by a 30% in 2050, and forestry crops will be responsible to cover this increment. Nowadays, 400 M m^3 of industrial wood are being produced by forestry plantations, and they are estimated to be 900 M m^3 in 2050. Correspondingly, forestry plantations cover is estimated to grow by approximately 80–90 M ha. Non-timber forest products (NTFPs) are also of great relevance within the forest sector, and they are under a strong growth; although statistical data are partial and incomplete, so global statistics cannot be herein provided. Within this global framework, new solutions and policies are urgently needed targeting current challenges for agriculture and forestry, such as growing welfare, changing diets, climate change mitigation, and food security of a growing global population needing 50% more food in 2050, as proposed by FAO (2017).

4.3.2 Strategic Keys to Implement Rhizoculture Models to Optimize Its Impact on the Mitigation of Greenhouse Gas Emissions, Pollution Associated to Fertilizers Use and Food Security

Considering this background, the main objectives of developing rhizoculture technology, both in OA and CA, would be to reduce fertilizer use, and to increase crop resilience, C sequestration, and even agricultural and forestry production, and N and P recycling, to be applied from the farm scale to much larger geographical units. These objectives would be directly related to the principles of organic agriculture and permaculture (Mollison and Reney-Mia 1991; FAO 2017), and therefore, rhizoculture would have to propose models for closing N, P and C cycling loops in agro ecosystems, which rely on sustainable ecosystem management rather than on external agricultural inputs.

Rhizocultural technology could be based on a set of experimental studies at the laboratory scale aiming to unravel the biological mechanisms (including molecular, biochemical and physiological ones) that are involved in root and rhizosphere intensification. On the other hand, and both at farm and landscape scales across different biogeographic regions (such as, Boreal, Atlantic, Continental and Mediterranean) and different continents (such as, Europe, Africa, Asia and Latin America), comparable pilot scenarios could be selected to implement rhizoculture by: (i) developing new applications of conventional OA techniques; (ii) designing new agricultural practices based on the observation of rhizosphere patterns of performance (such as liming and the control of organic matter and water inputs into the soil); and (iii) monitoring the closure of loops of N, P and C cycling, as well as diffuse contamination, GHG emissions and C sequestration in agricultural ecosystems. Moreover, rhizoculture could be integrated with (a) livestock management and feed, and aquaculture; (b) new technologies to enhance their implementation (i.e., recycling of agricultural residues); (c) socioeconomic and environmental assessment models, aiming to designing new prototypes on agro-ecosystems, and disseminating them via agriculture extension programs (García-Montero et al. 2015a, b).

Therefore, it is necessary to develop research projects unfolding strategic steps to implement rhizoculture in potential designing prototypes of sustainable agro-ecological systems. The main objectives in the design of these experiments could be:

(I) To develop agricultural, both organic and conventional, technologies to intensify root development and rhizosphere activity within both agricultural and forest management practices.
(II) To design rhizoculture strategies based on OM management models (e.g. livestock management and others) aiming to increase the ratio of symbiotic/saprophytic soil organisms in order to achieve these objectives:

   (a) To intensify biological NPK fertilisation (increasing NfBs and mycorrhization, fungal rock-eating activity, etc.).
   (b) To reduce CO_2 and GHG emissions and increase C stocks.
   (c) To increase plant resilience to climate change.

(III) To develop rhizocultural strategies increasing the presence levels of Ca, and its salts, in the soil by liming and the management of biological activity (earthworms, oxalotrophic bacteria, etc.) in order to intensify the symbiosis of plants with mycorrhizas and NfBs (using the *Ca-induced-chlorosis* model).
(IV) To implement rhizocultural strategies based on the management of fauna and organic waste recycling (including aquiculture, livestock, wild herbivorous, social insects, earthworm culture, and others) aiming to increase soil nutrients availability (by organic amendment, biological fertilisation, etc.) and the quality of soil properties (soil structure and aeration, cation exchange capacity, etc.).
(V) To design strategies integrating the use of chemical fertilisers and pesticides with biological fertilisation and crop protection based on rhizoculture.
(VI) To develop new information technologies that would favour the implementation of rhizoculture and its monitoring in agricultural and forestry systems. Some of these new technological developments may be:

   (a) Technological applications aiming to increase symbiotic activities among plants, mycorrhizas, and bacteria.
   (b) Technological applications to understand and intensify calcium roles in the symbiotic processes between plants, mycorrhizas and bacteria.
   (c) Molecular monitoring of biomass and diversity of soil organisms.
   (d) Remote sensing technologies monitoring fungal and soil organism activity.

(VII) To develop new models integrating rhizoculture in agriculture and forestry with consumer behaviour, food security, resources use and green economy.

In order to generate a set of explanatory and management models on rhizoculture on different scales, Table 4.1 shows a set of hypothetical scenarios of high agricultural and forest value at both the global and European scales, based on 15 crop types (including 4 tree species) and 15 agroforestry systems (including 3 livestock species), which have been selected because: (i) they together represent 71.73% of the European and 63.85% of the global crop areas, in basis to FAOSTAT productivity data (FAO 2017); (ii) a bibliographic review shows that these crops have valuable root systems and a good response to calcareous amendments (Tables 4.2 and 4.3); (iii) prototypes of rhizoculture based on this set of crops would allow comparisons

Table 4.1 Set of hypothetical scenarios for designing prototypes of sustainable agro-ecological systems based on rhizoculture

Regions	Countries	Annual crops to combining in some rotations: alternating crops and rhizoculture			Tree crops to be integrated with annual-rotations to stimulate rhizoculture	Intensive farming to integrate with N, P and C cycling in crops and rhizoculture	Agroforestry to integrate with N, P and C cycling in crops and with rhizoculture (farm and landscape levels)			Sources of P and N to intensify (circular economy)	C sinks to intensify (green economy)
		Crop type	Deeper roots	Shallower roots			Livestock and crops, and natural grasslands	Small forest, dehesa (savannah) and crop mosaics			
Boreal EU	Finland	*Cover*	Wheat; millet	Barley	Apple	Cow; pig	Extensive cow cattle	*Pinus sylvestris; Populus* spp. *Alnus glutinosa*	Improving soil N and P availability and C sinks	Aquaculture waste treatments (high-tech) and aquaponics	Rhizoculture and soil C sinks
		Rows	Tomato	Potato							
		N-fixers	Alfalfa	Green peas							
Atlantic EU	Germany France	*Cover*	Wheat; maize; millet	Barley	Apple; walnut	Cow; pig; aquaculture	Extensive cow sheep cattle	*P. sylvestris; Quercus robur*; *Populus* spp; *Alnus glutinosa*			Rhizoculture and ocean C sinks
		Rows	Tomato	Potato							
		N-fixers	Alfalfa	Green peas; soybeans							
Continental EU	Germany Czech Republic	*Cover*	Wheat; millet	Barley	Apple						Livestock and soil C sinks
		Rows	Tomato	Potato							
		N-fixers	Alfalfa	Green peas							
Mediterranean EU	Spain; Italy	*Cover*	Wheat; maize; millet	Barley	Apple; walnut; hazelnut; olive		Extensive cow, sheep, pig cattle; wild herbivores; rice-fishing	*Q. ilex*; *P. sylvestris*; *Populus* spp.; truffle culture; *A. glutinosa*; *Ceratonia siliqua*; soil social insects			Aquaculture and soil C sinks
		Rows	Tomato	Potato							Aquaculture and ocean C sinks
		N-fixers	Alfalfa; dry beans	Green peas; soybeans							
		Wetlands	Rice (paddy)								
Crops with global impact on N, P and C cycling and international trades	Argentina Brazil	*Cover*	Maize; millet	Barley	–	–	–			Aquaculture waste treatments (soft-tech) and aquaponics	
		Rows	Tomato	Potato							
		N-fixers	Dry beans	Soybean							
	Kenya	*Cover*	Maize; millet	Barley			Extensive cow sheep cattle; wild herbivores	Soil fauna; social insects	Improving soil N and P availability and C sinks		
		Rows	Tomato	Potato							
		N-fixers	Dry beans	Green peas							
	Bangladesh Philippines	*N-fixers*	*L. sativus*; *S. rostrata*			Aquaculture	Rice-fishing	–			
		Wetlands	Rice (paddy)								

Table 4.2 Root patterns and positive effects of liming on 15 representative crops that have been selected for designing comparative experiments on rhizoculture

15 selected crops	Root pattern in the soil	Liming effect	Impact	AM species
Wheat[1]	Both spring and winter wheat are annuals with deeply penetrating, widely spreading, and profusely branching, fine, fibrous roots; the roots of the spring varieties are less extensive[3]. Effective root zone[4] = 50–100 cm; 61–68% of their roots in < 30 cm; 50% >15 cm (avg. 16.8); 95% <103.8 cm; max depth = 150.4 cm[5]. 2–year-old crested wheat grass plant possessed over 500 m of roots occupying about 2.5 m^3 of soil[13]	Liming increased the efficiency of P use by wheat because of biological causes. Liming caused a significant increase in grain and straw yield of wheat (26.8 – 18.6%, respectively) over the no lime treatment, irrespective of the sources and levels of lime.	+	–
Barley[2]	Barley roots often occur nearer the surface than those of wheat; when grown in rich deep soil, has a root habit very similar to that of spring wheat; the fineness of the roots, degree of branching, and lateral spread often being intermediate. Effective root zone[4] = 50–100 cm ; 67–76% of their roots in < 30 cm; 50% <13 cm (avg. 11.5); 95% <99.6 cm; max depth = 146.1 cm[5]	Small positive effect on mycorrhizal colonization of barley and on propagule numbers.	≃	–

Maize[1]	Corn has a remarkably widely spreading, deeply penetrating, and profusely branching root system[3]. Effective root zone[4] = 50–100 cm; 50% of their roots <14.4 cm in average; 95% <88.9 cm; max depth = 118.3 cm[5]. In prairie soils, corn roots regularly penetrate to 2 m[13]	AM inoculation increased the yield when pH is intermediate.	+	–
		AM increases the productivity of limed maize	+	*Glomus etunicatum*
		Liming favoured specific AM species and many non-AM fungi. Higher richness of AM in non-limed soil	+	*Funneliformis*
		AM improved maize growth in limed soil and higher AM infection rates were achieved	+	*G. margarita*
		Increase plant height in limed soil. Liming and fertilizers (NPKS) increased maize yields as compared to the control plots which received NPK only.	+	–
		Liming increased the AMF colonisation of corn. Lower AM germination in non-limed and weak correlation with shoot dry weight	+	*G. mosseae*
Olive[2]	While other trees send their roots deep into the ground, olive trees feature shallow root systems. This allows olive roots to collect water from soil that typically dries fast, ensuring the tree gets enough moisture to stay hydrated	P and liming increased the leaf-P levels. Liming to many Australian soils, will improve the health, growth and crop of olive trees		
Potato[2,]	Potatoes have more superficial roots than many crops such as corn, winter wheat and most legumes. After extending horizontally 30–60 cm, majority of roots penetrate < 60 cm (exceptionally < 115 cm)	Liming had positive effects on yield, protein content, ash, starch, and Ca^{2+} of potato. However, Zn, Cu, Fe and P decreased with increased application of lime. AM inoculation increased yield when pH is intermediate	+	–

(continued)

Table 4.2 (continued)

15 selected crops	Root pattern in the soil	Liming effect	Impact	AM species
Apple[1]	Apple trees roots consist of a deep taproot (analogous to the carrot) and lateral fibrous roots. Lateral fibrous roots can extent to more than twice the spread of the canopy. Fine roots develop from the fibrous roots. During the third year, the maximum lateral spread reached 9 m and the maximum depth reached was 5 m. This greatly exceeded the lateral spread of three-year-old tops, which was about 2 m, and the height of the trees, which was 2 to 2,4 m[6]. Roots of 18-year-old apple trees penetrated to a depth of at least 10 m and fully occupied the soil between the rows, which were about 10 m apart[13]	Tree growth and apple fruit yield were significantly increased. Moreover, apple tree nutrition was benefited from liming.		
Rice[2]	Rice has a relatively shallow and compact (short and thick) root system compared with maize and wheat. Rarely grow deeper than 40 cm; about 90% of the total root system is restricted to the top 20 cm of the soil layer. Mean of root length of no fertilizer plants is 23.44 shorter than NPK fertilizer plants[7, 8, 9]	Upland rice dry matter and grain yields increased up to 32 and 19%, respectively, with lime addition	+	–

[1][For annual crops[1]: deeper rooting depth than other crops, or with 95% of roots < 100–138 cm, and max depth of 146–172 cm; or Effective Root Zone[4] = 50–100 cm / For tree crops: deeper root systems than other trees]; [2][For annual crops: shallowest rooting depth than other crops, with 95% of roots 64–85 cm, and max depth of 64–85 cm / For tree crops: shallower root systems than other trees]; [3][Weaver 1926]; [4][Effective root zone is the depth within which most crop roots are concentrated;]; [5][Fan et al. 2016]; [6][Yocum 1937]; [7][Kirk 1994]; [8][Morita and Keisuke 1995]; [9][Jeon ,2006]; [10][Weaver and Bruner, 1927]; [11][Baron 1985]; [12][Olsen 2013]; [13][Kramer and Boyer 1995]

Table 4.3 Root patterns and positive effects of liming on 15 representative crops that have been selected for designing comparative experiments on rhizoculture

15 selected crops	Root pattern in the soil	Liming effect	Impact	AM species	References
Soybean[2]	67–76% of their roots in < 30 cm; 50% of total root amount <13 cm (avg. 10.9); max depth = 111.3 cm[5]	Liming increased the ability of AMFs to enhance the efficiency of P use by soybean	+	–	
Tomato[1]	Mature plants have a wonderfully extensive root system. Usually 15 to 20 major branches spread widely (maximum, 168 cm), have very numerous large branches, which usually penetrate deeply, and finally, turning downward, extend into the 90, 120, and sometimes the 150 cm of soil. All are so profusely furnished with masses of much rebranched rootlets that the absorbing area is extremely intricate and extensive even beyond the working level of 106 cm[10].	Tomato fruit yield, plant growth and dry matter were significantly increased by liming at a pH between 3.3 to 5.2 while at a pH of 5.7 and above liming had little effects. In a soil of pH 4.2–5.1, liming 2 tones ha^{-1} of dolomite, and fertilizer (N–P–K–S–Zn), were optimum for better fruit yield tomato	–	–	
Green peas[2]	Plants 1.5 months old have a root depth of 60 cm. The soil at a depth of 5 to 20 cm is well filled with a network of nearly horizontal roots and their laterals to a distance of 46 cm on all sides of the plant. But in the deeper soil little absorbing area occurs[10]. Effective root zone[4] = 60–70 cm; 61–68% of their roots in < 30 cm; 50% of total root amount >15 cm (avg. 18.2); max depth = 150.4 cm[5]	The number of rhizobia that occurred naturally in non-inoculated plots increased rapidly in high–lime plots	+	–	
Beans (dry)[1]	The kidney bean rapidly develops a deeply penetrating taproot. When the plants are nearly mature, the soil to 60 cm on all sides is well ramified to a working level of 90 cm feet and numerous roots extend 120 cm deeper. Thus the general root habit of the kidney bean is not greatly unlike that of the pea although the lateral spread and depth of penetration are somewhat greater and the deeper soil, just beneath the plant, somewhat more thoroughly occupied[10].	Growth parameters and yield of *P. vulgaris* were significantly increased with liming (40% and 45%, respectively) or growing in vermicompost amended soil. Nodulating were significantly related with liming, largely due to changes in the predominance of the rhizobial species groups	+	–	

(continued)

Table 4.3 (continued)

15 selected crops	Root pattern in the soil	Liming effect	Impact	AM species	References
Hazelnut[2]	Hazelnuts are more shallow-rooted than most fruit trees, growing well in the marginal locations for walnuts[11]. Most of a hazelnut tree's roots are found in the first 0.6 m of soil; suitable soils allow trees to develop active root systems to depths of 1.8–3 m[11, 12]	Hazelnut uptake of N, K, Mg and Ca increase with liming, which also reduces Al and Mn uptake, but that liming is probably only cost effective up to pH 5.6 or 5.8. Liming increased very significantly the number of root tips	+	–	García-Montero et al. (García-Montero et al. 2012)
Walnut[1]	Walnut has an intricate extensive root system with the taproot penetrating to 3.7 m deep and many lateral roots which are mainly located in the horizon B at depth 60–80 cm, and sometimes extended to a distance exceeding the tree height and/or width. These trees excrete toxins to keep other plants with deep roots systems from growing[13]	Liming improved the yield of walnut very significantly, and improved also soil porosity, increased leave nitrogen content and leave total potassium	+	–	
Alfalfa[1]	Alfalfa is a long-lived, very deeply rooted perennial[3]. Effective root zone[4] = 120 cm ; 61–68% of their roots in < 30 cm; 50% of total root amount >15 cm (avg. 19.8); max depth = 176.8 cm[5]. In prairie soils, alfalfa roots have been found at depths of 10 m[13]	Higher rhizobium nodule numbers and total nodule weight. Higher shoots' N concentration without AM, while lower P concentration with AM	+	*Glomus intraradices*	Guo et al. (Guo et al. 2010)
Pearl millet	Root system development is characterized by a fast growing primary root that quickly colonizes deeper soil horizons (> 2 m)	Liming provided increments in the dry biomass production and in the accumulation of nutrients (N, P, K, Ca, Mg and S) by millet plants	+	–	
Mushrooms	Liming produces positive effects in many mycorrhizal fungi and truffles with agroforestry interest; see reviews of García-Montero et al. (García-Montero et al. 2009, García-Montero et al. 2012), Monfort-Salvador et al. (Monfort-Salvador et al. 2015)				

[1][For annual crops[1]: deeper rooting depth than other crops, or with 95% of roots <100–138 cm, and max depth of 146–172 cm; or Effective Root Zone[4] = 50–100 cm/For tree crops: deeper root systems than other trees]; [2][For annual crops: shallowest rooting depth than other crops, with 95% of roots 64–85 cm, and max depth of 64–85 cm/For tree crops: shallower root systems than other trees]; [3][Weaver 1926]; [4][Effective root zone is the depth within which most crop roots are concentrated;]; [5][Fan et al. 2016]; [6][Yocum 1937]; [7][Kirk 1994]; [8][Morita and Keisuke 1995]; [9][Jeon Jeon 2006]; [10][Weaver and Bruner 1927]; [11][Baron 1985]; [12][Olsen 2013]; [13][Kramer and Boyer 1995]

of "cover vs. row crops, annual vs. perennial, tall vs. short crops, deep vs. shallow roots, N-fixing and catch crops, species suited for companion planting and polyculture (i.e. beans, corn, peas, potato, tomato, alfalfa walnut)"; and (iv) other prototypes of rhizoculture based on a set of agricultural, forestry and other land uses would allow the analysis of arable land vs. adjacent forests, field margins vs. hedges and trees, tree species with N fixation capacity, and others scenarios such as buffer strips and mountain farming.

4.3.3 *Basis to Integrating the Rhizoculture and the Soil Organic Matter Management by Livestock, Biomass and Bioenergy Uses*

Livestock constitutes a powerful management tool at the landscape level. Apart from its proven capacity to influence biodiversity (Teillard et al. 2015), the high proportion of biomass consumed by livestock on grasslands (up to 85% according to estimates by Haynes and Willams (1993) and the dominance of domestic livestock in terrestrial grazed ecosystems (Smith et al. 2015) show how determinant this component can be. While livestock funnels a large proportion of the grassland biomass, most of the ingested components are excreted in an organic form and only mineralized by soil microorganisms once buried (Haynes and Williams 1993; Rufino et al. 2006). The interdependence of the soil ecosystem and the grazing ecosystem becomes therefore very relevant, especially in grazed ecosystems that represent more than half of the continental lands (Manzano 2015). It is then unsurprising that large mammalian herbivores influence the presence and activity of soil nematodes (Veen et al. 2010) or mycorrhizae (Murray et al. 2010).

At the interface between the grazing and the soil ecosystems, dung-burying invertebrates become critical in putting systems together and closing the nutrient cycle. While earthworms have been considered determinant in temperate humid climates, as well as dung beetles in semiarid Mediterranean systems (Lumaret et al. 1992), the influence of the latter seems to be restricted to the most humid months both in temperate and tropical semi-arid and arid areas. Harvester ants have been observed to be much more relevant in terms of dung burial during dry seasons in the Mediterranean area (Manzano et al. 2010), with similar roles attributed, in principle, to termites in the tropics (Freymann et al. 2008; West 1991). While such studies give a hint on the potential importance of social insects in dung burial, very little is known on which factors favour or limit the process or how it affects the soil community and the processes therein, such as carbon fixation or $CaCO_3$ production. Such research is particularly crucial when considering that a very large proportion of the continental land consists of drylands (up to 41.3% according to the Millennium Ecosystem Assessment), which have particularly been disregarded by research.

The management of grazing lands has also a very strong influence on their land degradation and biodiversity status, which in turn has a direct effect on C storage potential., Studies on rangeland governance have shown that security in land tenure is very determinant for keeping soils in grazing lands under a satisfactory conservation status (Herrera et al. 2014). Grazing management as a tool, including adequate land tenure systems, therefore seems a cheap and far-reaching management tool to promote carbon-fixing processes in the soil, provided that concluding research on how the different regimes influence it would be enabled.

Therefore, livestock and biomass use management can be major factors to develop soil organic matter management strategies aiming to increase the ratio of symbiotic/saprophytic soil organisms in order to achieve the main objectives of rhizoculture (described above) (García-Montero et al. 2015a, b).

In this regard, the need to mitigate climate change effects reducing GHG emissions, the growing interest on reducing crude oil dependence, and several other beneficial effects are major motivations for increasing the use of biomass-derived bioenergy (Long et al. 2015; UNTACD 2016). Biomass use has undergone a significant transformation along the last decade, that is, biomass is not only used for energy generation, but also for biomaterial production (food, feed, and fibre) thanks to the application of more innovative biotechnological processes, leading to develop what it is currently known as the bioeconomy (UNCTAD 2016). In 2014, the 10.1% of the world's Total Primary Energy Supply (TPES = 1,384 Mtoe = 57.8 EJ) was produced from biomass (solid and liquid biofuels; IEA 2016). Biofuel production in 2015 was 74.85 Mtoe, whereas 239.4 thousand million tonnes of oil were produced from geological reserves. The three largest biofuel producers were USA, Brazil and Germany, accounting for the 41.5%, 23.6% and 4.2% of the total worldwide outcome, respectively (BP 2016). The bulk of the production of the USA and Brazil is ethanol from corn and sugarcane, respectively; whereas the major part of the German production is biodiesel from rapeseed (BP 2016; Long et al. 2015). In the USA, a 30% of the average annual production of corn grain from 2006 to 2010 was used for ethanol production Three oil crops supply the vast majority of biodiesel: soybean (7.0 Mt), oil palm (6.3 Mt) and rapeseed/canola (6.0 Mt). Most soybean biodiesel is produced and consumed in the major centres of soybean cultivation, which are Brazil, USA and Argentina (Long et al. 2015). Bioenergy and biofuels produced from biomass have been considered carbon neutral in most Life Cycle Assessment (LCA) studies of biomass systems. It is assumed that the carbon released during the combustion of biomass is sequestered back into equivalent growing biomass. This convention is currently being discussed because the timing difference between the release and sequestration of carbon from forest biomass leads to a situation where part of the carbon remains in the atmosphere until it is fully integrated back into the growing forest. This would result in a warming impact if sequestration lags emission. Thus, carbon neutrality over a forest rotation period is not equal to climate change neutrality (Cherubini et al. 2011; Helin et al. 2013).

In biomass production (crop and harvest stages), other GHG emissions (i.e. N_2O and CH_4) should also be taken into account; and special attention should be paid to the role of fertilizers (Sastre et al. 2016). Net GHG savings are really highly uncertain and depend, among others, on the N-fertilization level and assumed N_2O emissions factors (Erisman et al. 2010). Recent studies point out that typical fertilization doses for bioenergy crops could produce nitrogen deficits in soil stocks, compromising soil sustainability and future crop fertility. However, raising nitrogen fertilization above the typical dosage aiming to compensate soil nitrogen deficit could compromise the achievement of GHG savings, as well as it could generate excessive effects in other relevant impact categories (i.e. eutrophication) (Sastre et al. 2016). In connection with rhizoculture, different crop management techniques are suggested to overcome this problem: (i) crop rotation with legumes; (ii) no-tillage farming; (iii) optimization of the crop collection time and/or the use of low N content species; (iv) introducing *gluconacetobacter* bacteria; and/or (v) using slow release fertilizers (Erisman et al. 2010; Sastre et al. 2016).

The cultivation of bioenergy crops is viewed as controversial because of this uncertainty with respect to net GHG-savings and its potential competition with biodiversity and food production for land use (Erisman et al. 2010). Research is currently focused on a second-generation biofuels (i.e. produced from lignocellulosic biomass). Typical resources for these fuels are short rotation forestry crops (poplar, willow and eucalyptus), perennial grasses (*Miscanthus* spp., switch grass and reed canary grass), and residues from the wood industry, forestry, and agriculture. Most of these second-generation biofuels are in their early stages of pilot production phase and commercialization (UNCTAD 2016).

4.3.4 Basis to Integrating the Rhizoculture with Alternative Fertilizers Based on Waste Recycling and Aquaculture

In 2012 and 2013, the average world's fertilizers consumption was about 120 kg/ha of cropland a year. Just considering developed countries with intensive agriculture, these figures turn to exceed 170 kg/ha yearly (i.e., the UE addressed an average of 177.026 and 176.419 kg/ha, respectively). These figures include N, P and K fertilizing fertilising products, but traditional organic compost and manure from plant and animal origins are not included (FAO 2016; World Bank 2016). The cost and environmental issues associated to these high levels of fertilizer application are promoting new alternative N and P sources, based on waste recycling, aquaculture and rhizoculture.

The world's production of aquatic animals from aquaculture summed up to 73.8 million tons in 2014, and 27.3 million tons of aquatic plants were also cultivated, accounting for an estimated first-sale value of €101,000 million, even overpassing the world fisheries captures of 93.2 million tons (FAO 2016). The total European aquaculture production reached 2,350,278 tonnes in 2015.

Cold-water marine species now represent the 71.4% of the total production, fresh water species sum up to the 15.1%, and marine Mediterranean species address the rest 13.5%. In 2015, the 94% of bred species were salmon, trout, seabream, seabass, and carp. Within the aquaculture sector, Integrated Multi-Trophic Aquaculture (IMTA) is a multi-culture system where several species from different trophic levels are bred in proximity. While the theoretical potential for the growth of biomass in the ocean is evidently very large, the focus on this kind of solution has been surprisingly low. As long as it is performed within ecologically sustainable limits, IMTA offers potential solutions to an increased and profitable production of local, resource-efficient, and climate-friendly food and biomass for energy purposes whilst capturing CO_2.This system is similar to a natural ecosystem, so the waste or excess of nutrients produced from fish being farmed (e.g. faeces and waste metabolites, and uneaten feed) become nourishment resources for lower trophic species, such as shellfish or seaweeds that are housed within the same system or environment.

This combination contributes to an ecosystem-based production that, in addition to recycling nutrients, naturally binds CO_2. Particularly within these systems, "aquaponics" is an innovative and sustainable food production system integrating aquaculture with hydroponic vegetal crops. Aquaponics may hold a key role to play in food provision and tackling global challenges such as water scarcity, food security, urbanization, and reductions in energy use and food miles.

On the other hand, membrane technologies, or membrane separation processes (MSP), are considered as highly efficient alternatives for the removal of nitrates and phosphates from water, in addition to other technological advantages such as the significant reduction of equipment size, energy savings, minimized environmental impact, and its easy staggering and automation (Drioli et al. 2011). Thus, some researchers have addressed this issue by applying different types of membrane and other separation processes, showing promising results (removal of nitrates and phosphates > 99%), such as electrocoagulation (Bektaş et al. 2004), microfiltration (MF) (Wasik et al. Wąsik et al. 2001), membrane bioreactors (MBR) (Wenhai et al. Wenhai et al. 2017), nanofiltration (NF), or reverse osmosis (RO) or direct one (DO) (Van Voorthuizen et al. 2005). In many of these studies, different combinations of processes are also proposed, i.e. MBR+RO (Wenhai et al. Wenhai et al. 2017), MBR+NF (Kangmin and Jaeweon 2012) or an integrated MF-NF-RO system, depending on the requirements of final water quality or the components to be removed. Nowadays, several companies are developing the application of these processes to the extraction of N and P from waste and agricultural waste recycling to produce alternative fertilizers, which is on the base of a circular economy compatible with rhizoculture.

4.3.5 Basis for Integrating Rhizoculture and Liming in Agriculture and Forestry: Impact of Mycorrhizae and Calcium on the Physiology of Plants

Calcium availability in the soil is crucial to many biological processes, so Ca^{2+} scarcity, whether caused by natural causes or human activity, may block certain plant physiological processes demanding leaf exchangeable Ca. In forest ecosystems, forest decay has partially been related to low soil Ca^{2+} contents (Borer et al. 2005;

Ting-Wu et al. 2011). Correspondingly, tree species generally react in a positive way to an increment of soil Ca^{2+} availability. Forestry extractions and acid deposition may cause substantial Ca^{2+} depletion in forest soils if rock weathering, atmospheric deposition or fertilization application do not compensate the losses (Huntington 2005), thus addressing a negative effect on the health of forests (Halman et al. 2014). Consequently, those forest species growing on soils with a higher soil Ca^{2+} availability show higher tolerance to root diseases, better leaf cold tolerance (Halman et al. 2008), higher leaf growth and biomass (Littke and Zabowski 2007), and increased leaf nutrients contents, as well as tree tops vitality is enhanced, stress metabolic indicators are lower (Minocha et al. 2010), and it promotes a higher relative growth of the basal area, in particular, and the overall plant growth, in general (Moore and Ouimet 2010; Huggett et al. 2007).

From a plant physiological and nutritional point of view, calcium is considered as an immobile essential element for plants, that is, it is an intrinsic component of their structure and metabolism; thus, its deficit may cause severe disorders to plant growth and development; as well as it limits the ability of plants to adaptively respond to environmental changes (Borer et al. 2005; Taiz and Zeiger 2015). Furthermore, attending to its relative abundance within plant tissue, it is considered as a macronutrient. In particular, the average content of Ca^{2+} in plant tissue is about the 0.5% (Epstein 1972). In addition, Ca^{2+} is considered as a nutrient that remains in its ionic form according to its biochemical functions (Mengel and Kirkby 2001), which comprises being a constituent of the middle lamella of cell walls, its participation as a cofactor of some enzymes of ATP hydrolysis and phospholipids, as well as it performs second messenger roles in metabolic regulation and serve to regulate the osmotic potential., Furthermore, being a divalent cation provides Ca^{2+} the ability to modify the permeability of plant membranes, which may serve, as well as cytosolic acidic pH values induced by flooding (Tournaire-Roux et al. 2003), to reduce the expression of aquaporins (e.g. Gerbeau et al. 2002), thus reducing the cell-to-cell water transport through roots tissues and, ultimately, water losses. Finally, the structural function of calcium on cell-walls is a major determinant of the xylem vulnerability to cavitation (Herbette and Cochard 2010).

On the other hand, AMFs symbiosis generally results in increased plant transpiration and stomatal conductance rates (Auge, Augé 2001). Among other factors, both AMFs and ECMFs may increase water uptake and transport thanks to the action of extraradical hyphae. Correspondingly, mycorrhizal plants have been reported to withstand drought better than non-mycorrhizal ones (Ruiz-Sanchez et al. 2010). Water use efficiency (WUE) was found to increase as the result of increasing assimilation rates in higher proportion than transpiration ones in mycorrhizal *B. papyrifera* seedlings under irregular watering (Birhane et al. 2012). Although similar results have been reported in other species (Shamshiri et al. Shamshiri et al. 2006, Kumar et al. 2016), the impact of mycorrhizae on WUE may depend on the plant species (Querejeta et al. 2003). Mycorrhizae may enhance shoot growth in host plants by means of improved nutrition and soil water acquisition, but below-ground biomass would also change, as far as the combined growth of plant roots and fungi mycelia are considered. As a result, alterations in plant allometry as a result of the mycorrhizal infection may have an impact on long-term WUE. Moreover, any modification of plant WUE at the global scale may affect the hydrological cycle. In fact, it should be taken into consideration that plant transpiration does not only play a role in the hydrological cycle, but it is also a meaningful way to cool the atmosphere. A better understanding of the effects of the proposed liming and rhizocultural practices on the physiology and use of soil water by mycorrhizal plants may, therefore, contribute to achieve better solutions for current agriculture, forestry and environmental concerns.

4.4 Conclusions

In summary, agricultural and forestry systems must support the global increasing demands of food and fibre, facing at the same time delicate and outstanding issues as critical as fertilizers availability, C sequestration, sustainable development, and climate change. Therefore, new nature-based solutions are required to perform the transition towards a circular economic system and a green society. Rhizoculture would lead to decrease fertilization and increase soil C storage, resilience of plants, and even agricultural and forest production, through a set of techniques focused on intensifying the abundance, biomass, depth, and activity of roots and the symbiotic microorganisms and fauna associated to the rhizosphere. The hypotheses supporting this rhizocultural approach are based on imitating the promising possibilities of some fungal behaviour that alter their environment to favour symbiotic relationships, such as the hypotheses of the "*calcium(pH)–induced cholorosis*" and the "*Ca^{2+} salts paradox*". Therefore, a deeper understanding of calcium salts management impacts, along with fungi and soil fauna management, and roots biology, will help (if discussed simultaneously) developing new rhizocultural technologies. They would also be based on the management of livestock and biomass (to increase ratios of symbiotic/saprophytic communities in the soil), agricultural and aquaculture residues (applying new recycling technologies), and new socioeconomic systems, that may significantly reduce the need for fertilizers. We believe that this approach will open the field for discussions on a new management of terrestrial ecosystems for the benefit of safe and sustainable food and forest production in a more environmentally friendly way.

Acknowledgements The authors would like to thank the following people for their corrections, suggestions and discussions on the concepts and models of Rhizoculture: Vicente Monleón, Dave Myrold, Thom Kuyper, Jim Trappe, José Miguel Barea, Domingo Moreno, John Baham, Jennifer Parke, Martin Lukac, Paola Grenni, Mauro Gamboni, Mike Castellano, Maribel Hernando, Francisco Pérez-Alfocea, Jesús Pastor, Miguel Quemada, Kira Hontoria, Rosa Mosquera, Marta Conde, Ana Rincón, José Luis Hernanz, Ruben Valbuena, Paloma Díaz, José Luis García-Manjón, Gabriel Moreno, Sergio Sánchez, Dan Luoma, Anssi Pekkarinen, Danilo Mollicone, Alfonso Sánchez-Paus, José Antonio Bonet, Dani Oliach, Conchi Azcón, Charles Lefevre, José Antonio Domínguez, Salvador Rivas-Martínez, Leo García Sancho, Aziz Türkoglu, Ayhan Oral, Susana Martín-Fernández, Eugenio Martínez-Falero, Antonio García-Abril, Ángel Martín, Fernando García-Robredo and Javi Rejos. Authors also acknowledge Pedro Cifuentes and Escuela Superior de Ingenieros de Montes, Forestal y del Medio Natural for their financial support.

References

Adesemoye, A. O., & Kloepper, J. W. (2009). Plant-microbes interactions in enhaced fertilizer-use efficiency. *Applied Microbiology and Biotechnology, 85*, 1–12.

Akeem, G. (2012). *Biodiversity, conservation and utilization in a diverse world*. Rijeka: Intech.

Andersen, M. M., Landes, X., Xiang, W., Anyshchenko, A., Falhof, J., et al. (2015). Feasibility of new breeding techniques for organic farming. *Trends in Plant Science, 20*, 426–434.

Augé, R. M. (2001). Water relations, drought and vesicular-arbuscular mycorrhizal symbiosis. *Mycorrhiza, 11*, 3–42.

Baron, L. C. (1985). *Growing hazelnuts in Oregon. Oregon State University Extension Service Circular, 1219*, 2–19.

Baum, C., El-Tohamy, W., & Gruda, N. (2015). Increasing the productivity and product quality of vegetable crops using arbuscular mycorrhizal fungi: a review. *Scientia Horticulturae, 187*, 131–141.

Bektaş, N., Akbulut, H., Inan, H., & Dimoglo, A. (2004). Removal of phosphate from aqueous solutions by electro-coagulation. *Journal of Hazardous Materials, 106*, 101–105.

Birhane, E., Sterck, F. J., Fetene, M., Bongers, F., & Kuyper, T. W. (2012). Arbuscular mycorrhizal fungi enhance photosynthesis, water use efficiency, and growth of frankincense seedlings under pulsed water availability conditions. *Oecologia, 169*, 895–904.

Borer, C. H., Schaberg, P. G., & DeHayes, D. H. (2005). Acidic mist reduces foliar membrane–associated calcium and impairs stomatal responsiveness in red spruce. *Tree Physiology, 25*, 673–680.

BP. (2016). *Statistical review of world energy*. Available at https://www.bp.com/content/dam/bp/pdf/energy-economics/statistical-review-2016/bp-statistical-review-of-world-energy-2016-full-report.pdf.

Briones, M. J. I., Ostle, N. J., & Piearce, T. G. (2008a). Stable isotopes reveal that the calciferous gland of earthworms is a CO_2–fixing organ. *Soil Biology and Biochemistry, 40*, 554–557.

Briones, M., López, E., Méndez, J., Rodríguez, J., & Gago-Duport, L. (2008b). Biological control over the formation and storage of amorphous calcium carbonate by earthworms. *Mineralogical Magazine, 72*, 227–231.

Callot, G., Byé, P., Raymond, M., Fernandez, D., Pargney, J. C., Parguez-Leduc, A., Janex-Favre, M. C., Moussa, R., & Pagès, L. (1999). *La truffe, la terre, la vie* (192 p). Paris: INRA Editions.

Canti, M. (2007). Deposition and taphonomy of earthworm granules in relation to their interpretative potential in Quaternary stratigraphy. *Journal of Quaternary Science, 22*, 111–118.

Canti, M. (2009). Experiments on the origin of ^{13}C in the calcium carbonate granules produced by the earthworm *Lumbricus terrestris*. *Soil Biology and Biochemistry, 41*, 2588–2592.

Canti, M. G., & Piearce, T. G. (2003). Morphology and dynamics of calcium carbonate granules produced by different earthworm species. *Pedobiologia, 47*, 511–521.

Castrignano, A., Goovaerts, P., Lulli, L., & Bragato, G. (2000). A geostatistical approach to estimate probability of occurrence of *Tuber melanosporum* in relation to some soil properties. *Geoderma, 98*, 95–113.

Chan, K. (2003). Using earthworms to incorporate lime into subsoil to ameliorate acidity. *Communications in Soil Science and Plant Analysis, 34*, 985–997.

Cherubini, F., Peters, G. P., Berntsen, T., Strømman, A. H., & Hertwich, E. (2011). CO_2 emissions from biomass combustion for bioenergy: atmospheric decay and contribution to global warming. *Global Change Biology Bioenergy, 3*, 413–426.

Clark, R., & Zeto, S. (1996). Growth and root colonization of mycorrhizal maize grown on acid and alkaline soil. *Soil Biology and Biochemistry, 28*, 1505–1511.

Coleman, D. C., Crossley, D., & Hendrix, P. F. (2004). *Fundamentals of soil ecology*. New York: Academic Press.

Courty, P. E., Buée, M., Diedhiou, A. G., Frey-Klett, P., Le Tacon, F., et al. (2010). The role of ectomycorrhizal communities in forest ecosystem processes: new perspectives and emerging concepts. *Soil Biology and Biochemistry, 42*, 679–698.

Crannell, W. K., Tanaka, Y., & Myrold, D. D. (1994). Calcium and pH interaction on root nodulation of nursfry-grown red alder (*Alnus rubra* bong.) seedlings by *Frankia*. *Soil Biology and Biochemistry, 26*, 607–614.
Cromack, K., Sollins, P., Todd, R., Fogel, R., Todd, A., et al. (1977). The role of oxalic acid and bicarbonate in calcium cycling by fungi and bacteria: some possible implications for soil animals. *Ecological Bulletins, 25*, 246–252.
Drioli, E., Stankiewicz, A. I., & Macedonio, F. (2011). Membrane engineering in process intensification—An overview. *Journal of Membrane Science, 380*, 1–8.
Epstein, E. (1972). *Mineral nutrition of plants: principles and perspectives*. New York: John Wiley and Sons Inc..
Erisman, J. W., van Grinsven, H., Leip, A., Mosier, A., & Bleeker, A. (2010). Nitrogen and biofuels; an overview of the current state of knowledge. *Nutrient Cycling in Agroecosystems, 86*, 211–223.
Eurostat. (2015). Common agricultural policy context indicators 2014–2020. In European Union (Ed.), *Statistical book on agriculture, forestry and fishery*. Brussels: European Commission.
Fan, J., McConkeya, B., Wanga, H., & Janzen, H. (2016). Root distribution by depth for temperate agricultural crops. *Field Crops Research, 189*, 68–74.
FAO. (2016). http://www.fao.org/
FAO. (2017). *What is organic agriculture?*. http://www.fao.org/organicag/oa-faq/oa-faq1/en/.
Farmer, B. H. (1986). Perspectives on the 'Green Revolution' in South Asia. *Modern Asian Studies, 20*, 175–199.
Fitter, A., & Garbaye, J. (1994). Interactions between mycorrhizal fungi and other soil organisms. *Plant and Soil, 159*, 123–132.
Fitter, A., & Sanders, I. (1992). Interactions with the soil fauna. In M. F. Allen (Ed.), *Mycorrhizal Functioning* (pp. 333–354). New York: Chapman and Hall.
Fraser, A., Lambkin, D., Lee, M., Schofield, P., Mosselmans, J., & Hodson, M. (2011). Incorporation of lead into calcium carbonate granules secreted by earthworms living in lead contaminated soils. *Geochimica et Cosmochimica Acta, 75*, 2544–2556.
Freymann, B. P., Buitenwerf, R., Desouza, O., & Olff, H. (2008). The importance of termites (Isoptera) for the recycling of herbivore dung in tropical ecosystems: a review. *European Journal of Entomology, 105*, 165–173.
Gago-Duport, L., Briones, M., Rodríguez, J., & Covelo, B. (2008). Amorphous calcium carbonate biomineralization in the earthworm's calciferous gland: pathways to the formation of crystalline phases. *Journal of Structural Biology, 162*, 422–435.
García-Montero, L. G., Valverde-Asenjo, I., Díaz, P., Pascual, C., & Menta, C. (2008). New data on impact of earthworms activity on black truffle soils. In D. Donnini (Ed.), *Proceedings of 3° Congresso Internazionale di Spoleto sul Tartufo* (130 pp). Spoleto: Comunità Montana dei Monti Martani e del Serano.
García-Montero, L. G., Quintana, A., Valverde-Asenjo, I., & Díaz, P. (2009). Calcareous amendments in truffle culture: A soil nutrition hypothesis. *Soil Biology Biochemistry, 41*, 1227–1232.
García-Montero, L. G., Valverde-Asenjo, I., Moreno, D., Díaz, P., Hernando, I., et al. (2012). Influence of edaphic factors on edible ectomycorrhizal mushrooms: new hypotheses on soil nutrition and C sinks associated to ectomycorrhizae and soil fauna using the *Tuber* brûlé model. In A. Zambonelli & G. M. Bonito (Eds.), *Edible Ectomycorrhizal Mushrooms: Current Knowledge and Future Prospects* (pp. 83–104). Berlin Heidelberg: Springer.
García-Montero, L. G., Valverde-Asenjo, I., Grande-Ortíz, M. A., Menta, C., & Hernando, I. (2013). Impact of earthworm casts on soil pH and calcium carbonate in black truffle burns. *Agroforestry Systems, 87*, 815–826.
García-Montero, L. G., García-Robredo, F., Ortuño, S., Ayuga, E., Pinto, L., et al. (2015a). Basis on rhizoculture: management of "underground agroforestry systems" to improve their climate change resilience. In E. Tielkes (Ed.), *Tropentag 2015. Management of Land Use Systems for Enhaced Food Security: Conflicts, Controversies and Resolutions*. Göttingen: Cuvillier Verlag. 330 pp.

García-Montero, L. G., Monleon, V., Myrold, D., Kuyper, T. W., Trappe, J., et al. (2015b). Could mycorrhizae perform "ecosystem engineering" on the host plants? Basis on rhizoculture of agroforestry systems to improve their resilience. In: P. Grenni and A. Bevivino (Eds.), *Soil biological communities and aboveground resilience.* Proceedings of the 3rd Annual Meeting, Cost Action FP1305 "Biolink–Linking belowground biodiversity and ecosystem function in European forests", Rome, p. 26. http://www.bio-link.eu/wp-content/uploads/2014/10/AbstractBook_Biolink_Rome2015.pdf

Gerbeau, P., Amodeo, G., Henzler, T., Santoni, V., Ripoche, P., & Maurel, C. (2002). The water permeability of *Arabidopsis* plasma membrane is regulated by divalent cations and pH. *The Plant Journal, 30*, 71–81.

Gist, C. S., & Crossley, D. A. (1975). A model of mineral cycling for an arthropod foodweb in a Southeastern hardwood forest litter community. In F. G. Howell & M. H. Smith (Eds.), *Mineral Cycling in Southeastern Ecosystems* (pp. 84–106). Washington DC: ERDA Symposium Series.

Guo, Y., Ni, Y., & Huang, J. (2010). Effects of rhizobium, arbuscular mycorrhiza and lime on nodulation, growth and nutrient uptake of lucerne in acid purplish soil in China. *Tropical Grasslands, 44*, 109–114.

Guo, Y., Ni, Y., Raman, H., Wilson, B., Ash, G., et al. (2012). Arbuscular mycorrhizal fungal diversity in perennial pastures; responses to long-term lime application. *Plant and Soil, 351*, 389–403.

Halman, J. M., Schaberg, P. G., Hawley, G. J., & Eagar, C. (2008). Calcium addition at the Hubbard Brook Experimental Forest increases sugar storage, antioxidant activity and cold tolerance in native red spruce (*Picea rubens*). *Tree Physiology, 28*, 855–862.

Halman, J. M., Schaberg, P. G., Hawley, G. J., Hansen, C. F., & Fahey, T. J. (2014). Differential impacts of calcium and aluminum treatments on sugar maple and American beech growth dynamics. *Canadian Journal of Forest Research, 45*, 52–59.

Hamilton, C. E., Bever, J. D., Labbé, J., Yang, X., & Yin, H. (2016). Mitigating climate change through managing constructed–microbial communities in agriculture. *Agriculture, Ecosystems and Environment, 216*, 304–308.

Hanlon, R., & Anderson, J. (1979). The effects of Collembola grazing on microbial activity in decomposing leaf litter. *Oecologia, 38*, 93–99.

Hansen, B., Alrøe, H. F., & Kristensen, E. S. (2001). Approaches to assess the environmental impact of organic farming with particular regard to Denmark. *Agriculture, Ecosystems and Environment, 83*, 11–26.

Haynes, R. J., & Williams, P. H. (1993). Nutrient cycling and soil fertility in the grazed pasture ecosystem. *Advances in Agronomy, 49*, 119–199.

Helin, T., Sokka, L., Soimakallio, S., Pingoud, K., & Pajula, T. (2013). Approaches for inclusion of forest carbon cycle in life cycle assessment – a review. *Global Change Biology Bioenergy, 5*, 475–486.

Herbette, S., & Cochard, H. (2010). Calcium is a major determinant of xylem vulnerability to cavitation. *Plant Physiology, 153*, 1932–1939.

Herrera, P. M., Davies, J., & Manzano Baena, P. (2014). *The Governance of Rangelands: Collective Action for Sustainable Pastoralism.* London: Routledge.

Huggett, B. A., Schaberg, P. G., Hawley, G. J., & Eagar, C. (2007). Long-term calcium addition increases growth release, wound closure, and health of sugar maple (*Acer saccharum*) trees at the Hubbard Brook Experimental Forest. *Canadian Journal of Forest Research, 37*, 1692–1700.

Huntington, T. (2005). Assessment of calcium status in Maine forests: review and future projection. *Canadian Journal of Forest Research, 35*, 1109–1121.

IEA. (2016). *Key renewables trends statistics.* International Energy Agency. Available at http://www.iea.org/publications/freepublications/publication/KeyRenewablesTrends.pdf

Jeon, W. T. (2006). Rice root distribution and rice-based cropping systems for sustainable soil rhizosphere management. In: *International Workshop on Sustained Management of the Soil-Rhizosphere System for Efficient Crop Production and Fertilizer Use.* Land Development Department, Bangkok.

Johnson, D., Leake, J., & Read, D. (2005). Liming and nitrogen fertilization affects phosphatase activities, microbial biomass and mycorrhizal colonisation in upland grassland. *Plant and Soil, 271*, 157–164.

Jongmans, A., Pulleman, M., & Marinissen, J. (2001). Soil structure and earthworm activity in a marine silt loam under pasture versus arable land. *Biology and Fertility of Soils, 33*, 279–285.

Killham, K. (1994). *Soil Ecology*. Cambridge: Cambridge University Press.

Kirk, G. J. D. (1994). *Rice roots: nutrient and water use*. Los Baños: International Rice Research Institute.

Kluber, L. A., Tinnesand, K. M., Caldwell, B. A., Dunham, S. M., Yarwood, R. R., et al. (2010). Ectomycorrhizal mats alter forest soil biogeochemistry. *Soil Biology and Biochemistry, 42*, 1607–1613.

Kramer, P., & Boyer, J. (1995). *Water relations of plants and soils*. San Diego: Academic Press.

Kumar, A., Choudhary, A. K., & Suri, V. (2016). Influence of AM fungi, inorganic phosphorus and irrigation regimes on plant water relations and soil physical properties in okra (*Abelmoschus esculentus* L.)–pea (*Pisum sativum* L.) cropping system in Himalayan acid alfisol. *Journal of Plant Nutrition, 39*, 666–682.

Lambkin, D. C., Gwilliam, K. H., Layton, C., Canti, M. G., Piearce, T. G., & Hodson, M. E. (2011). Production and dissolution rates of earthworm–secreted calcium carbonate. *Pedobiologia, 54*, S119–S129.

Laxminarayana, K. (2016). Response of mycorrhiza, organic sources, secondary and micro nutrients on soil microbial activities and yield performance of colocasia (*Colocasia esculenta* L.) in Alfisols. *Communications in Soil Science and Plant Analysis, 47*, 775–786.

Lee, M. R., Hodson, M. E., & Langworthy, G. (2008). Earthworms produce granules of intricately zoned calcite. *Geology, 36*, 943–946.

Littke, K., & Zabowski, D. (2007). Calcium uptake, partitioning, and sinuous growth in Douglas–fir seedlings. *Forest Science, 53*, 692–700.

Long, S. P., Karp, A., Buckeridge, M. S., Davis, S. C., Jaiswal, D., et al. (2015). Feedstocks for Biofuels and Bioenergy. In: G. Mendes, R. L. Victoria, C. A. Joly, L. M. Verdade (Eds.), *Bionergy sustainability: bridging the gaps*. Scope 72, Sao Paulo.

Lumaret, J. P., Kadiri, N., & Bertrand, M. (1992). Changes in resources: consequences for the dynamics of dung beetle communities. *Journal of Applied Ecology, 29*, 349–356.

Manzano, P. (2015). *Transporte de semillas por las ovejas trashumantes, y sus potenciales implicaciones en la vegetación*. PhD thesis, Universidad Autónoma de Madrid. Available at http://hdl. handle.net/10486/670899.

Manzano, P., Azcárate, F. M., Peco, B., & Malo, J. E. (2010). Are ecologists blind to small things? The missed stories on non–tropical seed predation on feces. *Oikos, 119*, 1537–1545.

Meier, M. S., Stoessel, F., Jungbluth, N., Juraske, R., Schader, C., & Stolze, M. (2015). Environmental impacts of organic and conventional agricultural products – Are the differences captured by life cycle assessment? *Journal of Environmental Management, 149*, 193–208.

Mello, A., Ding, G. C., Piceno, Y. M., Napoli, C., Tom, L. M., et al. (2013). Truffle brûlé have an impact on the diversity of soil bacterial communities. *PLoS One, 8*, e61945.

Mengel, K., & Kirkby, E. A. (2001). *Principles of plant nutrition* (Fifth ed.). Dordrecht: Kluwer Academic Publishers.

Menta, C., García-Montero, L. G., Pinto, S., Conti, F. D., Baroni, G., & Maresi, M. (2014). Does the natural "microcosm" created by *Tuber aestivum* affect soil microarthropods? A new hypothesis based on Collembola in truffle culture. *Applied Soil Ecology, 84*, 31–37.

Minocha, R., Long, S., Thangavel, P., Minocha, S. C., Eagar, C., & Driscoll, C. T. (2010). Elevation dependent sensitivity of northern hardwoods to Ca addition at Hubbard Brook Experimental Forest, NH, USA. *Forest Ecology and Management, 260*, 2115–2124.

Mollison, B., & Reney-Mia, S. (1991). *Introduction to Permaculture*. Sisters Creek: Tagari Publications.

Monfort-Salvador, I., García-Montero, L. G., & Grande, M. A. (2015). Impact of calcium associated to calcareous amendments on ectomycorrhizae in forests: a review. *Journal of Soil Science and Plant Nutrition, 15*, 217–231.

Moore, J. D., & Ouimet, R. (2010). Effects of two Ca fertilizer types on sugar maple vitality. *Canadian Journal of Forest Research, 40*, 1985–1992.
Morita, S., & Keisuke, N. (1995). Morphology and anatomy of rice roots with special reference to coordination in organo– and histogenesis. In F. Baluska, M. Ciamporova, O. Gasparíková, & P. W. Barlow (Eds.), *Structure and Fuction of Roots* (pp. 75–86). Dordrecht: Kluwer Academic Publisher.
Mosse, B. (1981). Vesicular–arbuscular mycorrhiza research for tropical agriculture. *Research Bulletin of Hawaii Institute of Tropical Agriculture and Human Resources, 194*, 82.
Munns, D. (1986). Acid soil tolerance in legumes and rhizobia. *Advances in Plant Nutrition, 2*, 63–91.
Murray, T. R., Frank, D. A., & Gehring, C. A. (2010). Ungulate and topographic control of arbuscular mycorrhizal fungal spore community composition in a temperate grassland. *Ecology, 91*, 815–827.
Nara, K. (2008). Community developmental patterns and ecological functions of ectomycorrhizal fungi: implications from primary succession. In: A. Varma (ed.), *Mycorrhiza: State of the Art, Genetics and Molecular Biology, Eco-Function, Biotechnology, Eco-Physiology, Structure and Systematics*, Springer, Berlin/Heidelberg, 581–599.
Newbould, P., & Rangeley, A. (1984). Effect of lime, phosphorus and mycorrhizal fungi on growth, nodulation and nitrogen fixation by white clover (*Trifolium repens*) grown in UK hill soils. In *Biological Processes and Soil Fertility* (pp. 105–114). Berlin/Heidelberg: Springer.
Norton, R. A., & Behan-Pelletier, V. M. (1991). Calcium carbonate and calcium oxalate as cuticular hardening agents in oribatid mites (Acari: Oribatida). *Canadian Journal of Zoology, 69*, 1504–1511.
Olsen, J. (2013). *Nut Growers handbook*. http://oregonhazelnuts.org/growerscorner/resources
Peleg, Z., Reguera, M., Tumimbang, E., Walia, H., & Blumwald, E. (2011). Cytokinin–mediated source/sink modifications improve drought tolerance and increase grain yield in rice under water–stress. *Plant Biotechnology Journal, 9*, 747–758.
Phillips, R. P. (2007). Towards a rhizo–centric view of plant–microbial feedbacks under elevated atmospheric CO_2. *New Phytologist, 173*, 661–663.
Plattner, I., & Hall, I. (1995). Parasitism of non–host plants by the mycorrhizal fungus *Tuber melanosporum. Mycological research, 99*, 1367–1370.
Querejeta, J. I., Barea, J. M., Allen, M. F., Caravaca, F., & Roldán, A. (2003). Differential response of δ13C and water use efficiency to arbuscular mycorrhizal infection in two aridland woody plant species. *Oecologia, 135*, 510–515.
Redecker, D., von Berswordt-Wallrabe, P., Beck, D. P., & Werner, D. (1997). Influence of inoculation with arbuscular mycorrhizal fungi on stable isotopes of nitrogen in *Phaseolus vulgaris. Biology and Fertility of Soils, 24*, 344–346.
Ricard, J. M., Bergougnoux, F., Callot, R., Chevalier, G., Olivier, J., et al. (2003). *La Truffe: Guide Technique de Trufficulture*. Paris: Centre technique interprofessionnel des fruits et légumes.
Rineau, F., & Garbaye, J. (2010). Effects of liming on potential oxalate secretion and iron chelation of beech ectomycorrhizal root tips. *Microbial Ecology, 60*, 331–339.
Rineau, F., Maurice, J. P., Nys, C., Voiry, H., & Garbaye, J. (2010). Forest liming durably impact the communities of ectomycorrhizas and fungal epigeous fruiting bodies. *Annals of Forest Science, 67*, 110.
Rotheray, T. D., Chancellor, M., Jones, T. H., & Boddy, L. (2011). Grazing by collembola affects the outcome of interspecific mycelial interactions of cord–forming basidiomycetes. *Fungal Ecology, 4*, 42–55.
Rufino, M. C., Rowe, E. C., Delve, R. J., & Giller, K. E. (2006). Nitrogen cycling efficiencies through resource–poor African crop–livestock systems. *Agriculture, Ecosystems and Environment, 112*, 261–282.
Ruiz-Sánchez, M., Aroca, R., Muñoz, Y., Polón, R., & Ruiz-Lozano, J. M. (2010). The arbuscular mycorrhizal symbiosis enhances the photosynthetic efficiency and the antioxidative response of rice plants subjected to drought stress. *Journal of Plant Physiology, 167*, 862–869.

Sastre, C. M., Carrasco, J., Barro, R., González-Arechavala, Y., Maletta, E., et al. (2016). Improving bioenergy sustainability evaluations by using soil nitrogen balance coupled with life cycle assessment: A case study for electricity generated from rye biomass. *Applied Energy, 179*, 847–863.

Schneider, J., Klauberg Filho, O., Fontoura, S., & Alves, M. V. (2011). Influência de diferentes sistemas de manejo e calagem em experimento de longa duração sobre fungos micorrízicos arbusculares. *Ciência e Agrotecnologia, 35*, 701–709.

Shah, Z., Adamst, W., & Haven, C. (1990). Composition and activity of the microbial population in an acidic upland soil and effects of liming. *Soil Biology and Biochemistry, 22*, 257–263.

Shamshiri, M., Usha, K., & Singh, B. (2006). Carbon–isotope discrimination, water use efficiency, net photosynthesis and transpiration in mycorrhizal kinnow (*Citrus Nobilis* × *C. Deliciosa*) plants. In M. Lee (Ed.), *Proceedings of the XXVII International Horticultural Congress on Global Horticulture: Diversity and Harmony* (pp. 261–265). Leuven: *International Society for Horticultural Science*.

Smith, S. E., & Read, D. J. (2008). *Mycorrhizal symbiosis* (Third ed.). London: Academic press.

Smith, F. A., Doughty, C. E., Malhi, Y., Svenning, J. C., & Terborgh, J. (2015). Megafauna in the Earth system. *Ecography, 38*, 1–10.

Sparling, G., & Tinker, P. (1978). Mycorrhizal infection in Pennine grassland. *I. Levels of infection in the field. Journal of Applied Ecology, 15*, 943–950.

Taiz, L., & Zeiger, E. (2015). *Plant physiology and development* (Sixth ed.). Sunderland: Sinauer Associates Inc..

Te Pas, C. M., & Rees, R. M. (2014). Analysis of differences in productivity, profitability and soil fertility between organic and conventional cropping systems in the tropics and sub–tropics. *Journal of Integrative Agriculture, 13*, 2299–2310.

Teillard, F., Anton, A., Dumont, B., Finn, J., Henry, B., et al. (2015). *A review of indicators and methods to assess biodiversity: application to livestock production at global scale*. Rome: Livestock Environmental Assessment and Performance (LEAP) Partnership, FAO.

Ting-Wu, L., Wu, F.-H., Wang, W.-H., Chen, J., Li, Z.-J., et al. (2011). Effects of calcium on seed germination, seedling growth and photosynthesis of six forest tree species under simulated acid rain. *Tree Physiology, 31*, 402–413.

Tournaire-Roux, C., Sutka, M., Javot, H., Gout, E., Gerbeau, P., et al. (2003). Cytosolic pH regulates root water transport during anoxic stress through gating of aquaporins. *Nature, 425*, 393–397.

UNCTAD. (2016). *Second generation biofuel markets: state of play, trade and developing country perspectives*. United Nations Conference on Trade and Development. Available at http://unctad.org/en/Pages/Home.aspx

Van Voorthuizen, E. M., Zwijnenburg, A., & Wessling, M. (2005). Nutrient removal by NF and RO membranes in a decentralized sanitation system. *Water Research, 39*, 3657–3667.

Veen, G. F., Olff, H., Duyts, H., & van der Putten, W. H. (2010). Vertebrate herbivores influence soil nematodes by modifying plant communities. *Ecology, 91*, 828–835.

Verrecchia, E. P., Braissant, O., & Cailleau, G. (2006). The oxalate-carbonate pathway in soil carbon storage: the role of fungi and oxalotrophic bacteria. In G. M. Gadd (Ed.), *Fungi in Biogeochemical Cycles* (pp. 289–310). Cambridge: Cambridge University Press.

Versteegh, E. A., Black, S., & Hodson, M. E. (2014). Environmental controls on the production of calcium carbonate by earthworms. *Soil Biology and Biochemistry, 70*, 159–161.

Wąsik, E., Bohdziewicz, J., & Błaszczyk, M. (2001). Removal of nitrates from ground water by a hybrid process of biological denitrification and microfiltration membrane. *Process Biochemistry, 37*, 57–64.

Weaver, J. E. (1926). *Root development of field crops*. New York: McGrawhill Book Company, Inc..

Weaver, J. E., & Bruner, W. E. (1927). *Root development of vegetable crops*. New York: McGraw-Hill Book Company, Inc..

Wenhai, L., Phan, H. V., Xie, M., Hai, F. I., Price, W. E., et al. (2017). Osmotic versus conventional membrane bioreactors integrated with reverse osmosis for water reuse: Biological stability, membrane fouling, and contaminant removal. *Water Research, 109*, 122–134.

West, N. E. (1991). Nutrient cycling in soils of semiarid and arid regions. In J. Skujins (Ed.), *Semiarid Lands and Deserts: Soil Resource and Reclamation* (pp. 295–332). New York: Marcel Dekker.

Wiecek, C., & Messenger, A. (1972). Calcite contributions by earthworms to forest soils in Northern Illinois. *Soil Science Society of America Journal, 36*, 478–480.

World Bank. (2016). http://www.worldbank.org/

Yang, H., Xu, J., Guo, Y., Koide, R. T., Dai, Y., et al. (2016). Predicting plant response to arbuscular mycorrhizas: the role of host functional traits. *Fungal Ecology, 20*, 79–83.

Yocum, W. W. (1937). Root development of young delicious apple trees as affected by soils and by cultural treatments. *College of Agriculture University of Nebraska, Agricultural Experiment Station, Research Bulletin, 95*, 2–55.

Zhang, W., Hendrix, P. F., Dame, L. E., Burke, R. A., Wu, J., et al. (2013). Earthworms facilitate carbon sequestration through unequal amplification of carbon stabilization compared with mineralization. *Nature Communications, 4*, 2576.

Chapter 5
Impact of Agricultural Land Management on Soil Bacterial Community: A Case Study in the Mediterranean Area

Annamaria Bevivino and Claudia Dalmastri

Abstract Soil is a complex and dynamic ecosystem whose functionality is related to the equilibrium existing among chemical, physical and biological parameters and the resident microbial communities. Soil microorganisms play a central role in decomposing organic matter, in determining the release of mineral nutrients, and in nutrient cycling, and have direct and indirect effects on both crop growth and quality, as well as on the sustainability of soil productivity. In addition, soil microorganisms substantially contribute to the resistance and resilience of agro-ecosystems to abiotic disturbance and stress. Therefore, changes in microbial communities may directly affect soil ecosystem function since microbes can respond rapidly to environmental changes because of the vastness of microbial biomass and diversity. An increasing number of studies have shown how environmental impacts that cause modifications in microbial community structure and diversity ultimately affect soil biological processes. Agricultural land management is one of most significant anthropogenic activities that substantially alter soil characteristics, including physical, chemical, and biological properties. The present chapter gives a picture of the effect of different agricultural management practices on soil microbial community structure and function. A case study on the effects of tillage and nitrogen fertilization on soil bacterial community structure is also reported.

Keywords Microbial diversity • Culture-based methods • Function • Mediterranean area • Land management • Nitrogen fertilization • Soil • Microbial community structure • Unculture-based methods

A. Bevivino (✉) • C. Dalmastri
Territorial and Production Systems Sustainability Department, Biotechnologies and Agroindustry Division, Italian National Agency for New Technologies, Energy and Sustainable Economic Development, ENEA, Rome, Italy
e-mail: annamaria.bevivino@enea.it

M. Lukac et al. (eds.), *Soil Biological Communities and Ecosystem Resilience*, Sustainability in Plant and Crop Protection, DOI 10.1007/978-3-319-63336-7_5

5.1 Introduction

Soil is the outermost weathered layer of the earth's crust and supports all terrestrial life forms. It is a complex, heterogeneous and dynamic system where lives a large variety of microorganisms including bacteria, archaea, fungi, yeasts, microalgae and protozoa. The soil microbial community is highly heterogeneous with arguably the highest level of prokaryotic diversity of any environment (Delmont et al. 2011). The balance of chemical, physical and biological (including microbial) components contribute to maintaining soil health. In some instances, detectable changes in soil physical and chemical properties can follow changes in microbial populations and/or activity, thereby providing an early evidence/warning of soil improvement or degradation, and making soil microorganisms excellent indicators of soil health assessment (Nielsen and Winding 2002).

Soil biological activity is mainly concentrated in the topsoil, the depth of which may vary from a few to 30 cm. The biological components (plants roots and soil organisms) occupy a little fraction (<0.5%) of the total soil volume with soil microorganisms being responsible for a large part of biological activity. Soil microbes play an active role in maintaining soil fertility and recycling of nutrients, since they are responsible for the decomposition of the organic matter entering the soil and are involved in the main biogeochemical cycles, improving plant health and contributing to higher crop yield (Aislabie and Deslippeet 2013). In fact, certain soil microorganisms produce compounds that stimulate the natural defense mechanisms of plants, improving resistance to pathogens. The involvement of microorganisms in many soil processes makes them suitable to give an integrated measure of soil quality and health, which cannot be obtained with physical/chemical measures alone.

Anthropogenic activities such as those including agricultural management practices alter physical and chemical soil properties, which directly affect microbial life strategies and bacterial community composition. To detect the effects that differing land uses and management strategies undoubtedly have on soil microbial communities, several techniques are applied, ranging from culture-based to culture-independent methods that offer new insights into the phylogenetic and functional diversity of microbial assemblages (Tilston et al. 2010).

In this chapter, an emphasis is given to soil microbial composition and function, with a focus on the response of soil bacterial communities to different agricultural land management options. The ability of an ecosystem to withstand severe disturbances may partly depend on its microbial components. Then, the assessment of bacterial community composition and structure is at the basis of better understanding their resistance and resilience, and permits to predict the response of soil bacteria to disturbance and to manipulate ecosystem processes (Bevivino et al. 2014; Shade et al. 2012; De Vries and Shade 2013). Finally, a case study of changes in the eco-physiological diversity of soil bacteria communities in an agricultural soil under long-term tillage system and nitrogen fertilizer application is reported.

5.2 Composition and Function of Microbial Communities in Soil Ecosystems

Soils harbor highly diverse communities of microorganisms (microbiota) which include members of each of the three domains of life, *Bacteria* and *Archaea* (both Prokaryotes), and fungi among *Eukarya*. Soil microbes are the dominant and highly diverse form of life in the soil making up 75–90% of living biomass. It has been estimated that every gram of soil contains up to 10 billion microorganisms (Delmont et al. 2011) with an estimated value ranging from 10^3 to 10^8 (bulk soil) up to 10^{11} (rhizosphere) prokaryotic cells (Sikorski 2015). The numbers of *Archaea* may be one to two order of magnitude below the numbers reached by bacteria, which have so far been the most extensively studied.

Taxonomic diversity of soil microbial communities is mirrored by the diversity of their protein-encoded functions, encompassing a seemingly limitless array of physiologies and life history strategies (Fierer et al. 2012). The high heterogeneity of soil environment represents the primary factor driving the enormous diversity of soil microbial life; i.e., the different components of the solid fractions in soil (sand, silt, clay, and organic matter) provide myriads of different microhabitats (niches) that differentially select bacteria and/or fungal types. Also, horizon development with soil depth provides a readily observable change in important soil characteristics that are known to have a profound effect on soil properties and in microbial community composition (Michel and Williams 2011). Microbial communities change along the soil depth profile due to the different environment of each horizon concerning nutrient and water availability, soil structure, organic matter content, pH, temperature, and oxygen. Even if microbes exist throughout the soil profile, they are mostly abundant in surface soils, plants rhizosphere, and around macropores.

Bulk soil is characterized by oligotrophic, carbon-limited environments, whilst rhizosphere soil is rich in organic nutrients. The higher microbial biomass of rhizosphere soil in respect to bulk soil is likely due to the greater availability of substrates for microbial growth. In fact, actively growing roots secrete a diverse array of organic root exudates that stimulate the growth of microbial populations present at the soil-root interface (i.e., rhizosphere). Then, the structure of soil microbial community is distinct from that in the bulk soil, with an increased presence of bacteria establishing positive interactions with plant roots, such as plant growth promoting rhizobacteria (PGPR). The vast majority of soil microorganisms are heterotrophic bacteria which rely on organic matter for energy and nutrients (Aislabie and Deslippeet 2013). Based on life strategies, they can be divided into *r*-strategists or zymogenous (microorganisms that rapidly grow in response to availability of a resource/substrate, being able to colonize unstable ecosystem) and *K*-strategists or autochthonous (microorganisms that use the resources slowly but more efficiently, being persistent in stable/mature ecosystems) (Pianka 1970).

Basically, the *r*-strategists are Gram-negative bacteria and also considered to be opportunistic since they require access to readily available organic matter, while *K*-strategists, which are able to survive with a low supply of nutrients, are mostly Gram-positive bacteria (Van Elsas et al. 2007). It is well known that many oligotrophic microorganisms (such as many members of the *Acidobacteria*) may be *r*-strategists, while many copiothrophic microorganisms (such as many members of the *β-Proteobacteria* and *Bacteroidetes*) may be *K*-strategists. This classification is useful for understanding soil C dynamics but cannot be related directly to a particular taxon since microorganisms can switch growth strategies in a complex environment such the soil. The same soil bacteria can alternate between these two states, depending on the nutrient supply resulting in a reversible transition between the *r* and *K* states (Stenström et al. 2001). As suggested by Dorodnikov et al. (2009), changes in growth rates of the whole microbial community after amendment with easily available substrates can reflect the shift toward *r* or *K* types. To provide new insights into the taxonomic and functional diversity of soil microorganisms it is necessary to analyze in depth the population structure, determining the number of species (taxa) present (richness) and their relative abundance, i.e. the distribution of individuals within the various species (evenness), and their physiological role in connection with the environment and other microorganisms (Tiedje et al. 1999).

Currently, the majority of soil microorganisms cannot be cultured via traditional laboratory techniques and can only be investigated by using molecular methods. Before the exploring of the taxonomic diversity and composition of soil microbial communities using polymerase chain reaction (PCR)-based approaches, *Agrobacterium*, *Alcaligenes*, *Arthrobacter*, *Bacillus*, *Flavobacterium*, *Micromonospora*, *Nocardia*, *Pseudomonas* and *Streptomyces* were considered to be the principal genera of soil bacteria, based on cultivation studies. Following the application of molecular ecological methods, that have allowed cultivation-independent investigations of soil microbial communities, members of the above nine genera together make up only 2.5 to 3.2% of soil bacteria (Janssen 2006). The dominant phyla in the libraries were *Proteobacteria*, *Acidobacteria*, *Actinobacteria*, *Verrucomicrobia*, *Bacteroidetes*, *Chloroflexi*, *Planctomycetes*, *Gemmatimonadetes* and *Firmicutes*, with members of the two first phyla representing the most abundant soil bacteria (making-up an average of 39% and 20% of libraries derived from soil bacterial communities, respectively). Otherwise, members of *Bacteroidetes*, *Firmicutes*, and *Planctomycetes*, are less abundant, with an average of 7%, 5% and 2% respectively (Janssen 2006). The majority (79 to 89%) of 16S rRNA gene sequences were from bacteria that were not affiliated with known genera, and some of these were associated with well-studied lineages of bacteria, such as *Actinobacteridae*, *Flavobacteria*, *Sphingobacteria*, *Bacilli*, *Clostridia*, *a-Proteobacteria*, *β-Proteobacteria*, *γ-Proteobacteria* and *δ-Proteobacteria*.

Soil microbes play diverse and often critical functions in soil ecosystems. The roles of soil microbes are highlighted in the cycling of the main biological elements (C, N, P), in the recycling of wastes, and the detoxification of environmental pollutants. Soil microbes contribute to soil formation through nutrient cycling and organic matter production, can support plant growth through increasing nutrient availability and by outcompeting invading pathogens, maintain soil fertility trough recycling nutrients, modulate the carbon storage capacity by mineralizing soil carbon and nutrients, and serve as repositories of genetic information, providing ecosystem services which are fundamental for human persistence (Aislabie and Deslippeet 2013).

In the last decade, the application of metagenomic approaches permitted to analyze microbial diversity and identify metabolic pathways and catalytic potential of the complex soil microbial communities, providing insight into the long-standing questions of "who's there?", "what are they doing?", "how sensitive are soil microbial communities to changing agriculture management and/or land use?", and "what are the dynamics of microbial communities in space and time?" (Myrold et al. 2014; Nesme et al. 2016). However, the large majority of metagenomic studies often measures the abundance and diversity of functional groups or genes associated with few relevant functions but this information is not always accompanied by the characterization of the new species' niche spaces. Microbial systems are responsible for the provision of a wide range of crucial ecosystem services, but little is known about the role of diversity in maintaining this function (Jurburg and Salles 2015). Furthermore, the link between phylogeny and function is often cut for prokaryotes, where horizontal gene transfer allows for the acquisition of functions associated with adaptability to new environments. Indeed, the ecosystem function of soil microbial communities demands much attention.

5.3 Agricultural Land Management Land Use and Soil Microbiota

Agricultural land management includes some of the most significant anthropogenic activities that alter soil physical, chemical, and biological properties, profoundly affecting microbial metabolism and survivor. The different members of a defined microbial community respond to these stresses with changes in population structure that further influence the ecosystem so that, ultimately, modification will be permanent with repercussions on the whole system, including plants and mammalians. In this scenario, the knowledge of structure dynamics of soil microbiota makes them useful as early indicators of soil quality and ecosystem stability concerning different soil uses.

5.3.1 Different Land Uses and Agriculture Managements

Land-use intensification includes the main drivers of biodiversity changes in soil ecosystems. Different managements and land use changes, such as the conversion of forest and pasture into cropped land, have ecosystem-scale impacts on soil cycling of organic compounds, biodiversity and soil nutrient dynamics (Bevivino et al. 2014; Dupouey et al. 2002; Houghton and Goodale 2004; Parfitt et al. 2003) that result in shifts in composition and function of soil microbial communities. Agricultural practices regarding soil nutrient management such as fertilization, tillage, agronomic practices and land uses are below described.

Chemical fertilizers (nitrogen, phosphorus, and potassium) enhance crop yield but also may result in shifts in the functionality and quality of soils by directly or indirectly changing their physical and chemical properties, and structure and diversity of microbial populations. A strict connection between soil microbial communities and the levels of phosphorus and soil moisture as well as their enzymatic activities has been widely described. Luo et al. (2016) suggested that microbes play important roles in determining the fertility of nitrogen-free fertilizer rice soils. In addition, functional diversity and evenness of soil bacterial communities were significantly higher in organically cultivated land than in chemically fertilized soil and fallow grassland, suggesting an improvement in soil quality (Chaundry et al. 2012). Although previous studies have examined well the response of soil microbial communities to single chemical fertilizer, only very few studies have explored the impact of the combination of chemical fertilizers on soil microbial biomass and composition suggesting that both N and P additions had different effects on soil microbial community and, thus, probably altered ecosystem functioning (Li et al. 2015).

Tillage is a common agricultural practice affecting soil structure -increases in bulk density and decreases in porosity-, and therefore biological properties. So, intensive tillage reduced the bacterial diversity due to the interruption of physical diversity of the soil environment (Acosta-Martínez et al. 2011), with negative consequences on the recycling of nutrients and proper balance among organic matter, soil organisms and plant diversity, that are necessary components of a productive and ecologically balanced soil environment (Hendrix et al. 1990). The ecological impact of management practices has been demonstrated as being a consistent source of disturbance on soil ecosystems (Hobbs and Hueneke 1992). Cropping system profoundly affects soil microbial biomass. So, microbial C and N and enzyme activities can be more affected by crop rotation than tillage management, as observed with sorghum in sandy soil in a semi-arid region where dry-land cropping system was required for a sustainable agricultural production (Acosta-Martínez et al. 2011).

In general, continuous monoculture systems tend to reduce soil organic matter due to low organic inputs and disturbance from tillage practices while crop rotations have positive effects on soil properties related to the higher C inputs and diversity of plant residues returned to soils in comparison with continuous systems (Miller and Dick 1995). For instance, integrated cotton (*Gossypiumhirsutum*) cropping and livestock production systems in West Texas were shown to provide more sustainable alternatives to the traditional continuous cotton system and improve soil quality (Acosta-Martínez et al. 2004). Organic and reduced-tillage management systems are aimed at favoring greater increases in soil microbial biomass and more diverse microbial communities with higher substrate utilization efficiency than with conventional management (Ghimire et al. 2014). In general, the root exudation produced by crops used in rotation systems varies with plant age and genotype, and consequently specific microorganisms respond and interact with different host plants, so explaining the effects of crop rotation and change in land-use on soil on soil microbial communities (Bergsma-Vlami et al. 2005; Chiarini et al. 1998; Dalmastri et al. 1999; Di Cello et al. 1997; Ramachandran et al. 2011).

Changes in land-use practices impact soil microbial community structure (Lauber et al. 2013). When forests are converted to grasslands, and grasslands turned into agricultural lands, a significant effect on chemical and structural composition of soil organic matter, as well as a sharp switch from one type of soil microbial community to another one occurred. For instance, in a typical Mediterranean ecosystem, dominated by *Quercus suber L.*, the microbial composition was found to change in response to five different land uses along with seasonal changes (Bevivino et al. 2014). Change in land use not only alters the taxonomic structure of soil microbial communities, but also their functional gene composition, as also reported in a study on the Amazon rainforest (Paula et al. 2014). Forest to pasture conversion varied the diversity significantly across functional gene groups, with genes linked to carbon and nitrogen cycling mostly altered, raising concerns about impacts of land use change at an ecosystem scale. Then, land-use type and, in particular, differences in vegetation dynamics have a role in modulating the temporal variability in soil bacterial communities (Lauber et al. 2013). Microbial communities result to be very different between managed and unmanaged agricultural systems. Conversion of the tidal wetlands into agricultural land was followed by a significant increase in microbial biomass and changes in diversity patterns that were more pronounced than those in functional gene abundances (Bannert et al. 2011).

In summary, different land uses and agriculture managements can cause decreased microbial production and biodiversity changes in microbial community composition, which can be seen even decades after the anthropogenic disturbance (Atlas et al. 1991; Buckley and Schmidt 2003).

5.3.2 How Microbes Respond to Land Uses

Environmental disturbances in general, and in particular different land uses and agriculture treatments as seen above, alter the microbial community composition, and land-use change is considered one of the main drivers of biodiversity changes in grassland ecosystems. So, it is crucial to understand how soil microorganisms respond to anthropogenic disturbance.

Responses of microbial populations to disturbances may include resistance or resilience (Griffiths et al. 2001; Westergaard et al. 2001). When the microbial community is not altered after a disturbance, it is considered to be resistant. Otherwise, when the community changes, but recovers and returns to the original state, it is deemed to be resilient. Soil microbiota response to land-use changes is correlated to microbe growth strategy since *r*-strategists require access to readily available organic matter, while *K*-strategists are able to survive with a low supply of nutrients (Van Elsas et al. 2007).

When toxic stress occurs in an ecosystem, the functions can be maintained by the replacement of sensitive with tolerant populations. Indeed, microbial communities may change in composition without any effects on microbial processes, since different microbial groups carry out the same functions, a phenomenon known as "functional redundancy" (Allison and Martiny 2008). The fact that ecological functions of different bacteria may overlap in the community (Botton et al. 2006) favors the persistence of the same functions even if changes occur, thanks to the activity of microbes which are resistant to changes/stress. Indeed, it is not the species composition but the functional aspect of the microbial community that is critical to maintain (Øvreås 2000). Therefore, diversity represents a large resource in the soil as well as in any natural habitat, and diverse communities are considered to be more resistant to environmental disturbances. However, the great diversity does not necessarily ensure functional stability in microbial communities (Peterson et al. 1998). In most cases, perturbations can alter microbial community structure resulting in permanent changes in microbial processes like nutrient cycling, decomposition, and energy flow. Substantially, any shifts in population structure will have consequences on ecosystem function, when the tolerant microorganisms fail to compensate for biogeochemical functions usually carried out by inhibited or eliminated microbial groups (Widenfalk et al. 2008).

In general, bacterial diversity is reduced by agricultural and management practices such as continuous monoculture, which can reduce soil organic matter because of low organic inputs systems (Miller and Dick 1995), and tillage (Acosta-Martínez et al. 2010). Soil bacterial communities showed a more stable structure in soils subjected to low human inputs (cork-oak forest and pasture) than in those with high human inputs (vineyards and managed meadow) (Bevivino et al. 2014). Differences in class composition across the site were also observed suggesting that the microbial composition changes in response to land use. In fact, seven classes (α, β and γ-*Proteobacteria*, *Sphingobacteria*, *Flavobacteria*, *Actinobacteria* and *Bacilli*) were present in the vineyard, cork-oak forest and pasture soils, while all classes but α-*Proteobacteria* in tillage vineyard and all classes but *Sphingobacteria* in managed meadow were found.

Investigation of bacterial and archaeal communities involved in inorganic nitrogen turnover in a study concerning the conversion of tidal wetlands into agricultural land for rice cultivation, showed higher abundances of ammonia-oxidizing microbes in the tidal wetland whereas fifty years of paddy management resulted in an increase of nitrogen-fixing bacteria (Bannert et al. 2011). Conversely, the functional diversity of denitrifying strains isolated from various rice paddy soils was minimally affected by crop rotation (Tago et al. 2011). The genus *Pseudogulbenkiania* was dominant at all locations, suggesting that *Pseudogulbenkiania* denitrifiers are ubiquitous in various rice paddy soils. Similarly, potential denitrifying activity was similar among the strains, regardless of the differences in taxonomic position and soil of origin (continuous cultivation of rice vs. rotational cultivation of rice and soybean), indicating that soil and other environmental factors, excluding cropping systems, could select for N_2-producing denitrifiers.

Also, organic amendments were found to influence the functional diversity and community structure in the soil enhancing soil suppressiveness and crop yield (Bonanomi et al. 2014; Pane et al. 2013). Diverse groups of microorganisms such as *Proteobacteria*, *Bacteroidetes* and *Gemmatimonadetes* were activated by its use (Chaudhry et al. 2012). Furthermore, an increase in *Proteobacteria* and *Ascomycota* groups and a reduction in *Acidobacteria* and *Mortierellales* were observed in soil subjected to composted almond shells as organic amendment. Also, genes related to the carbon cycle and other important soil processes had a higher significant relative abundance for amended soils as revealed by the functional GeoChip analysis (Vida et al. 2016). Interestingly, a group of specific probes included in the "soil benefit" category was found to be present only in almond shells-amended soils, corresponding to specific microorganisms described as potential biocontrol agents, such as *Pseudomonas* spp., *Burkholderia* spp., or *Actinobacteria.*

Finally, a decrease in both methanotroph (CH_4-oxidizing bacteria) diversity and methane consumption was observed in soils managed for row-crop agriculture when compared with native deciduous forests and never tilled soils managed as grasslands (Levine et al. 2011). This supports the fact that one of the most important drivers of species loss in terrestrial ecosystems worldwide is the simplification of ecosystem structure due to intensified land use (Sala et al. 2000). Conversely, a similar relationship between soil respiration and bacterial richness was not found, consistent with the prediction that microbial diversity is more likely to be important in specialized metabolic processes rather than in broadly distributed types of metabolism and that specialized processes are better targets for microbial mediation.

5.3.3 Detection of Microbial Response

Since the ability of soil ecosystem to withstand serious disturbances partly depends on its microbial components, characterizing bacterial community composition and structure will help to better understand and manipulate ecosystem processes.

Different approaches can be followed for studying diversity and community structure and evaluating dynamics processes at a global level or at the level of distinct taxonomic groups, for identification and typing, and for functional characterization. First of all, we need to distinguish between culture-based techniques and molecular assays which do not require cultivation. Culture-based techniques such as traditional plate counting can be used to enumerate microbial cells in a sample but they are limited by the fact that around 99–99.9% of the total microscopically countable bacterial cells in one gram of soil is not cultivable by standard culturing techniques (Hugenholz and Tyson 2008; Torsvik and Øvreås 2002). This is what has been described as "The Great Plate Count Anomaly" (Staley and Konopka 1985). The bacterial colony development method was used to determine eco-physiological differences in microbial communities in soil (De Leij et al. 1994) by using the concept of *r/K* strategy that attempts to explain changes in microbial community structure, including microbial shifts from *r*-strategists (copiotrophs) to *K*-strategists (oligotrophs).

To characterize the microbial communities, the ecophysiological index (EP, a modification of the Shannon diversity index) was calculated as described by De Leij et al. (1994). Indeed, based on these techniques, some studies on factors affecting soil communities were previously performed (Bevivino et al. 2014; Krzyżak et al. 2013; Papaleo et al. 2015).

One of the widely used culture-dependent methods is the community-level physiological profiles (CLPP) by the BIOLOG system, i.e. the Biolog Eco-Plates™ to measure the carbon substrate utilization patterns of microbial communities. Even if this culture-based method is more useful for comparing the general structure and the functional potential of soil microbial communities than for community characterization (Garland 1997; Preston-Mafham et al. 2002), it provides an exciting opportunity to overcome the drawbacks of conventional time consuming culture-based analyses or biochemical tests.

To overcome problems associated with non-culturable bacteria, various methods have been developed (Kirk et al. 2004). Detection of soil microbiota changes is currently performed by means of genetic fingerprinting methods that generate a profile of microbial communities based on direct analysis of PCR products amplified from soil DNA (Rastogi and Sani 2011). These techniques include denaturing-gradient and temperature-gradient gel electrophoresis (DGGE/TGGE), single-strand conformation polymorphism (SSCP), random amplified polymorphic DNA (RAPD), amplified ribosomal DNA restriction analysis (ARDRA), terminal restriction fragment length polymorphism (T-RFLP), length heterogeneity PCR (LH-PCR), and ribosomal intergenic spacer analysis (RISA), and produce a community fingerprint based on either sequence or length polymorphism. They all make possible to assess the association between molecular community fingerprints and environmental changes, and to follow the microbial community responses to a defined stress. Molecular analysis performed on total 16S rDNA present in the population furnishes a picture that is characteristic of defined conditions and permits to detect rapid changes as response to disturbances. Similarly to ecophysiological indexes, molecular indexes (Richness index, R; Shannon-Weaver index of general diversity, H'; Simpson index of dominance, D) are applied to elaborate profiles obtained in the different situations (Bevivino et al. 2014). The recent developments in new sequencing chemistries, bioinformatics, and instruments technologies have revolutionized the field of soil microbial ecology, providing researchers with the ability to assess bacterial diversity at lower costs, and quicker turnaround than prior 16S rRNA and sequencing methods (Shange et al. 2012). Barcode sequencing of 16S rDNA performed on total microbial DNA from soil samples enables to evaluate the diversity based on Operation Taxonomic Units (OTUs) composition as well as to assess their taxonomic status, and to further individuate molecular markers of changing.

Understanding how microbial communities function in natural environments is a central goal in microbial ecology. Given that metabolic function of indigenous microbes are essential to guarantee soil wellness and resilience, it is fundamental their knowledge also in the perspective of a correct soil management in agriculture. In addition to the conventional biochemical analysis of metabolites/enzymes, investigation of functional gene composition can be performed in soil population (Paula et al. 2014). Indeed, the currently used enzymatic activity measurements are not able to provide either the taxonomic information from the microbial community or probe the function from a specific microbial group (Luo et al. 2016). Analysis of total RNA extracted from environmental samples enables to assess microbial populations reaction at the transcription level providing more valuable information than DNA in revealing active microbial communities versus dormant microbial communities (Torsvik et al., 2002). This kind of analysis put in evidence changes in the expression/production of metabolites/proteins but does not permit to assess if changes occurred at the taxonomic level. Therefore, different approaches are needed to discover if new/different bacteria (taxa) get the better of previous ones providing for maintaining functional processes (resilience).

Comparative analysis of results obtained by 16S rDNA and functional genes investigation will enable to assess shifts occurring in taxonomic composition and responsible of functional changes. The development in metagenomic approach (which investigate the collective microbial genomes retrieved directly from environmental samples) open the way to previously unknown scenarios to detect microbial activities in microbes without requiring their cultivation since gives the possibility to analyze the meta-community dynamics and to identify markers of soil microbiota changes in response to the different land uses and treatments (Hugenholtz and Tyson 2008). In spite of the trend to give up traditional methods for molecular approaches, their importance cannot be forgotten since a profound knowledge of microbial ecosystem requires the ability to cultivate the resident microorganisms for studying individual strains given their possible application as soil status indicators. So, data from microbiota analysis will be helpful also in the perspective to improve cultivation techniques.

5.4 A Case Study in the Mediterranean Area

5.4.1 Introduction

Tillage practice represents one of the main factor affecting soil microbial community structure and activity. The altered soil physical and chemical conditions under conservation tillage result in significantly higher soil organic matter contents compared with conventionally tilled soils, then increasing microbial population and activity as well as microbial biomass (Mathew et al. 2012).

In the present work, we aimed at assessing the impact of applying no tillage (NT) compared with conventional tillage (CT), along with two fertilizer treatments on silty clay soil bacterial communities in a rain-fed *Triticum durum* and *Zea mays* rotation under temperate sub-Mediterranean conditions in a silty clay soil, in which crop residues were left on the soil (NT) or incorporated (CT). We considered the relative abundance of *r*- and *K*-strategists to assess the resistance and resilience of microbial communities in respect to changes due to soil management. The central hypothesis was that long-term use of no-tillage practices along with nitrogen fertilizer application would cause shifts in soil microbial community structure relative to conventional tillage practices.

5.4.2 *Experimental Design and Methods*

Sampling was performed at the "Pasquale Rosati" experimental farm of the Polytechnic University of Marche in Agugliano (AN) (43° 32′ N, 13° 22′E, 100 m a.s.l.) in the coastal hills of Marche, Italy (De Sanctis et al. 2012; Seddaiu et al. 2016). The Agugliano soil is a Calcaric Gleyic Cambisol with 20% slope, which in the first 30 cm has an Ap horizon (that is the homogeneous layer due to plowing) and a clay-silt texture (Lagomarsino et al. 2009).

The soil sampling was carried out in the frame of a National Research Project (SOILSINK Project) and soils used in this study were collected under consent of the landowners.

A split-plot with randomized blocks was designed to compare CT system (40 cm deep ploughing) with NT (sod seeding with chemical desiccation and chopping) and two fertilizer treatments (70 Kg/ha P_2O_5 with or without 90 Kg/ha NH_4NO_3) (Fig. 5.1). The soil samples were collected during wheat (*Triticum durum* L.) in autumn (October) in rotation with maize (*Zea mays* L.) in late spring (June). Five randomly replicates at each season and for each soil type (each one as a composite sample of 5 soil cores) were performed for a total of forty soil samples (5 replicates × 4 soil types × 2 seasons). After removal of litter layer, soil core samples (50 to 100 g; diameter, 5 cm) were taken at 20 cm depth, using a 5-on-dice sampling pattern

Fig. 5.1 The experimental area in the "Pasquale Rosati" experimental farm in Agugliano, Italy (gently provided by Prof. Pier Paolo Roggero and colleagues in the frame of the SOILSINK Project)

Table 5.1 Physical soil characteristics

Wheat-maize rotation[a]	Sand (g/kg)	Silt (g/kg)	Clay (g/kg)	Soil texture	Soil (pH)	Field capacity (% vol)
NT-0	105	512	383	Silty-clay	8.4	41.9
NT-90	103	403	494	Clay	8.4	42.5
CT-0	248	373	379	Clay loam	8.1	41.1
CT-90	249	353	398	Clay	8.1	41.7

[a]*CT* conventional tillage, *NT* no tillage; 0, without nitrogen fertilization; 90, with nitrogen fertilization (Pastorelli et al. 2010; Landi et al. 2011)

Table 5.2 Chemical soil characteristics

Wheat-maize rotation	C org (g/kg)	N tot (g/kg)	Nitric N (mg/kg)	OM	C/N	CEC (Meq/100g)
NT-0	5.0	0.8	2.05	1.18	7.82	24.07
NT-90	7.1	0.9	2.43	1.13	8.09	22.51
CT-0	11.8	1.4	1.36	1.78	8.43	23.28
CT-90	12.7	1.4	1.41	2.01	8.67	21.7

C org total organic carbon, *N tot* total nitrogen, *OM* organic matter, *CEC* cation exchange capacity

with ca. 70 m distance between each sampling point. Soil samples were immediately sieved (<2 mm) to remove fine roots and large organic debris, air dried and transported to the labs for microbiological analysis. The soil physical and chemical characteristics of the experimental areas are reported in Tables 5.1 and 5.2.

All soils had similar texture (clay soil), pH soil (8.1-8.4) and low C/N ratio (7.82-8.67). Differences were present in organic matter and nitrogen amounts; in particular, NT plots had a higher amount of organic matter and N-tot (1.4 g/kg) than CT plots, while nitric N was higher in CT soils than NT soils.

Soil bacterial community analysis was performed on five soil replicates for each treatment. Bacterial cell extraction was performed according to the recommendations of Smalla et al. (2001) with minor modifications as reported in Bevivino et al. (2014).

To determine the changes in the structure of cultivable fraction of soil bacteria, the *r/K*-strategy concept proposed by De Leij et al. (1994) was used. Colonies were enumerated at 1, 2 and 6 days of growth on 0.1 Tryptic Soy Agar (TSA), generating three counts (classes) per sample (C1, C2, C3).

The biodiversity of soil bacterial populations was investigated by using the Eco-Physiological (EP) index, that is a measure of both richness (i.e. total number of species in the community) and evenness (i.e. how evenly individuals in the community are distributed over the different species) of groups of microorganisms with similar developmental characteristics (De Leij et al. 1994). Bacterial population data (CFU/g of soil) were log transformed and subsequently analysed by one-way ANOVA (GraphPad Prism 7 Software). Percentage data of EP index value were *logit*-transformed, as follows: Logit (p) = log [p/(1-p)] for the proportion *p*, and compared using one-way ANOVA (GraphPad Prism 7 Software).

5.4.3 Results and Discussion

The four differently managed have similar characteristics, such as clay texture, pH and low C/N ratio, while they differ in some parameters resulting from the field's agricultural use. In the two seasons examined, a high diversity of colony morphology was observed in no-tilled system (data not shown).

The total microbial density ranged from 7.89×10^5 to 1.16×10^7cfu g^{-1} of soil. No significant differences were found in total microbial density in respect to both season and soil management (Fig. 5.2). A significant variation of EP-Index was found between N-fertilized and unfertilized soil in conventional tillage (CT) system ($p < 0.05$) while no significant differences in EP-Index were found in respect to both season and soil management (tilled vs. no-tilled soils) ($p > 0.05$) (Fig. 5.2).

In June, conventional tillage system in fertilized soil seemed to favor slow growers (*K*-strategists, oligotrophic bacteria) whilst no-tillage treatment positively affected the fast growers (*r*-strategists, copiotrophic bacteria) (Fig. 5.3). Nitrogen fertilization

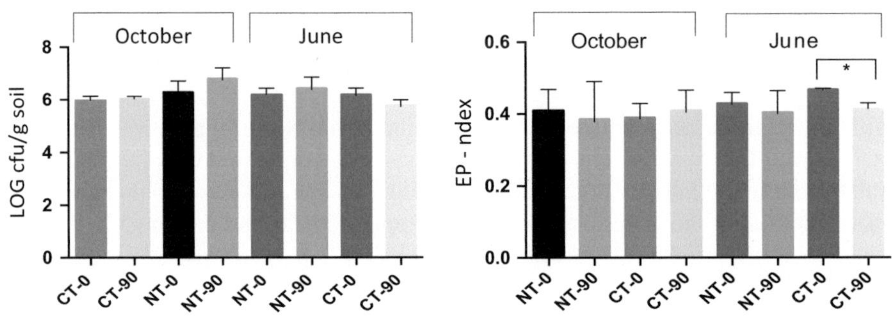

Fig. 5.2 Effect of tillage practices (CT, NT) and two fertilizer treatments [70 Kg/ha P_2O_5 with (90) or without (0) 90 Kg/ha NH_4NO_3)] on total bacteria (*left*) and EP index (*right*) in two seasons

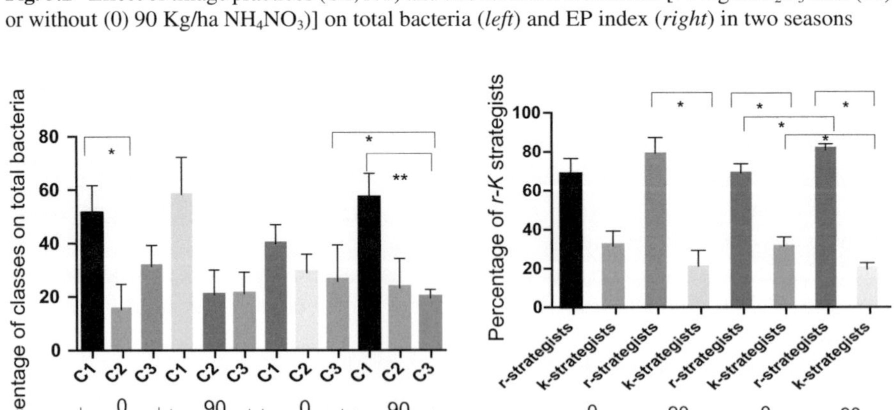

Fig. 5.3 Effect of tillage practices (CT, NT) and two fertilizer treatments (0, 90) on the percentage of C1, C2, C3 classes (left) and on the percentage of *r*-*K* strategists on total bacteria, in June

significantly affected the abundance of *r*-strategists in CT soils (P = 0.0002); in particular, N-fertilization reduced the abundance of *k*-strategists whilst increased abundance of *r*-strategists.

Overall, the investigation of the eco-physiological diversity of soil bacteria communities in relation to tillage system and nitrogen fertilizer revealed that only N-fertilization affected the relative abundance of *r-K* strategists as well as the evenness of population distribution and richness of the microbial community. We can conclude that N fertilization is able to increase the soil organic carbon level, particularly under conventional tillage, affecting soil microbial community structure. This study demonstrated that tillage systems along with different N fertilization rates influence soil microbial communities. Further research is needed to determine the impact of tillage-driven changes on soil microbial community composition and their dynamics.

5.5 Conclusions

In conclusion, diverse agricultural practices differently impact the microbial community resident in different soils. Given the role of soil microbes in soil safeness and function, there is the need to improve our knowledge about the ecology of microbial populations for proper agriculture management, at both levels of whole population structure and defined taxonomic/functional groups, such as those involved in inorganic nitrogen turnover and biogeochemical cycles. Clearly, a holistic approach taking into consideration all of potential factors and drivers is necessary when examining the structure–function relationships of soil microbial communities to improve understandings of controls and functioning of belowground processes.

Acknowledgements This study was supported by COST Action FP1305 "BioLink: Linking soil biodiversity and ecosystem function in European forests". The case-study in the Mediterranean Area was funded by MIUR (Integrated Special Fund for Research– FISR) in the frame of the Italian National Project SOILSINK "Climate change and agro-forestry systems, impacts on soil carbon sink and microbial diversity", and partially supported by MIUR (Research Department of Italian Government) in the framework of the Agreement Program ENEA-CNR (Art. 2, c. 44, Legge 23.12.2009 n. 191 – Legge Finanziaria 2010). The funders had no role in study design, data collection and analysis, decision to publish, or preparation of the manuscript. Authors greatly acknowledge the "Azienda Didattico-Sperimentale Pasquale Rosati" in Agugliano, the coordinator of SOILSINK Project Dr. Rosa Francaviglia (CRA-RPS, Rome), the responsible of the experimental site in Agugliano Prof. Pier Paolo Roggero (University of Sassari), and Dr. Roberto Orsini, Dr. Giuseppe Iezzi and Dr. Giuseppe Corti (Polytechnic University of Marche, Ancona, Italy) for providing data on soil physical properties and collecting soil samples.

References

Acosta-Martínez, V., Zobeck, T. M., & Allen, V. (2004). Soil microbial, chemical and physical properties in continuous cotton and integrated crop–livestock systems. *Soil Science Society of America Journal, 68*, 1875–1884.

Acosta-Martínez, V., Dowd, S. E., Bell, C. W., Lascano, R., Booker, J. D., Zobeck, T. M. M., & Upchurch, D. R. (2010). Microbial community composition as affected by dryland cropping systems and tillage in a semiarid sandy soil. *Diversity, 2*, 910–931.

Acosta-Martínez, V., Lascano, R., Calderón, F., Booker, J. D., Zobeck, T. M., & Upchurch, D. R. (2011). Dryland cropping systems influence the microbial biomass and enzyme activities in a semiarid sandy soil. *Biology and Fertility of Soils, 47*, 655–667.

Aislabie, J., & Deslippe, J. R. (2013). Soil microbes and their contribution to soil services. In J. R. Dymond (Ed.), *Ecosystem services in New Zealand – conditions and trends* (pp. 143–161). Lincoln: Manaaki Whenua Press.

Allison, S. D., & Martiny, J. B. H. (2008). Resistance, resilience, and redundancy in microbial communities. *Proceedings of the National Academy of Sciences, 12*, 11512–11519.

Atlas, R. M., Horowitz, A., Krichevsky, M., & Bej, A. K. (1991). Response of microbial populations to environmental disturbance. *Microbial Ecology, 22*, 249–256.

Bannert, A., Kleineidam, K., Wissing, L., Mueller-Niggemann, C., Vogelsang, V., Welzl, G., Cao, Z., & Schloter, M. (2011). Changes in diversity and functional gene abundances of microbial communities involved in nitrogen fixation, nitrification, and denitrification in a tidal wetland versus paddy soils cultivated for different time periods. *Applied Environmental Microbiology, 77*, 6109–6116.

Bergsma-Vlami, M., Prins, M. E., & Raaijmakers, J. M. (2005). Influence of plant species on population dynamics, genotypic diversity and antibiotic production in the rhizosphere by indigenous *Pseudomonas* spp. *FEMS Microbiology Ecology, 52*, 59–69.

Bevivino, A., Paganin, P., Bacci, G., Florio, A., Sampedro Pellicer, M., et al. (2014). Soil bacterial community response to differences in agricultural management along with seasonal changes in a Mediterranean region. *PLoS ONE, 9*, e105515.

Bonanomi, G., D'Ascoli, R., Scotti, R., Gaglione, S. A., Caceres, M. G., et al. (2014). Soil quality recovery and crop yield enhancement by combined application of compost and wood to vegetables grown under plastic tunnels. *Agriculture Ecosystems and Environment, 192*, 1–7.

Botton, S., van Heusden, M., Parsons, J. R., Smidt, H., & van Straalen, N. (2006). Resilience of microbial systems towards disturbances. *Critical Review in Microbiology, 32*, 101–112.

Buckley, D. H., & Schmidt, T. M. (2003). Diversity and dynamics of microbial communities in soils from agro-ecosystems. *Environmental Microbiology, 5*, 441–452.

Chaudhry, V., Rehman, A., Mishra, A., Chauhan, P. S., & Nautiyal, C. S. (2012). Organic and chemical amendments changes in bacterial community structure of agricultural land due to long-term organic and chemical amendments. *Microbial Ecology, 64*, 450–460.

Chiarini, L., Bevivino, A., Dalmastri, C., Nacamulli, C., & Tabacchioni, S. (1998). Influence of plant development, cultivar and soil type on microbial colonization of maize roots. *Applied Soil Ecology, 8*, 11–18.

Dalmastri, C., Chiarini, L., Cantale, C., Bevivino, A., & Tabacchioni, S. (1999). Soil type and maize cultivar affect the genetic diversity of maize root-associated *Burkholderia cepacia* populations. *Microbial Ecology, 38*, 273–284.

De Leij, F. A. A. M., Whipps, J. M., & Lynch, J. M. (1994). The use of colony development for the characterization of bacterial communities in soil and on roots. *Microbial Ecology, 27*, 81–97.

De Sanctis, G., Roggero, P. P., Seddaiu, G., Orsini, R., Porter, C. H., & Jones, J. W. (2012). Long-term no tillage increased soil organic carbon content of rain-fed cereal systems in a Mediterranean area. *European Journal of Agronomy, 40*, 18–27.

De Vries, F. T., & Shade, A. (2013). Controls on soil microbial community stability under climate change. *Frontiers in Microbiology, 4*, 1–16.

Delmont, T. O., Robe, P., Cecillon, S., Clark, I. M., Constancias, F., Simonet, P., Hirsch, P. R., & Vogel, T. M. (2011). Accessing the soil metagenome for studies of microbial diversity. *Applied Environmental Microbiology, 77*, 1315–1324.

Di Cello, F., Bevivino, A., Chiarini, L., Fani, R., Paffetti, D., & Tabacchioni, S. (1997). Biodiversity of a *Burkholderia cepacia* population isolated from maize rhizosphere at different plant growth stages. *Applied and Environmental Microbiology, 63*, 4485–4493.

Dorodnikov, M., Blagodatskaya, E., Blagodatsky, S., Fangmeier, A., & Kuzyakov, Y. (2009). Stimulation of r- vs. K-selected microorganisms by elevated atmospheric CO_2 depends on soil aggregate size. *FEMS Microbiology Ecology, 69*, 43–52.

Dupouey, J. L., Dambrine, E., Laffite, J. D., & Moares, C. (2002). Irreversible impact of past land use on forest soils and biodiversity. *Ecology, 83*, 2978–2984.

Fierer, N., Leff, J. W., Adams, B. J., Nielsen, U. N., Bates, S. T., et al. (2012). Cross-biome metagenomic analyses of soil microbial communities and their functional attributes. *Proceedings of the National Academy of Sciences, 109*, 21390–21395.

Garland, J. L. (1997). Analysis and interpretation of community-level physiological profiles in microbial ecology. *FEMS Microbiology Ecology, 24*, 289–300.

Ghimire, R., Norton, J. B., Stahl, P. D., & Norton, U. (2014). Soil microbial substrate properties and microbial community responses under irrigated organic and reduced-tillage crop and forage production systems. *PLoS ONE, 9*, e103901.

Griffiths, B. S., Bonkowski, M., Roy, J., & Ritz, K. (2001). Functional stability, substrate utilization and biological indicators of soils following environmental impacts. *Applied Soil Ecology, 16*, 49–61.

Hendrix, P. H., Crossley, D. A., & Coleman, D. C. (1990). Soil biota as components of sustainable agro-ecosystem. In C. A. Edwards, R. Lal, P. Madden, R. H. Miller, & G. House (Eds.), *Sustainable agricultural systems* (pp. 637–654). Soil Water Conservation Society: Iowa.

Hobbs, R. J., & Hueneke, L. F. (1992). Disturbance, diversity, and invasion: Implications for conservation. *Conservation Biology, 6*, 324–337.

Houghton, R. A., & Goodale, C. (2004). Effects of land-use change on the carbon balance of terrestrial ecosystems. In R. De Fries, G. Asner, & R. A. Houghton (Eds.), *Ecosystems and land use change, geophysical monograph series* (Vol. 153, pp. 85–98). Washington, DC: American Geophysical Union.

Hugenholtz, P., & Tyson, G. W. (2008). Metagenomics. *Nature, 455*, 481–484.

Janssen, P. H. (2006). Identifying the dominant soil bacterial taxa in libraries of 16S rRNA and 16S rRNA genes. *Applied and Environmental Microbiology, 72*, 1719–1728.

Jurburg, S. D., & Salles, J. F. (2015). Functional redundancy and ecosystem function – The soil microbiota as a case study. In: Y. H. Lo, J. A. Blanco, S. Roy (Eds.), *Agricultural and biological sciences. Biodiversity in ecosystems – Linking structure and function* (pp. 29–49).

Kirk, J. L., Beaudettea, L. A., Hartb, M., Moutoglisc, P., Klironomos, J. N., Lee, H., & Trevors, J. T. (2004). Methods of studying soil microbial diversity. *Journal of Microbiological Methods, 58*, 169–188.

Krzyżak, J., Wasilkowski, D., Płaza, G. A., Mrozik, A., Brigmon, R. L., & Pogrzeba, M. (2013). Culture methods as indicators of the biological quality of phytostabilized heavy metal-contaminated soil. *Environmental Biotechnology, 9*, 6–13.

Lagomarsino, A., Grego, S., Marhan, S., Moscatelli, M. C., & Kandeler, E. (2009). Soil management modifies micro-scale abundance and function of soil microorganisms in a Mediterranean ecosystem. *European Journal of Soil Science, 60*, 2–12.

Landi, S., Piccolo, R., Simoncini, S., & Pastorelli, R. (2011). Molecular characterization of the microbial community involved in the carbon cycle in different areas and types of soil management. *Environmental Quality, 6*, 29–38.

Lauber, C. L., Ramirez, K. S., Aanderud, Z., Lennon, J., & Fierer, N. (2013). Temporal variability in soil microbial communities across land-use types. *The ISME Journal, 7*, 1641–1650.

Levine, U. Y., Teal, T. K., Robertson, G. P., & Schmidt, T. M. (2011). Agriculture's impact on microbial diversity and associated fluxes of carbon dioxide and methane. *The ISME Journal, 5*, 1683–1691.

Li, J., Li, Z., Wang, F., Zou, B., Chen, Y., et al. (2015). Effects of nitrogen and phosphorus addition on soil microbial community in a secondary tropical forest of China. *Biology and Fertility of Soils, 51*, 207–215.

Luo, X., Fu, X., Yang, Y., Cai, P., Peng, S., et al. (2016). Microbial communities play important roles in modulating paddy soil fertility. *Nature Scientific Reports, 6*, 20326.

Mathew, R. P., Feng, Y., Githinji, L., Ankumah, R., & Balkcom, K. S. (2012). Impact of no-tillage and conventional tillage systems on soil microbial communities. *Applied Environmental Soil Sciences*, ID 548620. http://dx.doi.org/10.1155/2012/548620.

Michel, H. M., & Williams, M. (2011). Soil habitat and horizon properties impact bacterial diversity and composition. *Soil Science Society of America Journal, 75*, 1440–1448.

Miller, M., & Dick, R. P. (1995). Thermal stability and activities of soil enzymes influenced by crop rotations. *Soil Biology and Biochemistry, 27*, 1161–1166.

Myrold, D. D., Zeglin, L. H., & Jansson, J. K. (2014). The potential of metagenomic approaches for understanding soil microbial processes. *Soil Science Society of America Journal, 78*, 3–10.

Nesme, J., Achouak, W., Agathos, S. N., Bailey, M., Baldrian, P., et al. (2016). Back to the future of soil metagenomics. *Frontiers in Microbiology, 7*, 1–5.

Nielsen, M. N., & Winding, A. (2002). *Microorganisms as indicators of soil health* (Technical Report No. 388). National Environmental Research Institute, Denmark. Available at http://www.dmu.dk/1_viden/2_Publikationer/3_fagrapporter/rapporter/FR388.pdf

Øvreås, L. (2000). Population and community level approaches for analyzing microbial diversity in natural environments. *Ecology Letters, 3*, 236–251.

Pane, C., Piccolo, A., Spaccini, R., Celano, G., Villecco, D., & Zaccardelli, M. (2013). Agricultural waste-based composts exhibiting suppressivity to diseases caused by the phytopathogenic soil-borne fungi *Rhizoctonia solani* and *Sclerotinia minor*. *Applied Soil Ecology, 65*, 43–51.

Papaleo, M. C., Fondi, M., Maida, I., Perrin, E., Bevivino, A., et al. (2015). Analysis of a pool of small plasmids from soil heterotrophic cultivable bacterial communities. *The Open Microbiology Journal, 9*, 98–109.

Parfitt, R. L., Scott, N. A., Ross, D. J., Salt, G. J., & Tate, K. R. (2003). Land use change effects on soil C and N transformations in soils of high N status: Comparisons under indigenous forest, pasture and pine plantation. *Biogeochemistry, 66*, 203–221.

Pastorelli, R., Piccolo, R., Cocco, S., & Landi, S. (2010). mRNA recovery and denitrification gene expression in clay-soil bacterial communities under different agricultural managements. *Agrochimica, 54*, 179–192.

Paula, F. S., Rodrigues, J. L. M., Zhou, J., Wu, L., Mueller, R. C., et al. (2014). Land use change alters functional gene diversity, composition and abundance in Amazon forest soil microbial communities. *Molecular Ecology, 23*, 2988–2999.

Peterson, G., Allen, C. R., & Holling, C. S. (1998). Ecological resilience, biodiversity and scale. *Ecosystems, 1*, 6–18.

Pianka, E. R. (1970). On r and K selection. *The American Naturalist, 104*, 592–597.

Preston-Mafham, J., Boddy, L., & Randerson, P. F. (2002). Analysis of microbial, community, functional, diversity, using sole-carbon-source utilisation profiles: a critique. *FEMS Microbiology Ecology, 42*, 1–14.

Ramachandran, V. K., East, A. K., Karunakaran, R., Downie, J. A., & Poole, P. S. (2011). Adaptation of *Rhizobium leguminosarum* to pea, alfalfa and sugar beet rhizospheres investigated by comparative transcriptomics. *Genome Biology, 12*, R106–R106.

Rastogi, G., & Sani, R. K. (2011). Molecular techniques to assess microbial community structure, function, and dynamics in the environment. In I. Ahmad, F. Ahmad, & J. Pichtel (Eds.), *Microbes and microbial technology: Agricultural and environmental applications* (pp. 29–57). New York: Springer.

Sala, O. E., Chapin, F. S., Armesto, J. J., Berlow, E., Bloomfield, J., Dirzo, R., et al. (2000). Biodiversity-global biodiversity scenarios for the year 2100. *Science, 287*, 1770–1774.

Seddaiu, G., Icola, I., Farina, R., Orsini, R., Iezzi, G., & Roggero, P. P. (2016). Long term effects of tillage practices and N fertilization in rainfed Mediterranean cropping systems: durum wheat, sunflower and maize grain yield. *European Journal of Agronomy, 77*, 166–178.

Shade, A., Peter, H., Allison, S. D., Baho, D. L., Berga, M., et al. (2012). Fundamentals of microbial community resistance and resilience. *Frontiers in Microbiology, 3*, 1–19.
Shange, R. S., Ankumah, R. O., Ibekwe, A. M., Zabawa, R., & Dowd, S. E. (2012). Distinct soil bacterial communities revealed under a diversely managed agroecosystem. *PLoS ONE, 7*, e40338.
Sikorski, J. (2015). The prokaryotic biology of soil. *Soil Organisms, 87*, 1–28.
Smalla, K., Wieland, G., Buchner, A., Zock, A., Parzy, J., Kaiser, S., & Roskot, N. (2001). Bulk and rhizosphere soil bacterial communities studied by denaturing gradient gel electrophoresis: plant-dependent enrichment and seasonal shifts revealed. *Applied Environmental Microbiology, 67*, 4742–4751.
Staley, J. T., & Konopka, A. (1985). Measurement of in situ activities of non photosynthetic microorganisms in aquatic and terrestrial habitats. *Annual Reviews of Microbiology, 39*, 321–346.
Stenström, J., Svensson, K., & Johansson, M. (2001). Reversible transition between active and dormant microbial states in soil. *FEMS Microbiol Ecology, 36*, 93–104.
Tago, K., Ishii, S., Nishizawa, T., Otsuka, S., & Senoo, K. (2011). Phylogenetic and functional diversity of denitrifying bacteria isolated from various rice paddy and rice-soybean rotation fields. *Microbes and Environments, 26*, 30–35.
Tiedje, J. M., Asuming-Brempong, S., Nusslein, K., Marsh, T. L., & Flynn, S. J. (1999). Opening the black box of soil microbial diversity. *Applied Soil Ecology, 13*, 109–122.
Tilston, E. L., Sizmur, T., Dixon, G. R., Otten, W., & Harris, J. A. (2010). The impact of land-use practices on soil microbes. In G. R. Dixon & E. L. Tilston (Eds.), *Soil microbiology and sustainable crop production* (pp. 273–295). New York: Springer.
Torsvik, V., & Øvreås, L. (2002). Microbial diversity and function in soil: from genes to ecosystems. *Current Opinion in Microbiology, 5*, 240–245.
Van Elsas, J. D., Torsvik, V., Hartmann, A., Øvreås, L., & Jansson, J. K. (2007). The bacteria and archaea in soil. In J. D. Van Elsas, J. K. Jansson, J. T. Trevors (Eds.), *Modern soil microbiology* (2nd ed, pp. 23–54). Boca Raton: CRC Press.
Vida, C., Bonilla, N., de Vicente, A., & Cazorla, F. M. (2016). Microbial profiling of a suppressiveness-induced agricultural soil amended with composted almond shells. *Frontiers in Microbiology, 7*, 1–14.
Westergaard, K., Muller, A. K., Christensen, S., Bloem, J., & Sørensen, S. J. (2001). Effects of tylosin as a disturbance on the soil microbial community. *Soil Biology and Biochemistry, 33*, 2061–2071.
Widenfalk, A., Bertilsson, S., Sundh, I., & Goedkoop, W. (2008). Effects of pesticides on community composition and activity of sediment microbes – Responses at various levels of microbial community organization. *Environmental Pollution, 152*, 576–584.

Chapter 6
A Metagenomic Study on the Effect of Aboveground Plant Cover on Soil Bacterial Diversity

Mariantonietta Colagiero, Isabella Pentimone, Laura Cristina Rosso, and Aurelio Ciancio

Abstract To study the effect of cropping and plant cover on soil microbial diversity, three adjacent sites characterized by the same soil, but differing for land use and cover, were sampled at Carovigno (Brindisi). Soil samples were collected from a traditional greenhouse producing horticultural crops, a close olive grove and an adjacent Mediterranean (*Quercus ilex*) forest spot. The soil bacterial communities from replicated samples were identified with a metagenomic Next Generation Sequencing (NGS) approach. Total RNAs were extracted from 2 g soil subsamples and the V3-V4 hypervariable regions of the 16S rRNAs were sequenced with the Illumina MiSeq technology. A total of $2.46 \bullet 10^6$ reads was produced from 13 samples, of which 85% passed the quality threshold, yielding an average of $97–239 \bullet 10^3$ reads per sample. Almost all (99%) sequences belonged to the Kingdom Bacteria, and 30% were informative up to the species level. Using the Greengenes classification system, the average number of species per sample was around 10^3. Most represented phyla in all samples were *Proteobacteria*, *Actinobacteria*, *Firmicutes*, *Planctomycetes* and *Verrucomicrobia*, with α-, β- and γ-*Proteobacteria* as most represented classes. NGS data showed that samples from cultivated soils (olive and vegetables) had higher frequencies (5–10%) of Bacillales, which were under-represented in the Mediterranean forest. Data analysis at the species level identified changes in the bacterial composition at deeper taxonomic levels, related to agricultural practices.

Keywords Bacteria • Microbiome • Next generation sequencing • Rhizosphere • Roots ecology

M. Colagiero (✉) • I. Pentimone • L.C. Rosso • A. Ciancio
Istituto per la Protezione Sostenibile delle Piante, Consiglio Nazionale delle Ricerche, Via Amendola 122/D, 70126 Bari, Italy
e-mail: mariantonietta.colagiero@ipsp.cnr.it

M. Lukac et al. (eds.), *Soil Biological Communities and Ecosystem Resilience*, Sustainability in Plant and Crop Protection, DOI 10.1007/978-3-319-63336-7_6

6.1 Introduction

Soil biodiversity plays a fundamental role in the ecology of plants, underpinning a variety of ecosystem services. The identification of the soil microbial components and their interactions with other telluric organisms (mainly invertebrates) are useful to improve or forecast the effects of soil environment and ecosystem management. It is generally recognized today that thousands of microbial taxa may be found in a few g of soil. Bacteria represent the most numerous organisms inhabiting soil. A few g of soil may contain billion bacterial cells, including species inhabiting specific niches and present at very low densities (Paul and Clark 1996).

For many years, studies carried out on bacterial biodiversity and related metabolism were based on isolation and cultivation, usually performed on artificial substrates. However, by this way, only a small fraction of organisms present in a given ecosystem can be studied, and in particular only the cultivable ones (Pace et al. 1986). New molecular biology procedures facilitated nowadays the study of entire microbial communities from a wide range of niches, including soil and plant rhizosphere. By means of metagenomic analyses we can directly sequence the genetic material extracted from an environmental sample, which is representative of its species composition and abundance (Turnbaugh et al. 2009). A huge amount of data is produced by this approach, whose analysis and interpretation represent a real challenge. Extracting and sequencing all the nucleic acids present in soil is thus only the first step of metagenomics analyses, which have to be integrated by the application of advanced bioinformatic and statistical tools.

We have applied this approach to study the effect of cropping and plants on soil microbial diversity, in three adjacent sites only differing for land use and vegetation cover.

6.2 The Metagenomic Approach

Next-Generation Sequencing systems, such as Illumina™, are today standard tools of choice for in-depth analysis of microbiome, in various environments. Metagenomics allow the simultaneous analysis of one or more genes from all organisms (metagenome) present in a given sample or environment. The study of microbial diversity by these methods allows access to vastly more information than previously obtainable. Within the wide range of metagenomic study types, those based on sequencing of 16S ribosomal RNA variable regions (16S rRNA) represent the most frequent choice for taxonomic purposes (Pace et al. 1986). They allow the identification of the bacterial taxa present in a microbial community and of their relative abundance. Examples of metagenomic studies involve the microbial populations present in ocean waters (Huber et al. 2007; Sogin et al. 2006) or underground (Urich et al. 2008), as well as the comparative characterization of the microflora from the human oral cavity (Keijser et al. 2008) or gut (Turnbaugh et al. 2008).

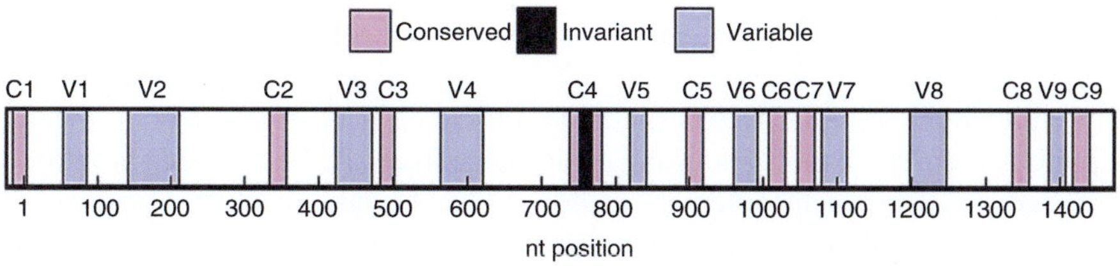

Fig. 6.1 A schematic representation of the bacterial 16S ribosomal RNA gene (16S rRNA). The variable regions are *highlighted* in *blue*, those preserved are shown in *pink*. The *black* region is an invariant motif, found in all Bacteria

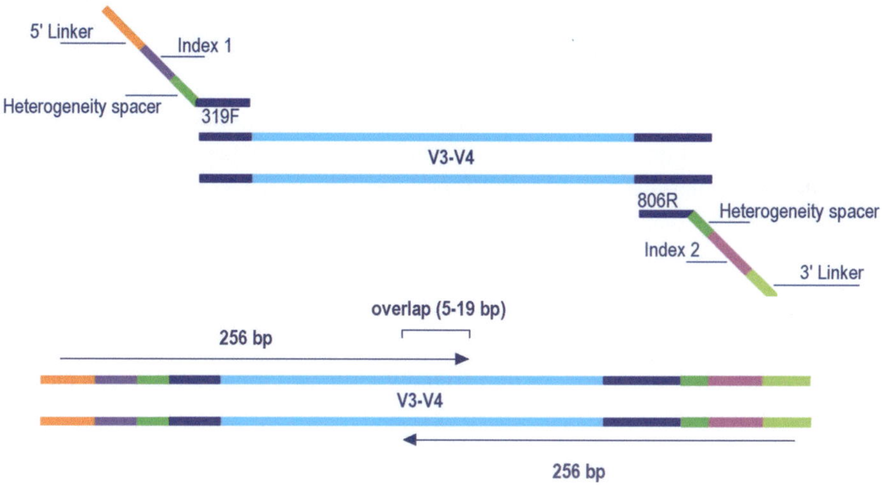

Fig. 6.2 Amplification scheme of the hypervariable regions V3-V4 of the 16S rRNA gene, on MiSeq™ platform (Adapted from Illumina™)

The 16S rRNA gene is generally used in molecular phylogenetic studies and biodiversity. Its structural organization based on 1542 nucleotides (Fig. 6.1), offers conserved and specific hypervariable regions.

The latter represent recognition sequences used to evaluate the genetic diversity within microbial communities and to determine the phylogenetic relationships along large evolutionary distances. Similarly, the conserved sequences allow to design universal primers, which are used in the DNA amplification reactions (Backer et al. 2003).

Using the MiSeq™ System Illumina platform (www.illumina.com/miseq) it is possible to sequence both ends of two adjacent hypervariable regions like the V3-V4 of the 16S rRNA gene, which is about 460 nt long. The region of interest is amplified using the primers 319F 5′-CCTACGGGNGGCWGCAG-3′ (forward) and 806R 5′-GGACTACHVGGGTWTCTAAT-3′ (reverse) to which unique adapters for each sample are anchored (Fig. 6.2). The sequencing (called "shot-gun") produces a high number of sequences that must be reassembled and analyzed by bionformatic tools.

6.3 Bioinformatic Tools

The study of the information content of DNA is the ultimate goal of bioinformatics. Application software tools, including open source freeware that has been developed and made available on the internet, can be used for the analysis of NGS sequencing data. The first analytical step involves the formation of contigs joining the single reads of the V3-V4 region of the 16S rRNA gene produced by the sequencing platform in two opposite directions (Pace et al. 1986). For this purpose, we used PandaSeq Assembler (https://github.com/neufeld/pandaseq) (Masella et al. 2012). Since the Illumina platform can generate sequences of amplicons in a paired-end format, the assembly requires a correspondence in the overlapping regions of two opposite reads, in order to produce the correct sequence (Bartram et al. 2011). PandaSeq uses as input the Illumina data in FASTA format, overlapping the reads for the reconstruction of the contig (the entirely assembled fragment from the original sequence). The mounting is extremely fast, and millions of paired-end sequences may be assembled quickly. The contigs formation is necessary to proceed further with the processing through a series of modules. The sequences are assembled for the final analysis that will lead to the taxonomic identification and evaluation of species biodiversity and distribution, depending on the different samples and treatments. For data analysis, we used the software QIIME (Quantitative Insights Into Microbial Ecology, http://qiime.org/), an open-source application that allows the analysis of data generated by most common sequencing platforms. The results may be easily interpreted also thanks to a friendly graphical interface implemented in this tool (Caporaso et al. 2010).

6.4 Soil Metagenomic Analysis

Soil microbial composition is influenced by several factors including the presence of the aboveground plants. In this view, we carried out a metagenomic study of three different vegetation covers in Apulia. We analysed the biodiversity present in a red soil (terra rossa) when subjected to different uses: (1) a greenhouse with horticultural productions, (2) an adjacent olive grove and (3) a nearby Mediterranean forest spot. The study area is located at Carovigno (Brindisi). The samples (three replicates for olive and greenhouse, and two for the forest environment) were taken from the different sites in summer 2014 and two g of soil from each sample were processed for extraction of total RNAs, using a specific kit (Mobio, CA, USA). The RNA, obtained accounted for the active, living cells present in soil, was retro-transcribed to cDNA amplifying the hypervariable V3-V4 region and sequenced by a commercial provider (IGA-Technology, Udine; http://www.igatechnology.com/) with an HiSeq Illumina platform.

Sequencing produced $2.46 \cdot 10^6$ valid sequences, later subjected to bioinformatic analyses concerning the number and distribution of species and to characterize the different sampling environments. QIIME produced 7492 Operational Taxonomic

Units (OTUs) differently distributed in the various samples. The OTU is an operational definition of a species or a taxonomic group of species, often used when only data from the genome sequence, based on their similarity, are available. Almost 99% of the sequences obtained were assigned to the Kingdom Bacteria with dominant groups represented by *Proteobacteria*, *Actinobacteria*, *Gemmatimonadetes*, *Planctomycetes* and *Verrucomicrobia*.

The soil microbial profile of the three environments was explored by evaluating the α and β diversities. These terms were proposed by Whittaker (1960) to indicate the species diversity within a given sample (α), and the that observed among different samples (β). The α-diversity was analyzed using different metrics: Chao-1, which provides a parameter indicative of the species richness; the observed species, describing the uniqueness of the OTUs present in each sample, and the PD-Whole tree, that relies on evolutionary distances, using a phylogenetic tree (Hughes et al. 2001).

The soil environment of the forest spot showed a greater number of OTUs, compared to the olive grove and the greenhouse soils (Fig. 6.3). This is likely due to the lack of intervention of anthropic and cultural activities that instead concerned the

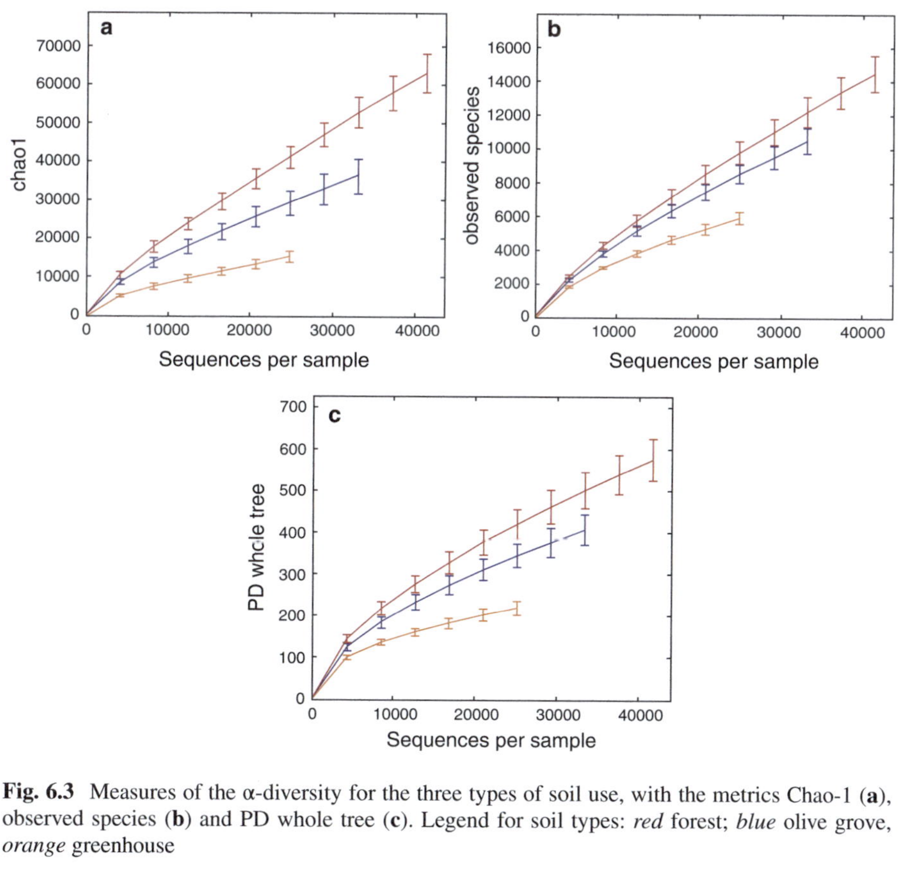

Fig. 6.3 Measures of the α-diversity for the three types of soil use, with the metrics Chao-1 (**a**), observed species (**b**) and PD whole tree (**c**). Legend for soil types: *red* forest; *blue* olive grove, *orange* greenhouse

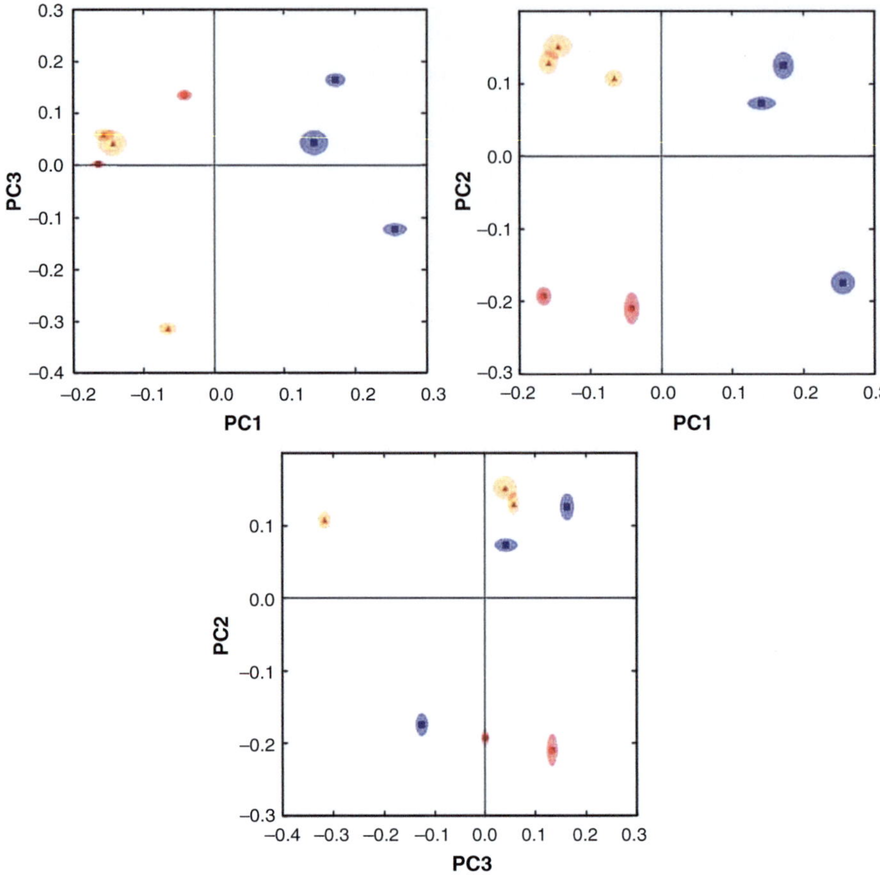

Fig. 6.4 Representation of the β-diversity among the different soil types (*red* forest, *blue* olive, *yellow* greenhouse), on the three PCA planes (the variation explained was: PC1 = 25.72%, PC2 = 19.92%, PC3 = 16.46%)

other two environments investigated, during long periods of time (estimated around 40–50 years, as inferred by the olive trees age). As shown in Fig. 6.4, the measure of the β-diversity, as shown by principal coordinate analysis (PCA), allowed a clear separation of the three soil micro-environments examined.

The OTUs identified at different taxonomic levels in the samples are shown through an interactive heatmap, which allows the fast display of taxonomies, the relative frequencies and the sample each OTU belongs to (Fig. 6.5).

At the Phylum level, the lineages identified were mainly represented by *Acidobacteria*, *Actinobacteria*, *Chloroflexi*, *Gemmatimonadetes*, *Proteobacteria*, *Planctomycetes* and *Verrucomicrobia* (Fig. 6.6).

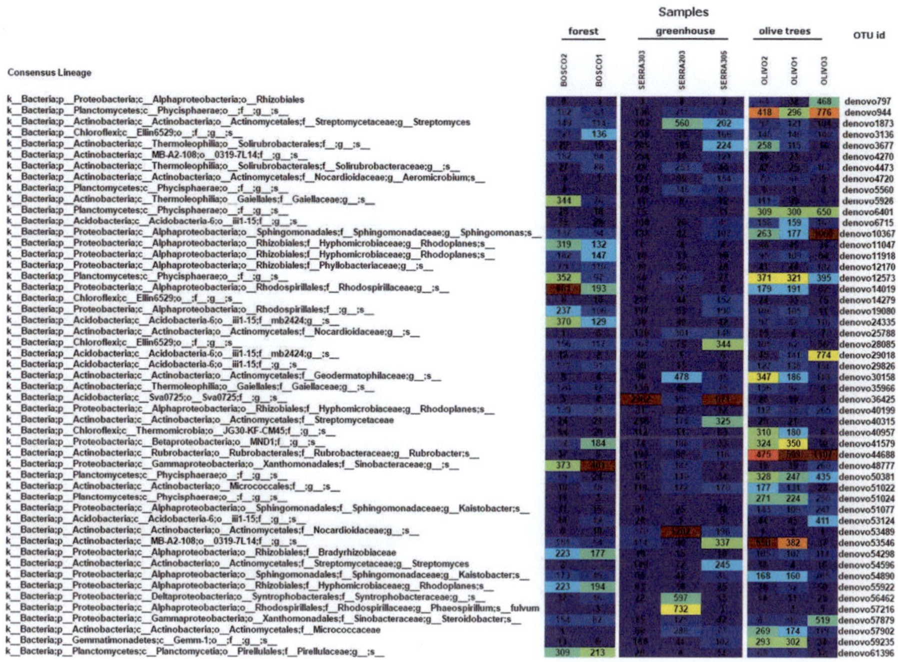

Fig. 6.5 OTUs (at last identified levels) and abundance per sample. For each column the colors vary from minima (*blue*) to maxima (*red*)

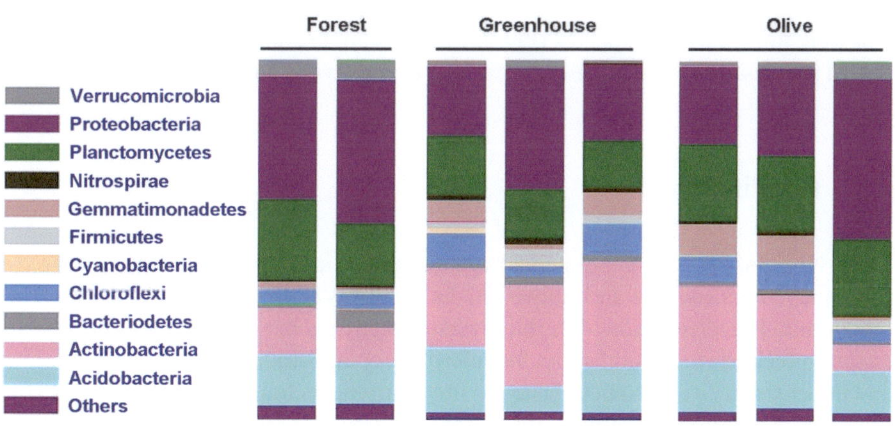

Fig. 6.6 Most represented Phyla (% of sequence higher than 1%), per sample

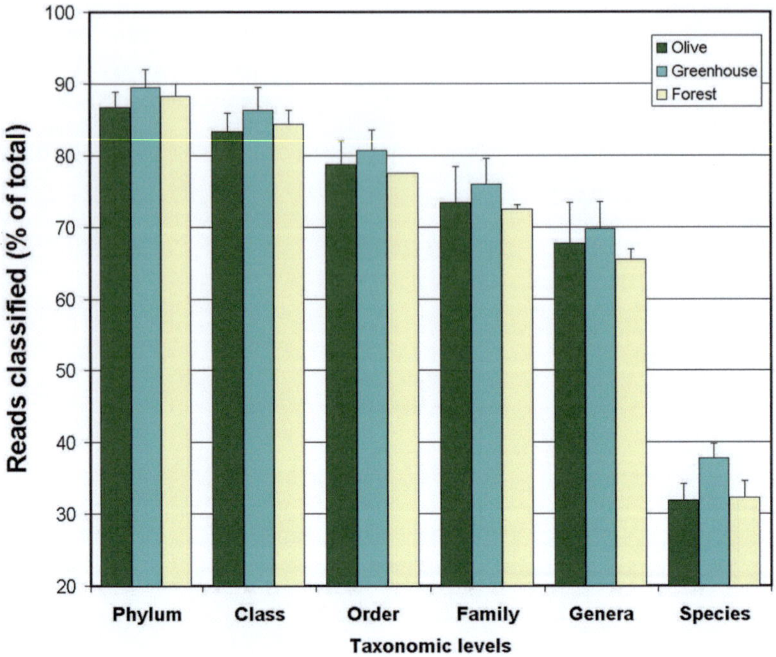

Fig. 6.7 Percent of reads classified for the three soil micro-environments (mean ± SD)

The majority (99%) of sequences belonged to the Kingdom Bacteria. Their taxonomic assignments progressively declined at finest levels, with only around 30% that were informative as species (Fig. 6.7). The sample composition for most represented species showed unique and common taxa as well (Table 6.1).

6.5 Conclusions

Many NGS assays have generated a huge amount of information that is challenging for management, storage and, above all, data analysis (Pop and Salzberg 2008). The taxonomic characterization of the biodiversity, analyzed in a metagenomic approach, ensures precise information about the microbial composition of soil at a certain time and under given circumstances. The sensitivity of the technique permits the description of each sample as a single entity, independent from the surrounding environment. This is supported by the variation observed in each replication, from the same environment. If integrated with other similar sampling plans, the data can be used as indicators of the functional state of soil and the effects caused by use and aboveground vegetation regimes.

Table 6.1 Percent of total reads for the top 8 species, in the three soil micro-environments sampled

Genus	Species	Olive			Greenhouse			Forest	
Aeromicrobium	*ponti*				1.13	3.19	1.05		
Asticcacaulis	*taihuensis*		0.85						
Bifidobacterium	*bombi*	0.94	0.74		1.27		1.08	1.19	1.37
Caldithrix	*paleochoryensis*	0.79							
Chthoniobacter	*flavus*			1.13				0.91	1.40
Cohnella	*soli*	0.80	1.03						0.73
Condromyces	*pediculatus*			2.80				2.02	1.20
Euzebya	*tangerina*		0.82		1.21		1.03		
Gemmatimonas	*aurantiaca*	1.79	1.78		1.09				
Kouleothrix	*aurantiaca*	0.74		1.29					
Kribbella	*ginsengisoli*						0.99	0.71	0.61
Megsphaera	*hominis*	1.38	1.17	1.45		0.89		1.06	1.28
Methylobaterium	*goesingense*			0.96					
Peptoniphilus	*coxi*					0.83			
Phaeospirillium	*fulvum*					1.10			
Rhodovibrio	*sodomensis*				1.42		1.05		
Runella	*limosa*							1.55	
Ca. "Scalindua	brodae"	2.02	1.56	2.90	1.13	1.23	1.66	1.19	2.17
Solirubrobacter	*soli*					0.83			
Sporotomaculum	*syntrophicum*				2.07	1.33	5.46		
Steroidobacter	*denitrificans*			1.82					

As reported in the literature, many microorganisms present in soil are still unknown, because they cannot be cultured in the laboratory by conventional techniques and therefore they were not described so far. The detailed knowledge of the microbial composition of soil is hence important to enlarge the range of species and identify the ecosystem services they provide.

References

Backer, G. C., Smith, J. J., & Cowan, D. A. (2003). Review and re-analysis of domain-specific 16S primers. *Journal of Microbiological Methods, 55*, 541–555.

Bartram, A. K., Lynch, M. D., Stearns, J. C., Moreno-Hagelsieb, G., & Neufeld, J. D. (2011). Generation of multimillion-sequence 16S rRNA gene libraries from complex microbial communities by assembling paired-end Illumina reads. *Applied and Environmental Microbiology, 77*, 3846–3852.

Caporaso, J. G., Kuczynski, J., Stombaugh, J., Bittinger, K., Bushman, F. D., et al. (2010). QIIME allows analysis of high-throughput community sequencing data. *Nature Methods, 7*, 335–336.

Huber, J. A., Mark Welch, D. B., Morrison, D. B., Huse, S. M., Neal, P. R., et al. (2007). Microbial population structures in the deep marine biosphere. *Science, 318*, 97–100.

Hughes, J. B., Hellmann, J. J., Ricketts, T. H., & Bohannan, B. J. M. (2001). Counting the uncountable: statistical approaches to estimating microbial diversity. *Applied and Environmental Microbiology, 67*, 4399–4406.

Keijser, B. J. F., Zaura, E., Huse, S. M., van der Vossen, J. M. B. M., Schuren, F. H. J., et al. (2008). Pyrosequencing analysis of the oral microflora of healthy adults. *Journal of Dental Research, 87*, 1016–1020.

Masella, A. P., Bartram, A. K., Truszkowski, J. M., Brown, D. G., & Neufeld, J. D. (2012). PANDAseq: PAired-eND assembler for Illumina sequences. *BMC Bioinformatics, 13*, 31.

Pace, N. R., Olsen, G. J., & Woese, C. R. (1986). Ribosomal RNA phylogeny and the primary lines of evolutionary descent. *Cell, 45*, 325–326.

Paul, E. A., & Clark, F. E. (1996). *Soil microbiology and biochemistry* (340 pp). San Diego: Academic.

Pop, M., & Salzberg, S. L. (2008). Bioinformatics challenges of new sequencing technology. *Trends in Genetics, 24*, 142–149.

Sogin, M. L., Morrison, H. G., Huber, J. A., Mark Welch, D. B., Huse, S. M., et al. (2006). Microbial diversity in the deep sea and the underexplored "rare biosphere". *Proceedings of the National Academy of Science, USA, 103*, 12115–12120.

Turnbaugh, P. J., Ridaura, V. K., Faith, J. J., Rey, F. E., Knight, R., & Gordon, J. I. (2009). The effect of diet on the human gut microbiome: A metagenomic analysis in humanized gnotobiotic mice. *Science Translational Medicine, 1*, 6ra14.

Turnbaugh, P. J., Bäckhed, F., Fulton, L., & Gordon, J. I. (2008). Diet-induced obesity is linked to marked but reversible alterations in the mouse distal gut microbiome. *Cell Host Microbe, 3*, 213–223.

Urich, T., Lanzén, A., Qi, J., Huson, D. H., Schleper, C., & Schuster, S. C. (2008). Simultaneous assessment of soil microbial community structure and function through analysis of the meta-transcriptome. *PLoS One, 3*, e2527.

Whittaker, R. H. (1960). Vegetation of the Siskiyou mountains, Oregon and California. *Ecological Monographs, 30*, 279–338.

Chapter 7
What Lies Beneath: Root-Associated Bacteria to Improve the Growth and Health of Olive Trees

David Ruano-Rosa, Antonio Valverde-Corredor, Carmen Gómez-Lama Cabanás, Rafael Sesmero, and Jesús Mercado-Blanco

Abstract During the last decades we have witnessed growing public concern on the abuse/misuse of agrochemicals to control plant pathogens. The fact that some relevant phytopathogens (for instance, the soil-borne fungus *Verticillium dahliae* Kleb.) are very difficult to control by methods alternative to chemical-based products, has urged researchers to seek effective measures within integrated disease management frameworks. Biological control, alone or in combination with other approaches, emerges as one of the most promising alternatives to confront plant pathogens in a sustainable, environment-friendly strategy. Effectiveness of biological control agents (BCA) largely depends on colonization and persistence capabilities in the ecological niches (e.g. root and/or rhizosphere) where their benefits are expected to be deployed. As a consequence, due to BCA-host specificity (or co-adaptation) the search of potential BCAs in their target environments seems an appropriate strategy. This chapter describes the isolation, identification and characterization of indigenous antagonist bacteria from the olive rhizosphere that can be eventually exploited as BCA against relevant pathogens affecting this woody crop, with emphasis on *V. dahliae*. The approach here implemented could be of interest for other pathosystems involving trees and soil-borne pathogens.

Keywords *Pseudomonas* spp. • *Paenibacillus* spp. • Biological control • Plant growth promoting rhizobacteria • *Verticillium dahliae* • Olive pathogens

D. Ruano-Rosa • A. Valverde-Corredor • C. Gómez-Lama Cabanás
R. Sesmero • J. Mercado-Blanco (✉)
Department of Crop Protection, Institute for Sustainable Agriculture (IAS), Agencia Estatal Consejo Superior de Investigaciones Científicas (CSIC), Avda. Menéndez Pidal s/n, Campus 'Alameda del Obispo', E-14004 Córdoba, Spain
e-mail: jesus.mercado@ias.csic.es

M. Lukac et al. (eds.), *Soil Biological Communities and Ecosystem Resilience*, Sustainability in Plant and Crop Protection, DOI 10.1007/978-3-319-63336-7_7

7.1 Introduction

Olive (*Olea europaea* L.) is a woody cultivated species of major importance for many countries in the Mediterranean Basin, where constitutes a fundamental economical pillar. In fact, more than 90% of global olive oil and table olive production is concentrated in this area (FAOSTAT 2016). The soil-borne fungal phytopathogen *Verticillium dahliae* Kleb, the causal agent of Verticillium wilt of olive (VWO), is one of the main biotic constrains affecting olive. Nowadays, the disease is considered one of the most serious threats for this crop. It has been described in many areas where olive is relevant, and reported to have spread alarmingly during the last two decades (López-Escudero and Mercado-Blanco 2011). At present, a single effective control measure against VWO is not available (Bubici and Cirulli 2011), and the management of the disease is a very complicated task. This difficulty can be explained by a combination of several causes, including the current lack of effective fungicides able to access the pathogen during its parasitic phase (i.e. within the xylem vessels) (López-Escudero and Mercado-Blanco 2011) even when applied in the trunk (Bubici and Cirulli 2012). Furthermore, *V. dahliae* produces resistant, dormant structures (microsclerotia), under adverse conditions, able to endure in soils for a prolonged period of time. The wide range of plants that can be infected by the pathogen, its genetic diversity, or inadequate agronomic practices are other factors reducing the chance to control VWO successfully (López-Escudero and Mercado-Blanco 2011; Tsror 2011). Therefore, the development of new, sustainable and alternative disease control measures is crucial to confront this pathogen. Approaches that should be considered as complementary within an integrated disease management framework.

The public concern on the abuse of agrochemicals (i.e. chemically-based fertilizers, fungicides, insecticides, etc.) increased in recent years, due to relevant negative effects on the environment, the ecosystems biodiversity and human or animal health. Consequently, several research teams have focused their efforts on the development of alternative, environment-friendly inputs aiming to increase crops yields and to control biotic threats towards relevant agro-ecosystems (Ruano-Rosa and Mercado-Blanco 2015). However, the implementation of control strategies replacing the traditional use of chemical biocides can be difficult in many situations. This is the case of trees and woody plants, although examples of successful biological control of diseases affecting these hosts are available (Cazorla and Mercado-Blanco 2016; and references therein).

In the olive-*V. dahliae* pathosystem, the use of biological control tools has been proposed as a complementary approach within the above-mentioned integrated disease management strategy, mainly as a preventive measure (López-Escudero and Mercado-Blanco 2011; Tjamos 1993). Biological control usually involves the application of microorganisms with antagonistic and/or plant growth promotion (PGP) effects. For instance, fungi such as *Trichoderma* spp. (*e.g.* Aleandri et al. 2015; Carrero-Carrón et al. 2016; Ruano-Rosa et al. 2016) or several bacterial genera have been reported as effective (Sanei and Razavi 2011) and/or with potential (Aranda et al. 2011) as biological control agents (BCA) against VWO. Regarding the use of bacteria to manage VWO, several studies showed that strains from *Pseudomonas fluorescens* and *Pseudomonas putida* (Maldonado-González et al. 2015b; Mercado-Blanco et al. 2004; Prieto et al. 2009), *Serratia plymuthica* (Müller et al. 2007) or *Paenibacillus alvei* (Markakis et al. 2016) performed well under different experimental conditions.

A reasonable strategy for BCA selection is to search for microorganisms residing in the same ecological niche in which they will be eventually applied (Knudsen et al. 1997; Ruano-Rosa and López-Herrera 2009). It is therefore expected that these BCA will be adapted to such a niche, being able to effectively colonize and endure in the target site (*e.g.* the roots) where they can deploy their beneficial effects (i.e. biological control activity). In this context, the aims of the present work were: (i) to generate a collection of cultivable bacteria originated from and adapted to the olive roots/rhizosphere; (ii) to screen for bacterial isolates showing antagonistic ability against *V. dahliae* and other phytopathogens; and (iii) to molecularly, biochemically and phenotypically characterize selected bacterial isolates showing traits traditionally associated with biological control and/or PGP. The final objective is to identify novel bacterial strains with potential to be used in future bioformulations effective against different olive diseases, with emphasis in VWO.

7.2 Materials and Methods

7.2.1 Sampling and Isolation of Cultivable Bacteria from Olive Roots

Root samples from one-year-old olive plants (cv. Picual) were collected from ten commercial nurseries located in Córdoba province (South Spain). Ten grams of fresh root tissue, thoroughly washed under tap water to remove most of the attached soil/substrate particles, and representative of the whole radical system of each plant (three per nursery), were cut and ground using a mortar and pestle with 10 mL of sterile distilled water. Aliquots (100 μl) of 10-fold serial dilutions of the macerates obtained from each sample were plated under aseptic conditions on different culturing media, namely potato dextrose agar (PDA; Oxoid, Basingstoke, UK), Luria Bertani agar (LB, 1% Bacto-tryptone, 0.5% yeast extract and 1% NaCl) and nutrient agar (NA, Oxoid). Plates were incubated at 28 °C in the dark at least for 72 h. Pure cultures originating from single colonies grown in these media were carefully selected and isolated attending to distinctive morphological characteristics, and cryopreserved in 2-ml vials containing 33% glycerol at −80 °C, until use.

7.2.2 *Molecular Identification of Bacteria*

A selection of bacteria (189 strains) from the generated collection was molecularly identified (genus level) by partial sequencing of the *16S rDNA* ribosomal gene. Then, a subset of 40 strains selected according to their potential as BCA, were further characterized by partial sequencing of the *gyrB* genes (coding for DNA gyrase, subunit B). For this aim, genomic DNA was obtained from 1 mL of each bacterial isolate grown in LB medium during 16 hours (28°C in the dark; 200 rpm) by using 'JETFLEX Genomic DNA Purification Kit' (Genomed, Löhne, Germany) according to the manufacturer's instructions. Amplifications were performed in a total volume of 25 µl containing 2.5 µl of 10× PCR buffer (50 µM KCl, 10 mM Tris–HCl pH 9 [25 °C], 1% v/v Triton X- 100), 1.5 (*16S rDNA*) or 2 (*gyrB*) mM $MgCl_2$, 0.2 µM each primer, 0.2 mM each dNTP, 1.25 U of BioTaq DNA Polymerase (Bioline Ltd, London, UK) and 100 ng of bacterial DNA. The *16S rDNA* sequence was amplified by PCR as follows: denaturation for 4 min at 94 °C, followed by 35 cycles of 45 s at 94 °C, 45 s at 55 °C and 1 min at 72 °C and a final extension step of 10 min at 72 °C. The primers pair F27 (5′AGAGTTTATCMTGGCTCAG3′) and R1492 (5′GRTACCTTGTTACGACTT3′) was used for *16S rDNA* amplification (1,465 bp amplicon size). PCR conditions for *gyrB* were: denaturation for 3.30 min at 94 °C, followed by 30 cycles of 1 min at 94 °C, 1 min at 53.8 °C and 2 min at 72 °C. The degenerated primers UP-1 (5′GAAGTCATCATGACCGTTCTGCAYGCNGGN GGNAARTTYGA3′) and UP-2r (5′AGCAGGGTACGGATGTGCGAGCCRTCN ACRTCNGCRTCNGTCAT3′) were used for amplification (amplicon size ≈ 1200 bp). Amplicons were purified using 'Favorgen GEL/PCR Purification Mini Kit' (Biotech Corp.; Pingtung, Taiwan) according to the manufacturer's specifications. DNA sequencing was performed by a commercial service (Sistemas Genómicos S.L., Valencia, Spain). For *gyrB,* primers UP-1S (5′GAAGTCATCA TGACCGTTCTGCA3′) and UP-2Sr (5′AGCAGGGTACGGATGTGCGAGCC3′) were used for sequencing as described by Yamamoto and Harayama (1995). DNA sequences were assembled in 'contigs' with CLC bio software (Aarhus, Denmark) and compared (http://www.ncbi.nlm. nih.gov/) with available databases to identify each bacterial isolate genus and, when possible, species.

7.2.3 **In vitro** *Antagonism Assays Against* **V. dahliae** *and Other Plant Pathogens*

The antagonistic activity of bacteria isolated from olive roots was assessed by *in vitro* dual culture assays in two stages. An initial screening of the whole collection (327 isolates) was performed attending to the antagonistic activity against V-937I, a highly-virulent, defoliating (D) isolate of *V. dahliae* previously characterized at the genetic, molecular and pathogenic level (Collado-Romero et al. 2006; Prieto et al. 2009; Maldonado-González et al. 2015a, b) (Table 7.1). A second screening round against a group of additional plant pathogens (Table 7.1) was then performed, using

Table 7.1 Plant pathogens used for *in vitro* assays to test the antagonistic activity of rhizobacteria isolated from olive roots

Specie	Isolate/strain	Reference/Source
Alternaria alternata	CN191, CN194	Dr. Longxian Ran, Agricultural University of Hebei, P.R. China
Colletotrichum nymphaeae	Col. 114	Moral et al. (2012)
Colletotrichum godetiae	Col. 516	Dr. A. Trapero Casas, Universidad de Córdoba, Córdoba, Spain
Fusarium oxysporum f. sp. *dianthi*	Fod 113, Fod108	Gómez-Lama Cabanás et al. (2012)
Phytophthora cinnamomi	CH1100	Dr. C. J. López-Herrera, Instituto de Agricultura Sostenible, CSIC, Córdoba, Spain
Pseudomonas savastanoi pv. savastanoi	PSS-3	Culture collection of the Plant-Microorganism Interactions Laboratory, IAS-CSIC
P. savastanoi pv. savastanoi	NCPPB 3335	Pérez-Martínez et al. (2007)
Rosellinia necatrix	Rn 320, Rn 400	López-Herrera and Zea-Bonilla (2007)
Verticillium dahliae (D)[a]	V-937I	Collado-Romero et al. (2006)
V. dahliae (D)[a]	Lebrija 1	Culture collection of the Plant-Microorganism Interactions Laboratory, IAS-CSIC
V. dahliae (ND)[a]	V-249I	Collado-Romero et al. (2006)

[a]*D* defoliating pathotype, *ND* non-defoliating pathotype

only those isolates that already showed positive antagonism against V-937I in the first screening (189 isolates). Assays for fungal pathogens were performed by placing combinations of the target pathogen (5-mm diameter mycelial agar plugs from 7-day-old cultures)/potential BCA (inoculated with a sterilized toothpick) at 2.5 cm distance on PDA (for fungal pathogens) and NA (for bacterial pathogens). Plates were incubated (25 °C in the dark) until the pathogen covered the distance separating both microorganisms in control plates (i.e. without antagonist) (Fig. 7.1a; upper left plate). For assays involving bacterial pathogens, potential antagonists were grown as described above over a pathogen lawn (OD_{600}: 0.1) on PDA and NA plates (28 °C in the dark) during 48 hours. Antagonistic activity (i.e. production of haloes and/or inhibition zones) was then scored. These experiments were repeated at least once for each phytopathogen/antagonist combination.

7.2.4 Phenotypic Characterization of Indigenous Bacteria from Olive Roots

To determine the presence of phenotypes associated to antagonism and/or PGP among isolates selected after the first screening, several activities were assessed, namely protease (Naik et al. 2013), catalase (Holt et al. 1994), phosphatase (Katznelson and Bose 1959), chitinase (Murthy and Bleakley 2012), siderophore production (Schwyn and Neilands 1987) and 2,3-butanediol (Methyl Red Voges Proskauer [MRVP] medium, according to manufacturer's specifications; Micro

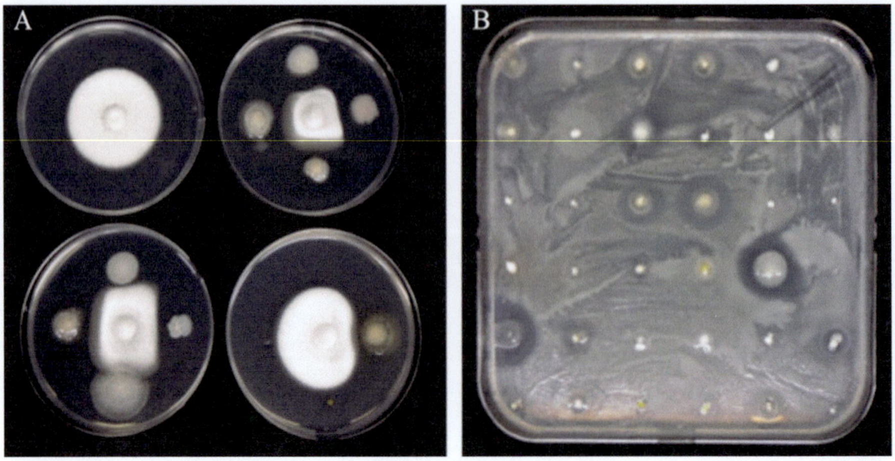

Fig. 7.1 *In vitro* antagonism of olive rhizosphere bacteria against two olive pathogens, on PDA. (**a**) antagonistic activity displayed by rhizobacteria against *V. dahliae* (D pathotype) (7-days-after inoculation). A comparison plate shows normal, full growth of *V. dahliae* in absence of bacteria (control, upper left plate). (**b**) antagonistic activity revealed by inhibition haloes around some bacterial colonies against *Pseudomonas savastanoi* pv. savastanoi (2-days-after inoculation)

Media, Nebotrade Ltd.; Budapest, Hungary). Additionally, in order to perform an in-depth evaluation of the nutritional requirements of a selected group of isolates (44, selected based on their antagonistic capability, phenotypic traits and suitability to be produced as BCA), we assessed their capability to metabolize 71 C sources and their sensibility to 23 chemical substances by using GEN III MicroPlate™ (Biolog; Hayward, CA). Evaluation was performed according to the manufacturer's specifications. All assays were repeated at least once.

7.3 Results

7.3.1 Generation of a Collection of Culturable Bacteria from Olive Roots and First In Vitro Antagonism Screening

A collection of 327 culturable bacterial isolates, natural inhabitants of roots from olive (cv. Picual) plants propagated under conditions usually implemented in nurseries at Southern Spain, was obtained following the sampling/isolation scheme described. Bacteria selected were representative of all colony morphologies found in the culturing media used. Overall, similar types of bacterial colonies were found regardless the media used (i.e. PDA, LB or NA). No major difference (i.e. prevalence of any colony morphology) was observed among nurseries either. After two rounds of growth to check for possible contaminants, all isolates selected were individually cryopreserved at -80°C until further use, and a code was assigned to each of them.

The first screening based on the antagonistic activity against *V. dahliae* V-937I (D pathotype) resulted in the selection of 121 isolates. Additionally, 68 isolates were included in the final selection according to different colony morphology, irrespective of their antagonistic capability against the D pathotype of *V. dahliae*. In the final selection, representative bacteria originating from all sampled nurseries were present.

7.3.2 Molecular Characterization of Cultivable Bacteria from Olive Roots

Sequencing of the *16S rDNA* (amplicon size 1,465bp) and *gyrB* (amplicon size ≈ 1,200bp) genes revealed the prevalence, in the collection, of taxa representative of the Proteobacteria (54.5%) followed by *Firmicutes* (34.4%) and *Actinobacteria* (9.5%) (Table 7.2). The genera distribution showed *Bacillus* spp. as the most abundant (33.3%), followed by isolates identified as *Pseudomonas* spp. (32.3%). Some isolates (1.57%) could not be unequivocally identified and were thus catalogued as unclassified (Table 7.2).

Table 7.2 Molecular identification of 189 bacteria isolated from roots of olive plants (var. Picual) based on analysis of the *16S rDNA* and *gyrB* gene sequences

Phyla	Class	Genera	Frequency (%)
Actinobacteria	*Actinobacteria*	*Cellulomonas*	1.06
		Curtobacterium	0.53
		Microbacterium	2.65
		Micrococcus	0.53
		Streptomyces	4.76
Firmicutes	*Bacilli*	*Bacilllus*	33.33
		Paenibacillus	1.06
Proteobacteria	*α-Proteobacteria*	*Agrobacterium*	1.59
		Ensifer	0.53
		Rhizobium	3.17
		Roseomonas	0.53
		Sinorhizobium	0.53
		Sphingomonas	0.53
	β-Proteobacteria	*Achromobacter*	1.59
		Cupriavidus	2.12
		Massilia	1.06
	γ -Proteobacteria	*Cronobacter*	3.70
		Enterobacter	2.65
		Pseudomonas	32.28
		Pseudoxhantomonas	3.17
		Serratia	1.06
Unclassified*			1.57

*Sequencing of *16S rDNA* and *gyrB* genes did not unambiguously resolve the genus

7.3.3 In vitro *Antagonism of Indigenous Olive Rhizobacteria Against Selected Plant Pathogens*

The *in vitro* antagonism against a representative of the *V. dahliae* D pathotype determined in the first screening round was confirmed in a second assay (Table 7.3). In general, differences (presence/absence of antagonism) were observed depending on the media used (with the exception of *V. dahliae*) (Fig. 7.1a), highlighting the cases of *C. godetiae* and *P. savastanoi* pv. savastanoi in which antagonism events were more frequent in NA and PDA, respectively (Table 7.3). For some rhizobacteria, differences in the percentage of effective antagonism between the two isolates used were found (*viz. R. necatrix*, *F. oxysporum* f. sp. *dianthi* and *P. savastanoi* p. var. savastanoi). The most dramatic case was for strain *P. savastanoi* NCPPB 3335 in PDA, which was inhibited by a significant lower number of rhizobacteria compared to antagonism observed against strain PSS-3 (25.4% and 9.0%, respectively). On the contrary, similar percentages were observed for *V. dahliae* and *Colletotrichum* spp. representatives, even though isolates here used belonged to different pathotypes (*V. dahliae*) or species (*Colletotrichum* spp.). Overall, *Paenibacillus* spp., *Pseudomonas* spp., *Rhizobium* spp. and *Bacillus* spp. representatives displayed a broad-spectrum antagonism against phytopathogens assayed. Remarkably, the best results were achieved for one strain of *Paenibacillus polymyxa*, which was able to inhibit the growth of 92% (in PDA) and 100% (in NA) of the pathogens assayed.

7.3.4 *Phenotypic Characterization of Selected Olive Rhizobacteria with Potential as Biological Control Agents*

Some phenotypes traditionally associated with biological control and/or PGP activities were evaluated and results are summarized in Table 7.4. All bacteria tested showed catalase activity, while phosphatase activity was found only in 2% of

Table 7.3 Percent of tested olive rhizobacteria showing antagonistic activity against phytopathogens

	Isolates (%)[a]	
Pathogen	PDA	NA
Alternaria alternata[b]	43.4	37.0
Colletotrichum godetiae[b]	15.3	49.7
Colletotrichum nymphaeae[b]	18.0	32.8
Fusarium oxysporum f. sp. *dianthi*	13.8	15.1
Phytophthora cinnamomi	33.9	19.0
Pseudomonas savastanoi pv. savastanoi[b]	17.2	1.4
Rosellinia necatrix[b]	23.6	37.0
Verticillium dahliae[b]	62.0	60.6

[a]Means from two assays
[b]Pathogens affecting olive

Table 7.4 Percentage of olive rhizobacteria showing phenotypes associated with biological control and/or plant growth promotion activities

Phenotypic characterization	Isolates[a] (%)
2,3-butanediol	30
Protease	29
Phosphatase	2
Siderophore	49
Catalase	100
Chitinase	20

[a]Means from two replications, for a total of 189 isolates selected after the first screening

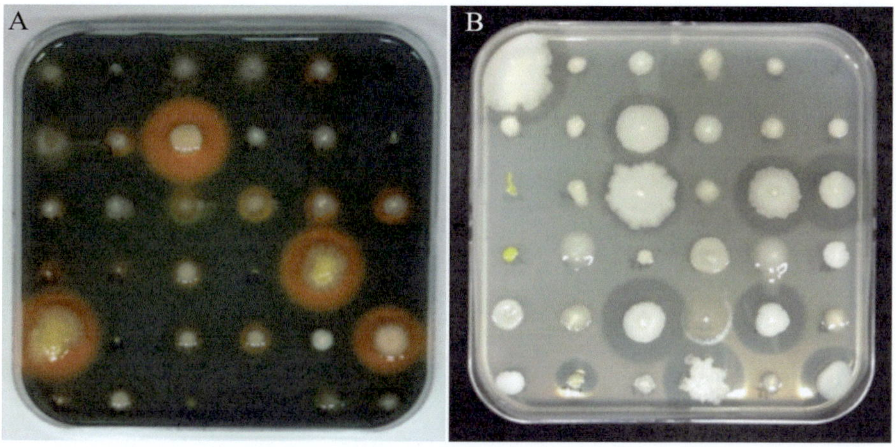

Fig. 7.2 Phenotypic characterization of olive rhizobacteria as potential biological control and plant-growth promotion agents. (**a**) siderophore production assay (according to Schwyn and Neilands 1987). (**b**) protease assay (according to Naik et al. 2013)

isolates (belonging to *Pseudomonas* and *Enterobacter*). A significant number of olive rhizobacteria displayed siderophore activity (49%), highlighting some *Pseudomonas* spp. isolates producing large halos when tested in CAS medium (Fig. 7.2a). The rest of activities tested showed a moderate-low presence. While catalase activity was present in representatives of all genera tested, production of 2,3-butanediol (30%), protease (29%) (Fig. 7.2b) and chitinase (20%) were associated almost exclusively to *Bacillus* spp. (Table 7.4).

GEN III MicroPlate™ assays provided a phenotypic fingerprinting of a selection of isolates (44). In general, high variability was observed among isolates regarding to the different substrates utilization, being *Proteobacteria* the most versatile phylum with some isolates able to metabolize up to 46 carbon sources (*Rhizobium* spp.). The capability to metabolize different C sources is summarized in Table 7.5. Regarding to chemical responsiveness, 98% of the assessed isolates grew in 1% NaCl and pH 6. It is worth mentioning the high percentage of isolates capable to grow in Na lactate 1% (93%), a well-known bactericidal used in alimentary industry, and the overall sensibility to minocycline, an antibiotic belonging to the tetracycline group (Table 7.6).

Table 7.5 Percentage of olive rhizobacteria showing the capability to metabolize different carbon sources according to the GEN III MicroPlate™ test

C source	Number of isolates (%)[a]	C source	Number of isolates (%)*
Acetic Acid	86	β-Hydroxy-D,L- Butyric Acid	35
α-D-Glucose	77	α-Keto-Glutaric Acid	30
Glycerol	70	D-Malic Acid	30
D-Fructose	67	Glycyl-L-Proline	26
Dextrin	65	N-Acetyl-D- Glucosamine	23
Pectin	65	myo-Inositol	23
L-Serine	58	Quinic Acid	21
Sucrose	53	D-Cellobiose	19
D-Fructose- 6-PO4	51	β-Methyl-D- Glucoside	19
L-Malic Acid	51	D-Mannose	19
Acetoacetic Acid	51	D-Salicin	16
L-Aspartic Acid	49	D-Galactose	16
L-Glutamic Acid	49	Glucuronamide	16
L-Histidine	49	Gentiobiose	14
Methyl Pyruvate	47	D-Turanose	14
D-Mannitol	44	D-Serine	14
L-Alanine	44	α-Keto-Butyric Acid	14
D-Gluconic Acid	44	L-Fucose	12
Citric Acid	44	Stachyose	9
D-Maltose	42	D-Raffinose	9
L-Pyroglutamic Acid	42	α-D-Lactose	9
D-Galacturonic Acid	42	D-Melibiose	9
L-Galactonic Acid Lactone	42	D-Sorbitol	9
Bromo-Succinic Acid	42	D-Fucose	7
Propionic Acid	42	L-Rhamnose	7
Gelatin	40	Inosine	7
L-Arginine	40	N-Acetyl-D- Galactosamine	5
D-Saccharic Acid	40	D-Arabitol	5
L-Lactic Acid	40	N-Acetyl-β-D- Mannosamine	2
γ-Amino-Butryric Acid	40	3-Methyl Glucose	2
Formic Acid	40	D-Aspartic Acid	2
D-Trehalose	37	p-Hydroxy- Phenylacetic Acid	2
D-Glucose- 6-PO4	37	D-Lactic Acid Methyl Ester	2
Mucic Acid	37	α-Hydroxy- Butyric Acid	2
Tween 40	37	N-Acetyl Neuraminic Acid	0
D-Glucuronic Acid	35		

[a]Means from two assays, for a total of 44 isolates selected based on their antagonistic capability, phenotypic characteristics and suitability to be produced as BCA

Table 7.6 Olive rhizobacteria capable to grow under diverse chemical substances and conditions according to the GEN III MicroPlate™ test

Chemical responsiveness	Number of isolates (%)[a]	Chemical responsiveness	Number of isolates (%)[a]
pH 6	98	Troleandomycin	42
1% NaCl	98	Niaproof 4	42
1% Sodium Lactate	93	Vancomycin	42
Potassium Tellurite	88	Tetrazolium Violet	42
Guanidine HCl	86	Tetrazolium Blue	42
Aztreonam	86	Fusidic Acid	30
Lithium Chloride	70	pH 5	26
4% NaCl	65	Nalidixic Acid	26
Sodium Butyrate	58	8% NaCl	19
D-Serine	49	Sodium Bromate	12
Rifamycin SV	49	Minocycline	0
Lincomycin	44		

[a]Means from two assays, for a total of 44 isolates selected based on their antagonistic capability, phenotypic characteristics and suitability to be produced as BCA

7.4 Discussion

Rhizosphere microbial communities are exposed to several biotic and abiotic factors strongly determining their structure and composition, being one of them the host plant (Marschner et al. 2004). Considering this scenario, it is not surprising that when a given microorganism is used as a BCA, one of the main problems encountered is its low adaptation to the target environment. This may result in the lack of success or the inconsistency frequently reported when implementing biological control approaches (Handelsman and Stabb 1996). However, this obstacle can be somehow overcome by searching, identifying and characterizing potential BCA from the same ecological niche where its beneficial action will be required (Knudsen et al. 1997; Ruano-Rosa and López-Herrera 2009). Taking this premise as starting point, we designed a strategy in which the first objective was to generate a collection of indigenous olive rhizobacteria as potential BCA against *V. dahliae*.

The molecular analysis of members of this collection revealed a high diversity and prevalence of phyla *Proteobacteria* and *Firmicutes*. This finding is in agreement with results earlier reported for wild olives (Aranda et al. 2011).

The selection process of candidate BCA was primarily driven by checking effective *in vitro* antagonism against the D pathotype of *V. dahliae*. However, since a broad spectrum of biological control activity is a desirable characteristic for a BCA (Whipps and Davies 2000), a number of phytopathogens affecting olive trees and other crops were evaluated as potential targets of newly-isolated olive rhizobacteria. We found several bacteria showing *in vitro* antagonism against tested pathogens. Because of the intensity and spectrum of the inhibition displayed, representatives of *Paenibacillus* spp., *Pseudomonas* spp. and *Bacillus* spp. are worth mentioning.

Members of these genera are frequently reported as per their involvement on biological control and/or PGP (Borriss 2015; Lugtenberg and Kamilova 2009; Mercado-Blanco 2015; Hayat et al. 2012; Santoyo et al. 2012; Rybakova et al. 2016).

In the olive rhizobacteria collection generated in this study, two strains of *Paenibacillus* spp. highlighted due to their strong antagonistic capability, both in intensity and range of pathogens inhibited. Species of this genus have been described as BCA, plant growth-promoting rhizobacteria and microorganism with high biotechnological potential (Lal and Tabacchioni 2009; Bhattacharyya and Jha 2012). Some of them are present in several commercial products against a broad range of phytopathogens (Rybakova et al. 2016). For instance, *P. polymyxa* has been previously reported as an excellent BCA against *V. dahliae* in cotton (Yang et al. 2013) and other fungi, oomycetes or bacteria (Raza et al. 2015; Hong et al. 2016; Xu and Kim 2016). Likewise, a strain of *Paenibacillus alvei* showed effective against *V. dahliae* in olive trees (Markakis et al. 2016). Whether the two *Paenibacillus* spp. strains here characterized behave as true BCA against VWO and/or the other pathogens tested still needs confirmation by conducting *in planta* bioassays.

Strains belonging to other genera also showed *in vitro* antagonism against tested phytopathogens, although to a lesser extent. The genus *Pseudomonas* is one of the most studied regarding to biological control and PGP activities. Important features such as broad colonization ability, production of a wide diversity of secondary metabolites and versatility in using different molecules as C sources make beneficial strains of this genus excellent tools in biological control strategies (Mercado-Blanco and Bakker 2007). Regarding to biological control of VWO, different strains of *Pseudomonas* spp. have been previously reported as effective BCA (Mercado-Blanco et al. 2004; Sanei and Razavi 2011). Another renowned genus used in biological control approaches is *Bacillus*. Mechanisms ranging from production of antifungal compounds to induction of systemic resistance in the host plants along with high colonization and sporulation capabilities and advantageous competition for nutrients, make them excellent BCAs as well (Borriss 2015; Santoyo et al. 2012; Kumar et al. 2011; Zheng et al. 2013). Isolates of *Bacillus* spp. have shown good performance against Verticillium wilt of cotton. As for the *Paenibacillus* spp. strains mentioned above, suitable BCA candidates assigned to the genera *Pseudomonas* and *Bacillus* could be expected from the olive rhizobacteria collection here generated. The next step yet to be taken is *in planta* experiments to verify effectiveness against VWO.

Besides assessment of antagonist ability, the first step towards the identification of candidate BCA, aim of this study was to evaluate the presence of bacterial traits related to biological control and/or PGP in indigenous olive rhizobacteria. Among the phenotypes detected, it is worth mentioning the outstanding number of isolates displaying siderophore production. These low-molecular-weight, iron-chelating molecules (Höfte 1993), present in nearly 50% of the isolates in our collection, are produced under iron-limiting growth conditions and have been frequently described as involved in competition and disease suppression for a number of BCA, including beneficial *Pseudomonas* spp. (Mercado-Blanco 2015). Additionally, since chitin, glucans and glycoproteins are the major components of the fungal cell wall (Bowman and Free 2006), it is also interesting to emphasize the presence of isolates displaying production of lytic enzymes such as chitinases and proteases. Both phenotypes are very appreciated during BCA selection processes, and pose relevance in industrial and medical biotechnology as well (Wang et al. 2009). While production of these exoenzymes is frequent in many rhizobacteria (Whipps 2001; Ajit et al. 2006), they were almost restricted to representatives of *Bacillus* spp. Although at low percentage, production of 2,3-butanediol was also found in our bacterial selection. This is an interesting volatile involved in growth stimulation (Ryu et al. 2005) and has been suggested to be implicated in induced systemic resistance responses (Han et al. 2006; Borriss 2015). Finally, substrate utilization and chemical responsiveness profiles obtained by the BIOLOG system provided relevant information for future formulations involving some of these bacteria, as well as to facilitate their identification (George et al. 2013).

7.5 Conclusions

A collection of bacteria originating from olive roots or rhizosphere of nursery-produced olive plants showing antagonism against several pathogens affecting olive trees, mainly *V. dahliae*, was generated. Furthermore, some of them displayed phenotypes related to biological control and/or PGP activities. Results indicate that the olive rhizosphere is actually a reservoir of candidate BCA against major olive diseases. These bacteria could therefore constitute the basis to develop bioformulations effective to control the pathogens evaluated in this work. Moreover, they offer the advantage to be adapted to the ecological niche where their benefits can be deployed. However, bioassays aiming at demonstrating their true biological control activity *in planta* are still needed, both under controlled and field conditions. This is now our current challenge.

Acknowledgements We are grateful to Drs. F. J. López-Escudero and A. Trapero (University of Córdoba, Spain), C. J. López-Herrera and E. Pérez-Artes (IAS-CSIC), C. Ramos (University of Málaga, Spain) and L. Ran (Agricultural University of Hebei, P.R. China) for their gifts of plant pathogens used in this study. Supported by grants P12-AGR667 (Junta de Andalucía, Spain) and RECUPERA 2020 (MINECO-CSIC agreement), both co-funded by ERDF from the EU.

References

Ajit, N. S., Verma, R., & Shanmugam, V. (2006). Extracellular chitinases of fluorescent pseudomonads antifungal to *Fusarium oxysporum* f. sp. *dianthi* causing carnation wilt. *Current Microbiology, 52*, 310–316.

Aleandri, M. P., Chilosi, G., Bruni, N., Tomassini, A., Vettraino, A. M., & Vannini, A. (2015). Use of nursery potting mixes amended with local *Trichoderma* strains with multiple complementary mechanisms to control soil-borne diseases. *Crop Protection, 67*, 269–278.

Aranda, S., Montes-Borrego, M., Jiménez-Díaz, R. M., & Landa, B. B. (2011). Microbial communities associated with the root system of wild olives (*Olea europaea* L. subsp *europaea* var. sylvestris) are good reservoirs of bacteria with antagonistic potential against *Verticillium dahliae*. *Plant and Soil, 343*, 329–345.

Bhattacharyya, P. N., & Jha, D. K. (2012). Plant growth-promoting rhizobacteria (PGPR): Emergence in agriculture. *World Journal of Microbiology and Biotechnology, 28*, 1327–1350.

Borriss, R. (2015). *Bacillus*, a plant-beneficial bacterium. In *Principles of Plant-Microbe Interactions* (pp. 379–391). Cham: Springer.

Bowman, S. M., & Free, S. J. (2006). The structure and synthesis of the fungal cell wall. *Bioessays, 28*, 799–808.

Bubici, G., & Cirulli, M. (2011). Verticillium wilt of olives. In *Olive diseases and disorders* (pp. 191–222). Kerala: Research Signpost.

Bubici, G., & Cirulli, M. (2012). Control of Verticillium wilt of olive by resistant rootstocks. *Plant Soil, 352*, 363–376.

Carrero-Carrón, I., Trapero-Casas, J. L., Olivares-García, C., Monte, E., Hermosa, R., & Jiménez-Díaz, R. M. (2016). *Trichoderma asperellum* is effective for biocontrol of Verticillium wilt in olive caused by the defoliating pathotype of *Verticillium dahliae*. *Crop Protection, 88*, 45–52.

Cazorla, F. M., & Mercado-Blanco, J. (2016). Biological control of tree and woody plant diseases: an impossible task? *BioControl, 61*, 233–242.

Collado-Romero, M., Mercado-Blanco, J., Olivares-García, C., Valverde-Corredor, A., & Jiménez-Díaz, R. (2006). Molecular variability within and among *Verticillium dahliae* vegetative compatibility groups determined by fluorescent amplified fragment length polymorphism and polymerase chain reaction markers. *Phytopathology, 96*, 485–495.

FAOSTAT. (2016). Available online at: http://faostat.fao.org/

George, P., Gupta, A., Gopal, M., Thomas, L., & Thomas, G. V. (2013). Multifarious beneficial traits and plant growth promoting potential of *Serratia marcescens* KiSII and *Enterobacter* sp. RNF 267 isolated from the rhizosphere of coconut palms (*Cocos nucifera* L.) *World Journal of Microbiology and Biotechnology, 29*, 109–117.

Gómez-Lama Cabanás, C., Valverde-Corredor, A., & Pérez-Artes, E. (2012). Molecular analysis of Spanish populations of *Fusarium oxysporum* f. sp. *dianthi* demonstrates a high genetic diversity and identifies virulence groups in races 1 and 2 of the pathogen. *European Journal of Plant Pathology, 132*, 561–576.

Han, S. H., Lee, S. J., Moon, J. H., Park, K. H., Yang, K. Y., et al. (2006). GacS-dependent production of 2R, 3R-butanediol by *Pseudomonas chlororaphis* O6 is a major determinant for eliciting systemic resistance against *Erwinia carotovora* but not against *Pseudomonas syringae* pv. tabaci in tobacco. *Molecular Plant-Microbe Interactions, 19*, 924–930.

Handelsman, J., & Stabb, E. V. (1996). Biocontrol of soilborne plant pathogens. *The Plant Cell, 8*, 1855–1869.

Hayat, R., Ahmed, I., & Sheirdil, R. A. (2012). An overview of plant growth promoting rhizobacteria (PGPR) for sustainable agriculture. In *Crop Production for Agricultural Improvement* (pp. 557–579). Dordrecht: Springer.

Höfte, M. (1993). Classes of microbial siderophores. In *Iron chelation in plants and soil microorganisms* (pp. 3–27). San Diego: Academic.

Holt, J. G., Krieg, N. R., Sneath, P. H. A., Staley, J. T., & Williams, S. T. (1994). In W. R. Hensyl (Ed.), *Bergey's manual of determinative bacteriology* (9th ed., 1094 pp). Baltimore: Williams & Wilkins.

Hong, C. E., Kwon, S. Y., & Park, J. M. (2016). Biocontrol activity of *Paenibacillus polymyxa* AC-1 against *Pseudomonas syringae* and its interaction with *Arabidopsis thaliana*. *Microbiological Research, 185*, 3–21.

Katznelson, H., & Bose, B. (1959). Metabolic activity and phosphate dissolving capability of bacterial isolates from wheat roots in the rhizosphere and non rhizosphere soil. *Canadian Journal of Microbiology, 5*, 79–85.

Knudsen, I. M. B., Hockenhull, J., Funck Jensen, D., Gerhardson, B., Hökeberg, M., et al. (1997). Selection of biological control agents for controlling soil and seed-borne diseases in the field. *European Journal of Plant Pathology, 103*, 775–784.

Kumar, A., Prakash, A., & Johri, B. N. (2011). *Bacillus* as PGPR in crop ecosystem. In *Bacteria in agrobiology: Crop ecosystems* (pp. 37–59). Berlin/Heidelberg: Springer.

Lal, S., & Tabacchioni, S. (2009). Ecology and biotechnological potential of *Paenibacillus polymyxa*: A minireview. *Indian Journal of Microbiology, 49*, 2–10.

López-Escudero, F. J., & Mercado-Blanco, J. (2011). Verticillium wilt of olive: A case study to implement an integrated strategy to control a soil-borne pathogen. *Plant Soil, 344*, 1–50.

López-Herrera, C. J., & Zea-Bonilla, T. (2007). Effects of benomyl, carbendazim, fluazinam and thiophanate methyl on white root rot of avocado. *Crop Protection, 26*, 1186–1192.

Lugtenberg, B., & Kamilova, F. (2009). Plant-growth-promoting rhizobacteria. *Annual Review of Microbiology, 63*, 541–556.

Maldonado-González, M. M., Bakker, P. A. H. M., Prieto, P., & Mercado-Blanco, J. (2015a). *Arabidopsis thaliana* as a tool to identify traits involved in *Verticillium dahliae* biocontrol by the olive root endophyte *Pseudomonas fluorescens* PICF7. *Frontiers in Microbiology, 6*, 266.

Maldonado-González, M. M., Schilirò, E., Prieto, P., & Mercado-Blanco, J. (2015b). Endophytic colonization and biocontrol performance of *Pseudomonas fluorescens* PICF7 in olive (*Olea europaea* L.) are determined neither by pyoverdine production nor swimming motility. *Environmental Microbiology, 17*, 3139–3153.

Markakis, E. A., Tjamos, S. E., Antoniou, P. P., Paplomatas, E. J., & Tjamos, E. C. (2016). Biological control of Verticillium wilt of olive by *Paenibacillus alvei*, strain K165. *BioControl, 61*, 293–303.

Marschner, P., Crowley, D., & Yang, C. H. (2004). Development of specific rhizosphere bacterial communities in relation to plant species, nutrition and soil type. *Plant and Soil, 261*, 199–208.

Mercado-Blanco, J. (2015). *Pseudomonas* strains that exert biocontrol of plant pathogens. In: *Pseudomonas. Volume 7: New aspect of Pseudomonas biology* (pp. 121–172). London: Springer.

Mercado-Blanco, J., & Bakker, P. (2007). Interactions between plants and beneficial *Pseudomonas* spp.: Exploiting bacterial traits for crop protection. *Antonie van Leeuwenhoek, 92*, 367–389.

Mercado-Blanco, J., Rodríguez-Jurado, D., Hervás, A., & Jiménez-Díaz, R. M. (2004). Suppression of Verticillium wilt in olive planting stocks by root-associated fluorescent *Pseudomonas* spp. *Biological Control, 30*, 474–486.

Moral, J., Jurado, J., & Trapero, A. (2012). Effect of temperature and relative humidity on mycelial growth, conidial germination and fruit infection by *Colletotrichum* spp. causing olive anthracnose. *IOBC-WPRS. Bulletin, 79*, 14.

Müller, H., Tejedor-González, E., Mercado-Blanco, J., Rodríguez-Jurado, D., Jiménez-Díaz, R. M., & Berg, G. (2007). Effect of the biological control strain *Serratia plymuthica* HRO-C48 on Verticillium wilt of olive trees cv. Arbequina. In: *Fundamental and practical approaches to increase biocontrol efficacy. IOBC/wprs Bulletin, 30*, 173–177.

Murthy, N., & Bleakley, B. (2012). Simplified method of preparing colloidal chitin used for screening of chitinase-producing microorganisms. *The Internet Journal of Microbiology 10*, 2. Available online at: http://ispub.com/IJMB/10/2/14186

Naik, D. N., Wahidullah, S., & Meena, R. M. (2013). Attenuation of *Pseudomonas aeruginosa* virulence by marine invertebrate–derived *Streptomyces* sp. *Letters in Applied Microbiology, 56*, 197–207.

Pérez-Martínez, I., Rodríguez-Moreno, L., Matas, I. M., & Ramos, C. (2007). Strain selection and improvement of gene transfer for genetic manipulation of *Pseudomonas savastanoi* isolated from olive knots. *Research in Microbiology, 158*, 60–69.

Prieto, P., Navarro-Raya, C., Valverde-Corredor, A., Amyotte, S. G., Dobinson, K. F., & Mercado-Blanco, J. (2009). Colonization process of olive tissues by *Verticillium dahliae* and its *in planta* interaction with the biocontrol root endophyte *Pseudomonas fluorescens* PICF7. *Microbial Biotechnology, 2*, 499–511.

Raza, W., Yuan, J., Ling, N., Huang, Q., & Shen, Q. (2015). Production of volatile organic compounds by an antagonistic strain *Paenibacillus polymyxa* WR-2 in the presence of root exudates and organic fertilizer and their antifungal activity against *Fusarium oxysporum* f. sp. *niveum*. *Biological Control, 80*, 89–95.

Ruano-Rosa, D., & López-Herrera, C. J. (2009). Evaluation of *Trichoderma* spp. as biocontrol agents against avocado white root rot. *Biological Control, 51*, 66–71.

Ruano-Rosa, D., & Mercado-Blanco, J. (2015). Combining biocontrol agents and organics amendments to manage soil-Borne phytopathogens. In *Organic amendments and soil suppressiveness in plant disease management* (pp. 457–478). Cham: Springer.

Ruano-Rosa, D., Prieto, P., Rincón, A. M., Gómez-Rodríguez, M. V., Valderrama, R., et al. (2016). Fate of *Trichoderma harzianum* in the olive rhizosphere: Time course of the root colonization process and interaction with the fungal pathogen *Verticillium dahliae*. *BioControl, 61*, 269–282.

Rybakova, D., Cernava, T., Köberl, M., Liebminger, S., Etemadi, M., & Berg, G. (2016). Endophytes-assisted biocontrol: novel insights in ecology and the mode of action of *Paenibacillus*. *Plant and Soil*. doi:10.1007/s11104-015-2526-1.

Ryu, C. M., Farag, M. A., Pare, P., & Kloepper, J. W. (2005). Invisible signals from the underground: bacterial volatiles elicit plant growth promotion and induce systemic resistance. *The Plant Pathology Journal, 21*, 7–12.

Sanei, S. J., & Razavi, S. E. (2011). Suppression of Verticillium wilt of olive by *Pseudomonas fluorescens*. *American Journal of Experimental Agriculture, 1*, 294–305.

Santoyo, G., Orozco-Mosqueda, M. C., & Govindappa, M. (2012). Mechanisms of biocontrol and plant growth-promoting activity in soil bacterial species of *Bacillus* and *Pseudomonas:* A review. *Biocontrol Science and Technology, 22*, 855–872.

Schwyn, B., & Neilands, J. B. (1987). Universal chemical assay for the detection and determination of siderophores. *Analytical Biochemistry, 160*, 47–56.

Tjamos, E. C. (1993). Prospects and strategies in controlling Verticillium wilt of olive. *Bulletin OEPP/EPPO, 23*, 505–512.

Tsror, L. (2011). Epidemiology and control of Verticillium wilt on olive. *Israel Journal of Plant Sciences, 59*, 59–69.

Wang, S. L., Chao, C. H., Liang, T. W., & Chen, C. C. (2009). Purification and characterization of protease and chitinase from *Bacillus cereus* TKU006 and conversion of marine wastes by these enzymes. *Marine Biotechnology, 11*, 334–344.

Whipps, J. M. (2001). Microbial interactions and biocontrol in the rhizosphere. *Journal of Experimental Botany, 52*, 487–511.

Whipps, J. M., & Davies, K. G. (2000). Success in biological control of plant pathogens and nematodes by microorganisms. In *Biological control: measures of success* (pp. 231–270). Dordrecht: Springer.

Xu, S., & Kim, B. S. (2016). Evaluation of *Paenibacillus polymyxa* strain SC09-21 for biocontrol of Phytophthora blight and growth stimulation in pepper plants. *Tropical Plant Pathology*, 1–7.

Yamamoto, S., & Harayama, S. (1995). PCR amplification and direct sequencing of *gyrB* genes with universal primers and their application to the detection and taxonomic analysis of *Pseudomonas putida* strains. *Applied and Environmental Microbiology, 61*, 1104–1109.

Yang, P., Sun, Z. X., Liu, S. Y., Lu, H. X., Zhou, Y., & Sun, M. (2013). Combining antagonistic endophytic bacteria in different growth stages of cotton for control of Verticillium wilt. *Crop Protection, 47*, 17–23.

Zheng, Y., Chen, F., & Wang, M. (2013). Use of *Bacillus*-based biocontrol agents for promoting plant growth and health. In *Bacteria in Agrobiology: Disease Management* (pp. 243–258). Berlin/Heidelberg: Springer.

Chapter 8
Norway Spruce Fine Roots and Fungal Hyphae Grow Deeper in Forest Soils After Extended Drought

Isabella Børja, Douglas L. Godbold, Jan Světlík, Nina E. Nagy, Roman Gebauer, Josef Urban, Daniel Volařík, Holger Lange, Paal Krokene, Petr Čermák, and Toril D. Eldhuset

Abstract Global warming will most likely lead to increased drought stress in forest trees. We wanted to describe the adaptive responses of fine roots and fungal hyphae, at different soil depths, in a Norway spruce stand to long-term drought stress induced by precipitation exclusion over two growing seasons. We used soil cores, minirhizotrons and nylon meshes to estimate growth, biomass and distribution of fine roots and fungal hyphae at different soil depths. In control plots fine roots proliferated in upper soil layers, whereas in drought plots there was no fine root growth in upper soil layers and roots mostly occupied deeper soil layers. Fungal hyphae followed the same pattern as fine roots, with the highest biomass in deeper soil layers in drought plots. We conclude that both fine roots and fungal hyphae respond to long-term drought stress by growing into deeper soil layers.

I. Børja (✉) • N.E. Nagy • H. Lange • P. Krokene • T.D. Eldhuset
Norwegian Institute of Bioeconomy Research, P.O.Box 115, N-1431 Ås, Norway
e-mail: isabella.borja@nibio.no

D.L. Godbold
BOKU Vienna, Gregor Mendel Str. 33, A-1180 Vienna, Austria

J. Světlík
Centre MendelGlobe – Global Climate Change and Managed Ecosystems, Mendel University in Brno, Zemědělská 3, 61300 Brno, Czech Republic

R. Gebauer • D. Volařík
Departement of Forest Botany, Dendrology and Geobiocoenology, Mendel University in Brno, Zemědělská 3, 61300 Brno, Czech Republic

J. Urban
Siberian Federal University, 79 Svobodny pr, 660041 Krasnoyarsk, Russia

Departement of Forest Botany, Dendrology and Geobiocoenology, Mendel University in Brno, Zemědělská 3, 61300 Brno, Czech Republic

P. Čermák
Department of Forest Protection and Wildlife Management, Mendel University in Brno, Zemědělská 3, 61300 Brno, Czech Republic

M. Lukac et al. (eds.), *Soil Biological Communities and Ecosystem Resilience*, Sustainability in Plant and Crop Protection, DOI 10.1007/978-3-319-63336-7_8

Keywords Fine root biomass • Hyphal biomass • Hyphal mesh • Minirhizotrons • *Picea abies*

8.1 Introduction

Predicted global warming will most likely increase the frequency of summer droughts and lead to soil water shortage in many forest regions of Europe (IPCC 2013). Drought is already considered to be a major driver for Norway spruce crown discoloration and tree mortality in SE Norway (Solberg 2004), and drought stress has been identified as a limiting factor for Norway spruce growth in this region (Andreassen et al. 2006). Adverse effects of drought on forest trees are also known from many other parts of the world (Allen et al. 2010; Bréda et al. 2006). Problems with drought-induced forest dieback will likely become exacerbated in the future, given that by 2100 the climate in Europe is estimated to be warmer with higher frequencies of hydrological droughts and other extreme climatic events (Roudier et al. 2016).

Maintained and sustainable tree growth requires that there is a balance between water uptake by the roots and transpiration losses in the canopy. Efficient water management that minimizes transpiration losses and maximizes water uptake is essential for tree survival during drought periods, and fine roots play a crucial role here as they are the organ that absorbs water and nutrients from the soil. The fine root system has an extended surface area that is almost always greater than the leaf area (Jackson et al. 1997). In trees with a diminished fine root system or fine roots that do not absorb water efficiently whole-tree water transport is compromised. This may lead to fine root adaptation, carbon starvation or root mortality (McDowell et al. 2008; McDowell and Sevanto 2010; Sevanto et al. 2014).

Although fine roots make up only a small fraction (2–5%) of the total tree biomass almost 50% of the carbon that is cycled in a forest may be used for fine root production (Jackson et al. 1997). Also, due to the rapid turnover of individual fine roots, about 66% of the annual photosynthate assimilated by a tree is spent on fine root production (Hendrick and Pregitzer 1993; Vogt et al. 1996). Because fine root production consumes such a large proportion of the total assimilated carbon in forest ecosystems fine root mortality is a major source of carbon input into forest soils (Aber et al. 1985; Steele et al. 1997). This means that any damage to the fine root system inflicted by e.g. drought affects not only on the health of individual trees, but the carbon and nutrient cycling of the entire forest ecosystem.

Fine roots are associated with an extensive network of ectomycorrhizal hyphae that provides the tree with extra water and nutrients and in return receives photosynthates from the tree crown. This mutualistic symbiosis may increase water uptake by the fine roots during periods of water shortage. Ectomycorrhizal fungi are probably modulating the adaptive responses of the fine root system to drought and may influence whether the fine root system is stimulated or inhibited by drought (Brunner et al. 2015; Ekblad et al. 2013).

Many studies have described effects of drought on aboveground tree organs such as stems, twigs, and leaves (e.g. Ditmarová et al. 2010; Gebauer et al. 2015; Sohn et al. 2012), but we still know little about how the root system of mature trees responds to water shortage. In particular, we know almost nothing about how drought affects the fine roots and their associated network of ectomycorrhizal hyphae. Because fine roots are responsible for water uptake they are in immediate contact with soil moisture and are thus likely to respond quicker and more directly to drought stress than leaves and other aboveground tree parts. For example, fine roots have an inherent growth plasticity and may adapt to local moisture conditions by changing their architecture or growing deeper in the soil (e.g. Gaul et al. 2008; Joslin et al. 2000; Moser et al. 2015; Pregitzer et al. 1993).

Although severe and recurrent droughts often have been identified as a main driver of tree decline and mortality (Bréda et al. 2006) we still lack evidence for the mechanisms that operate below ground, and especially for how drought affects fine roots and their associated fungal hyphae. Numerous drought experiments on potted saplings may give an indication of how fine roots respond to water shortage (e.g. Eldhuset et al. 2013; Schall et al. 2012), but little is known about fine root dynamics in trees growing in the field. More realistic data representing true forest conditions requires experiments that manipulate forest hydrology through precipitation exclusion in the field (e.g. Borken et al. 2006; Gaul et al. 2008; Joslin et al. 2000) or along the precipitation gradient (Meier and Leuschner 2008). However, the few existing studies of how water shortage affects the root system of larger trees report conflicting results, making it difficult to draw any firm conclusions about how the fine root system responds to drought. Some studies have reported a decrease in fine root biomass with decreasing rainfall (Joslin et al. 2000; Leuschner et al. 2004), whereas others have reported a stimulation of fine root growth in response to drought (Leuschner et al. 2001; Moser et al. 2015; Santantonio and Hermann 1985).

The study we report here investigated effects of water shortage on fine roots and fungal hyphae at different soil depths in a 20-year-old Norway spruce forest. Our aim was to determine the adaptive responses of fine roots and fungal hyphae in forest plots subjected to drought induced by precipitation exclusion. We hypothesized that water shortage in the upper soil layers would promote growth of fine roots and hyphae in deeper soil layers. To test this hypothesis, we evaluated fine root biomass at different soil depths using soil cores and nylon meshes, we determined the vertical distribution of fine root and fungal growth using minirhizotrons, and we quantified fungal hyphae in different soil layers by trapping hyphae in nylon meshes.

8.2 Materials and Methods

8.2.1 Study Site and Experimental Design

Our experimental plots were located in a 20-year-old Norway spruce stand (*Picea abies* L. Karst) at the Hoxmark experimental farm in SE Norway (59°40′14″N, 10°47′36″E, 90 m asl). The experiment was part of larger project described in

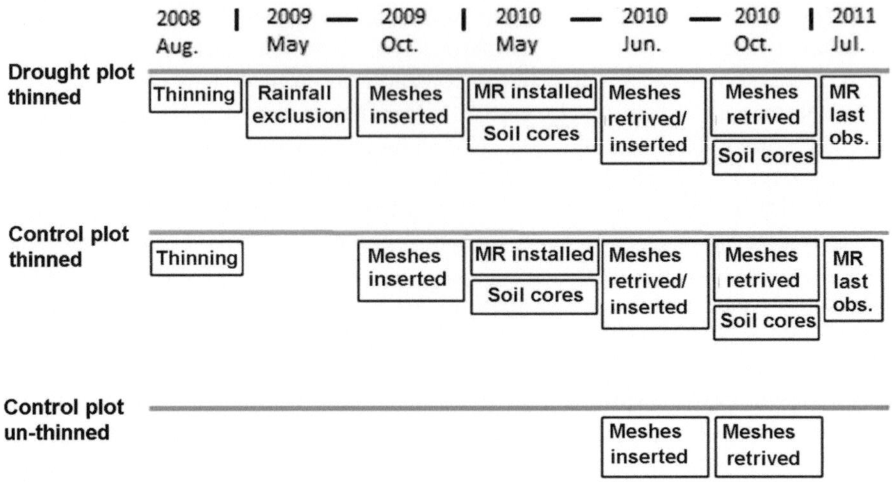

Fig. 8.1 Timeline for the experimental activities on a thinned drought plot and a thinned and un-thinned control plots (*MR* minirhizotrons; last observation in 2011)

Gebauer et al. (2011, 2012). Briefly, trees from three different full-sib families were growing in a 1 m × 1 m grid on former agricultural soil, described as albeluvisol, with mean tree height 9 m, stand density 10.000 trees ha^{-1}, and basal area 50 m^2 ha^{-1}. Two study plots of about 150 m^2 each were established within the stand. Both plots were thinned in August 2008 by removing the tree rows flanking our experimental trees, reducing stand density by 50%. On one plot, the drought plot (D), a plastic roof was set up under the tree canopy in May 2009 and remained in place until the end of the experiment in July 2011. A 30 cm deep trench was dug around the drought plot to intercept the precipitation run off. The other plot was left as it was and served as the control plot (C). A third plot, situated about 30 m away from the other plots, was established later on as an un-thinned control plot (C0) where additional set of nylon meshes were placed in the soil (see below) in June-Oct 2010 (Fig. 8.1). Both control plots (C, C0) received natural rainfall. Mean annual precipitation in the study area is 785 mm and mean annual temperature is 5.3 °C (data from the meteorological station in Ås, 3.9 km from the study site). The three full-sib families were the result of controlled crosses of parent trees from the following locations and altitudes in S Norway: family 14: mother no. 2598 from Krødsherad, Buskerud, 440 masl and father no. 5262 from Gol, Buskerud, 395 masl; family 27: mother 2526, Flesberg, Buskerud, 455 masl and father 5262, Gol, Buskerud, 395 masl; family 29: mother 2627, Brandbu, Oppland, 430 masl and father 2767, Stor-Elvdal, Hedmark, 400 m asl.

In plots C and D we selected one tree from each full-sib family, i.e. a total of three trees per plot. The use of pairs of similar-sized trees from the same full-sib families reduced phenotypic variation between plots and facilitated the comparison of results between the C and D plots.

8.2.2 Measurement of Soil Hydrology, Sap Flow and Climatic Variables

Triplicate measurements of soil water potential and soil humidity were made in plot C and D throughout the experiment. Soil water potential was measured using gypsum blocs (Delmhorst, Inc., USA) placed at 10 cm, 30 cm and 50 cm soil depth, and soil humidity was measured in the upper 30 cm of the soil using CS-616 TDR probes (Campbell, USA). Data were acquired every 10 minutes and stored in a data logger (ModuLog 1037, EMS Brno, Czech Republic). All soil characteristics were measured from May to November in 2009 and 2010. Sap flow was monitored in all six experimental trees in plot C and D from May to November in 2009 and 2010 using EMS 51 sap flow meters (EMS Brno, Czech Republic) working under the trunk heat balance principle (Čermák et al. 1973, 2004). A more detailed description of measurements of sap flow, tree and soil characteristics is given in Gebauer et al. (2011). A climate station was set up in an open field 100 m from the experimental plots to measure global radiation, air temperature and relative air humidity using EMS11 and EMS33 (EMS Brno, Czech Republic), as well as precipitation and wind speed (MetOne Instruments, Grants Pass, Oregon, USA). Data were measured every minute and 10-minute means were stored in a data logger.

8.2.3 Fine Root Growth and Biomass

8.2.3.1 Fine Root Growth Estimated from Minirhizotrons

To estimate fine root growth at different soil depths we installed minirhizotrons in the C and D plots (n=3) on May 20 2010. The minirhizotrons consisted of 1 m long transparent acrylic tubes made of Escaril® with 50 mm inner and 60 mm outer diameter. Minirhizotrons were installed about 1 m from each experimental tree in holes made at 45° angle into the soil using a minirhizotron drill (Fig. 8.1). Fine roots and fungal hyphae at different soil depths were photographed through the minirhizotrons seven times during 2010 and 2011 using a custom-made digital camera with associated image capture software BTC I-CAP (both from Bartz Technology Corporation, Santa Barbara, CA, USA). All individual roots in each image were assigned a specific id, manually traced to determine root tip numbers, root length and diameter, and root volume and surface area were automatically calculated using the software Rootfly (Version 2.0.2, Clemson University, SC, USA). The images along the 45° angle were recalculated to vertical soil depth and analysed to a soil depth of about 60 cm.

8.2.3.2 Fine Root Biomass Estimated from Soil Cores

The 60 mm diameter soil cores that were removed to install the minirhizotrons in May 2010 were used to collect fine roots at different soil depths. Segments along the 45° soil cores were recalculated to vertical depths. On October 25 2010, additional vertical soil cores were taken close to the other soil core holes and subdivided into different vertical sections (0–10, 10–20, 20–30 and 30–40 cm depth). All core sections from the vertical and slanted soil cores were air dried at 25 °C and all fine roots (<2 mm in diameter) were collected, sorted according to colour and tissue strength, and stored in tap water at 4 °C for up to three weeks until further processing. The fine roots were finally dried at 65 °C and the total dry weight of roots from each sample was determined and calculated as gram dry weight per soil volume (g dm^{-3}).

8.2.3.3 Fine Root Growth Estimated from Nylon Meshes

Fine root growth was also quantified using a parallel method involving 6–7 cm wide and 25 cm long pieces of woven nylon mesh (Normesh, Lancashire, UK, product code N1000/59105R) with 1 mm pore size. Each mesh was inserted vertically into the soil to 20 cm depth by using a sharp steel plate and a hammer, as described by Lukáč and Godbold (2010). In October 2009, four nylon meshes were placed 10 cm apart about 1 m from the trunk of each experimental tree in plot C and D (n = 12 meshes per plot). The meshes were removed in June 2010 and replaced with a new set of meshes. At this time, meshes were also installed in the un-thinned control plot (C0; n = 12). These meshes were removed in October 2010 by cutting the soil 5 cm from each side of the mesh, lifting the soil block and carefully removing the mesh as described in Lukáč and Godbold (2010) and Hirano et al. (2009). Fine roots trapped in the mesh were removed and fine root biomass per mesh was calculated as described in Lukáč and Godbold (2010). After root retrieval, the meshes were stored at −20 °C until further processing for estimation of hyphal biomass (below).

8.2.4 Hyphal Growth and Biomass

8.2.4.1 Hyphal Biomass Estimated from Nylon Meshes

The biomass of fungal hyphae trapped in the nylon meshes was quantified using a newly developed method of mesh image analysis according to Børja et al. (unpublished results). The method involves an algorithm that estimates hyphal biomass per soil volume based on the proportion of mesh pores that are filled with fungal hyphae. Briefly, after the fine roots had been removed the meshes were cleaned for soil and scanned at 1200 dpi resolution using a color image scanner (CanoScan 9000F, Canon Inc., USA). The scanned images were analyzed by first determining the "grayness" range associated with each component in the image (the mesh grid itself, fungal hyphae, soil particles) using the software GIMP (www.gimp.org/). The image processing software ImageJ (http://imagej.net/) was then used to generate a "grayness histogram" for each mesh, the histogram data was exported to Excel and converted to biomass of hyphae per soil volume unit using the algorithm we developed. Each scanned nylon mesh was divided into four layers corresponding to soil depths of 0–5, 5–10, 10–15 and 15–20 cm that were analyzed separately.

8.2.4.2 Hyphal Growth Estimated from Minirhizotrons

To assess hyphal growth in the soil we used the minirhizotron images as described above. Each image was subjectively, visually evaluated for percentage of hyphal coverage in each image. Roots and hyphae were discerned on basis of their diameter and distinct looks. We used the software Rootfly (Version 2.0, Clemson University, SC, USA) to keep track of image id and analysed images down to about 60 cm soil depth.

8.2.5 Statistical Analysis

We used analysis of variance (ANOVA) to compare fine root biomass in drought versus control plots. To achieve normality and homoscedasticity, the weight data for living fine roots was log-transformed. For data that were normally distributed (Shapiro-Wilks analysis) and homoscedastic (Levene analysis), we used one-way ANOVA and Tukey's post hoc comparison analysis to test for significant treatment differences. For data with non-normal distribution and heteroscedasticity, we used nonparametric ANOVA (Kruskal-Wallis analysis). The level of significance in all tests was set at $p < 0.05$. All statistical tests were done in either R or JMP Pro12 software (SAS, Cary, NC, USA).

8.3 Results

8.3.1 Climatic Conditions

The summers 2009 and 2010 had an even distribution of precipitation with no long drought periods, except in the spring 2009 (Fig. 8.2). Temperatures were higher in 2009 than in 2010, due to the warm and dry spring in 2009. From May 1 to September 30, mean air temperature in 2009 and 2010 was 14.5 and 13.1 °C respectively, total precipitation was 409 and 493 mm, and mean vapour pressure deficit was 471 and 278 Pa.

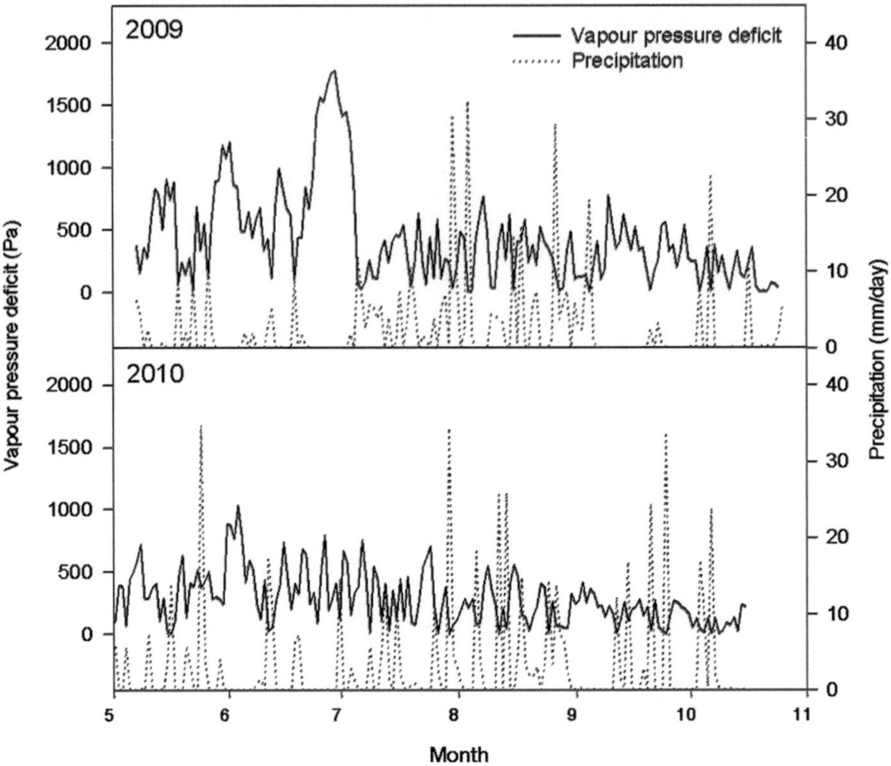

Fig. 8.2 Daily means of vapor pressure deficit and precipitation at the Hoxmark experimental site in SE Norway from May to November, in 2009 and 2010

Precipitation exclusion in the drought plot gradually led to severe drought in the upper 30 cm of the soil, with moderate drought conditions occurring down to 50 cm soil depth (Fig. 8.3). Soil water potential (SWP) values showed that water availability began to differ between the control and drought plot in July 2009 at 10 and 30 cm soil depth, and in August 2009 the plots also differed at 50 cm soil depth (Fig. 8.3). During the following two years, the control plot experienced only occasional brief drought spells in the soil. The drought plot, in contrast, experienced continuously severe drought conditions in the upper 30 cm of the soil throughout the 2-year-period. From August 2009, SWP remained below −1.1 MPa at 30 cm soil depth, the lower detection limit of our instruments, and from June 2010 medium dry conditions were detected even at 50 cm soil depth (Fig. 8.3).

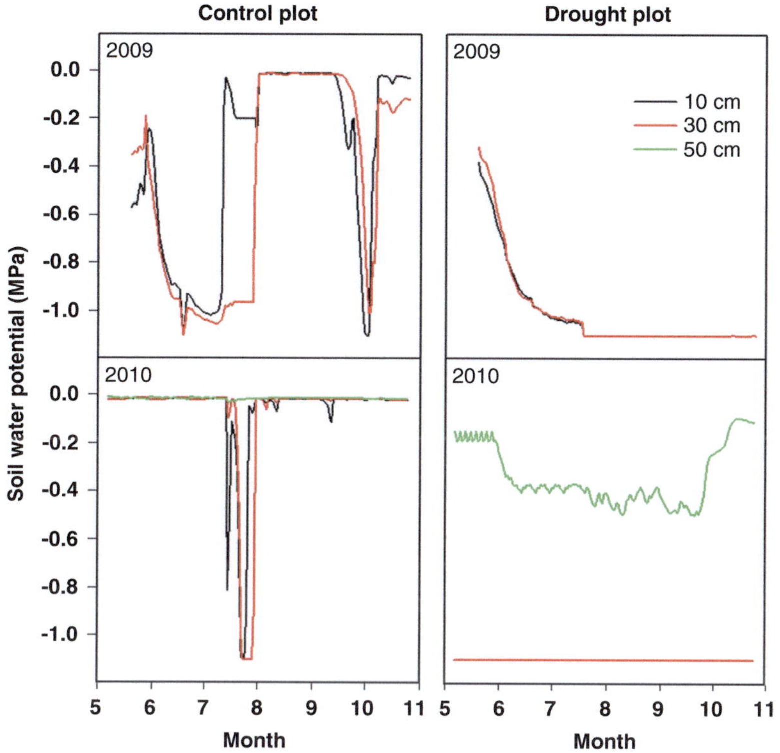

Fig. 8.3 Mean daily values of soil water potential in 2009 and 2010 at different soil depths in a thinned control plot and a plot subjected to drought

8.3.2 *Fine Root Growth and Biomass*

8.3.2.1 Growth Estimated from Minirhizotrons: Shift in Root Growth to Deeper Soil Layers Following Drought

Fine root growth, expressed as number of root tips per minirhizotron image, was increasing in the control plot throughout the 14-month observation period (Fig. 8.4). Fine roots were present both in upper (0–20 cm) and lower soil layers (20–40 cm), and at the end of the study the control plot had more than twice as many root tips in the lower soil layer as in the upper layer (Fig. 8.4). In contrast, no root growth was detected in the upper soil layer of the drought plot throughout the experiment, and new root tips only appeared at the last observation in July 2011, but in low numbers (Fig. 8.4).

Different full-sib families responded differently to drought. Family 14 had the most vigorous fine root growth of all families in the control plot, whereas in the

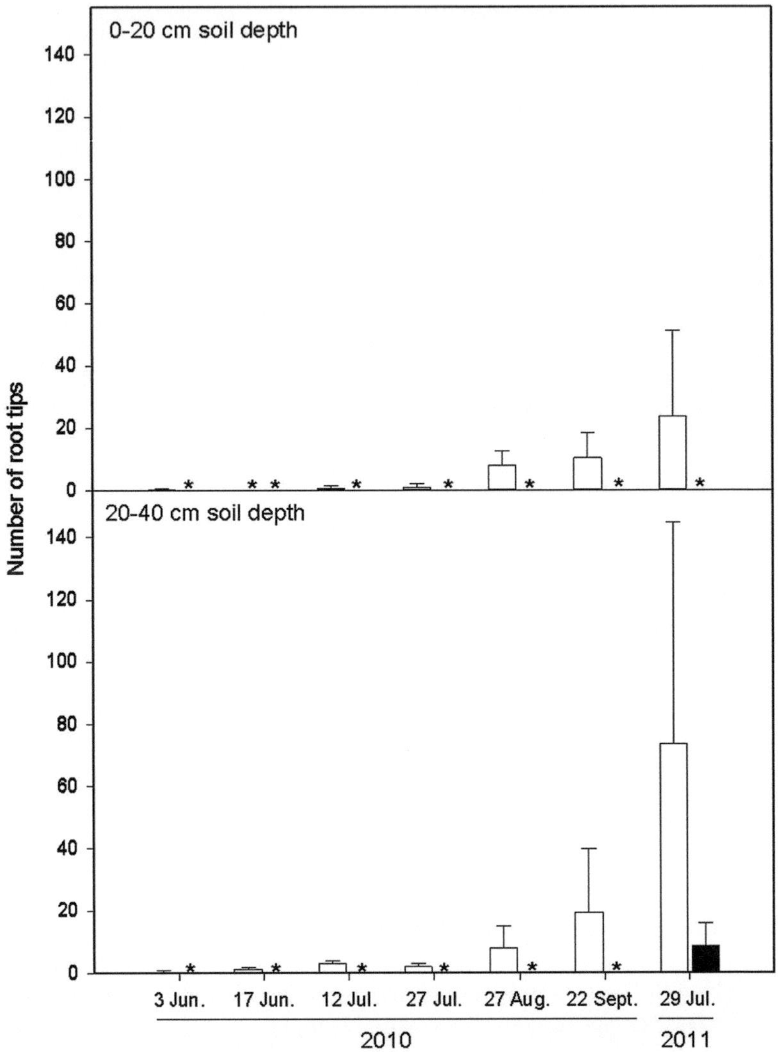

Fig. 8.4 Number of fine root tips at different soil depths in a thinned control plot (*white bars*) and a drought plot (*black bars*), following the start of precipitation exclusion in the drought plot in May 2009. Each bar shows the mean root tip number in three minirhizotrons per plot (error bar = 1 standard error, samples with zero values are marked with*)

drought plot it had almost no root growth (Fig. 8.5). Family 27 had intermediate fine root growth in the control plot and strongly reduced growth in the drought plot (Fig. 8.5). Finally, family 29 had the lowest fine root growth of all families in the control plot and was relatively unaffected by drought (Fig. 8.5).

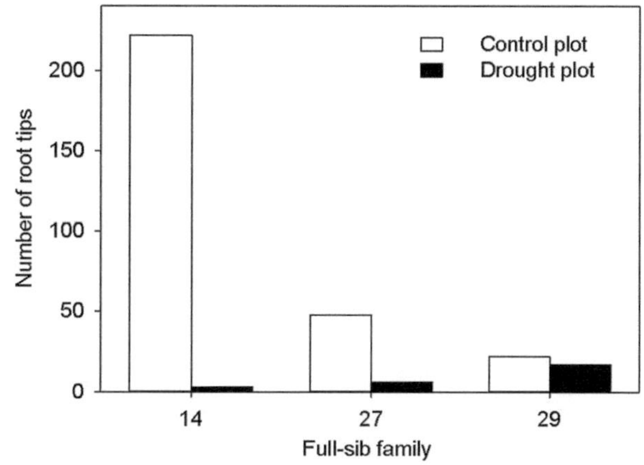

Fig. 8.5 Total number of fine root tips in Norway spruce trees from three different full-sib families in a thinned control plot (*white bars*) and a thinned drought plot (*black bars*). Root tip numbers were estimated from minirhizotron images taken over a 14-month-period following the onset of drought in May 2009 (n = 1 minirhizotron per tree)

8.3.2.2 Biomass Estimated from Soil Cores: Initial Stimulation of Root Growth by Drought, Followed by a Decrease

Fine root biomass was higher in the drought plot than in the control plot in May 2010, one year after the onset of drought (Fig. 8.6). The drought plot had higher root biomass than the control plot at all soil depths, but the difference was significant only for the 10–20 cm soil layer ($p = 0.022$). In both plots root biomass decreased with depth, but the only significant difference between soil layers was that between the uppermost soil level in the control plot and all deeper layers ($p < 0.026$).

In October 2010, 17 months after the onset of drought, the control plot had a similar vertical distribution of fine root biomass as in May, with a gradual decrease in root biomass with increasing soil depth (Fig. 8.6). The uppermost soil level had significantly more root biomass than the 20–30 cm layer ($p = 0.0008$). The drought plot had lower root biomass in all soil layers compared to May, and root biomass in the uppermost soil layer was significantly lower than in the control plot ($p = 0.032$).

8.3.2.3 Growth Estimated from Nylon Meshes: No New Root Growth in Upper Soil Layers After Drought

No fine roots were growing through the nylon meshes between October 2009 and June 2010 in either of the plots (data not shown). From June 2010 to October 2010 fine root growth was detected in both the thinned and un-thinned control plot, but not in the drought plot. Fine root biomass was twice as high in the thinned control plot as in the un-thinned control plot (0.038 ± 0.00 vs. 0.017 ± 0.00 g DW cm^{-3}, respectively (mean ±SD), $p = 0.05$).

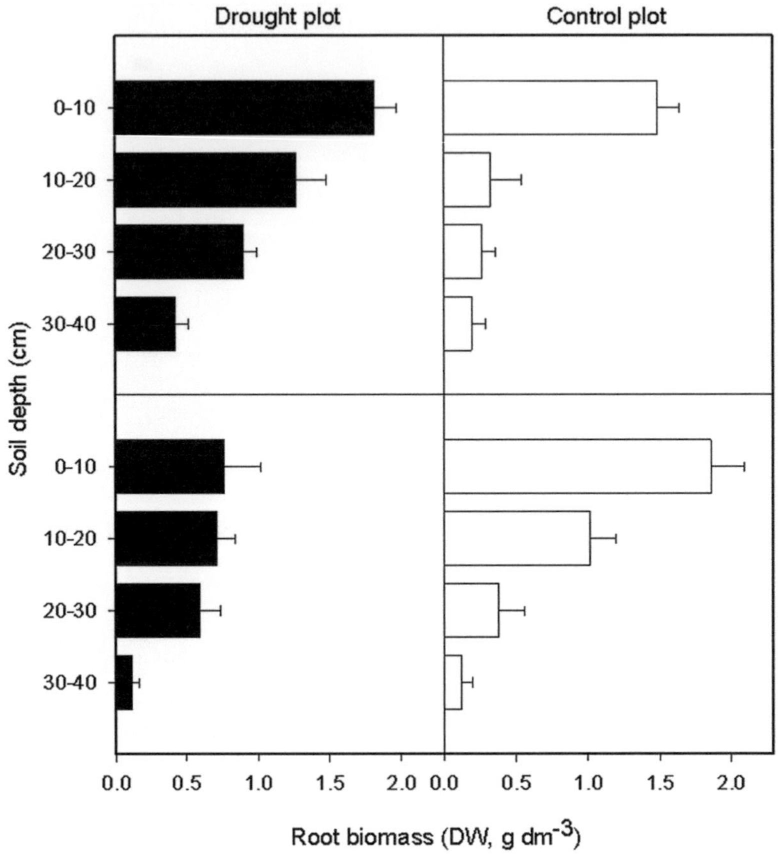

Fig. 8.6 Biomass of Norway spruce fine roots (diameter <2 mm) in soil cores collected in a thinned drought and control plot in May (*upper panels*) and October (*lower panels*) 2010, following the onset of drought in May 2009. Each bar shows the mean root biomass in individual soil cores taken near three Norway spruce trees in each plot (n = 3). *Bars* = 1 standard error

8.3.3 Hyphal Growth and Biomass

8.3.3.1 Biomass Estimated from Nylon Meshes: Hyphae Disappeared from Upper Soil Layers After Drought

The control plot had more hyphal biomass in the upper 10 cm of the soil than the drought plot (Fig. 8.7). Biomass values in the control were higher between autumn and spring (October 2009 to June 2010) than between spring and autumn (June 2010–October 2010) (Fig. 8.7). The drought plot had significantly lower hyphal biomass than the control plot, and the vertical distribution of hyphal biomass was constant over time (Fig. 8.7). The most striking difference between plots was in the vertical distribution of hyphal biomass in the soil (Fig. 8.7). In the thinned and

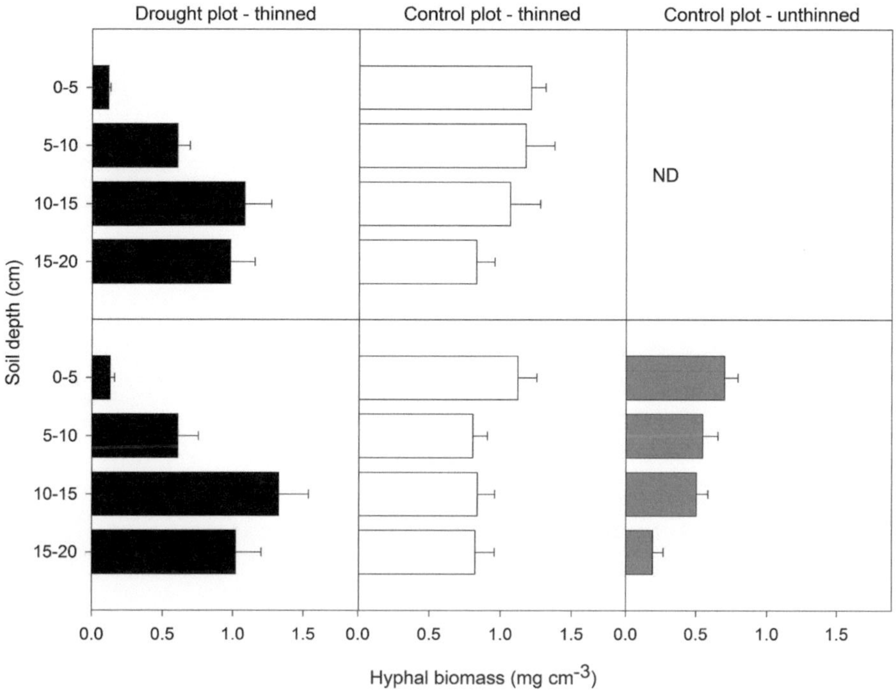

Fig. 8.7 Biomass of fungal hyphae at different soil depths in three experimental plots inside a Norway spruce stand. Hyphae were trapped in nylon meshes inserted into the soil (n = 12 per plot) during two time periods (*upper panels*: October 2009–June 2010; *lower panels*: June 2010–October 2010). Drought treatment started in May 2009 (bars = 1 standard error; *ND* not detected, meshes not installed in this period)

un-thinned control plots hyphal biomass was relatively evenly distributed between soil layers, with slightly higher levels in the upper five centimeters (Fig. 8.7). In the drought plot, however, hyphal biomass was much higher in the deeper soil layers (10–20 cm), with very little hyphae in the upper soil layer (0–5 cm) (Fig. 8.7). In fact, only 8% of the total hyphal biomass in the drought plot was located in the upper soil layer, compared with 30% in the two control plots. Interestingly, as for fine root growth estimated from nylon meshes hyphal biomass was about twice as high in the thinned control plot as in the un-thinned plot (Fig. 8.7).

8.3.3.2 Growth Estimated from Minirhizotrons: More Hyphae in Deeper Soil Layers After Drought

Analysis of hyphal coverage in minirhizotron images revealed a similar pattern as for hyphal biomass estimates obtained from the nylon meshes (Figs. 8.7 and 8.8). The control plot had much more fungal hyphae in upper soil layers (0–20 cm) than the drought plot, and hyphal coverage in the control plot increased through the

Control plot

Depth (cm)	3/6/10	17/6/10	12/7/10	27/7/10	27/8/10	22/9/10	29/7/11
0	7	30	38	37	38	38	25
2.3	8	33	51	53	55	55	43
4.5	10	20	52	52	55	57	58
6.8	12	25	38	42	45	52	67
9.1	7	15	37	38	43	52	68
11.3	8	17	37	35	43	50	65
13.6	12	18	32	28	40	47	62
15.8	12	15	28	22	25	28	40
18.1	12	10	15	15	20	20	35
20.4	10	10	15	20	33	33	35
22.6	10	10	13	13	27	27	27
24.9	10	8	12	12	25	25	23
27.2	8	7	8	10	23	25	25
29.4	7	7	8	8	23	23	23
31.7	7	5	7	7	12	17	22
33.9	10	12	13	13	23	27	17
36.2	10	12	13	13	18	17	17
38.5	10	12	13	13	13	13	13
40.7	10	12	15	15	15	15	15
43.0	13	12	15	15	15	15	15
45.3	13	13	17	22	22	22	22
47.5	15	13	22	23	22	22	22
49.8	15	12	20	23	23	23	28
52.0	13	10	22	23	25	25	27
54.3	13	8	22	23	25	25	25
56.6	0	0	0	0	0	0	0
58.8	0	0	0	0	0	0	0

Drought plot

Depth (cm)	3/6/10	17/6/10	12/7/10	27/7/10	27/8/10	22/9/10	29/7/11
0	0	0	0	0	0	0	0
2.3	0	0	0	0	0	0	0
4.5	0	0	0	0	0	0	0
6.8	0	0	0	0	0	0	0
9.1	0	0	0	0	0	0	0
11.3	0	0	0	0	0	0	0
13.6	0	0	0	0	0	1	1
15.8	0	1	1	1	1	1	1
18.1	0	2	3	5	7	9	10
20.4	0	2	3	5	7	12	13
22.6	0	1	5	5	8	12	13
24.9	0	1	3	3	8	12	17
27.2	0	1	3	5	10	12	15
29.4	0	1	3	5	12	12	17
31.7	0	1	2	4	7	12	17
33.9	0	1	2	3	5	8	14
36.2	1	1	1	2	4	5	13
38.5	1	1	2	3	4	5	12
40.7	1	1	2	5	6	7	10
43.0	1	1	9	11	14	17	20
45.3	2	1	9	13	13	17	25
47.5	2	2	18	22	27	30	35
49.8	2	6	23	27	32	35	38
52.0	2	6	23	25	32	35	40
54.3	2	4	18	20	28	30	40
56.6	1	4	10	12	17	20	30
58.8	0	0	2	2	7	13	15

Fig. 8.8 Percent coverage of fungal hyphae in minirhizotron images taken over a 14-month period at different soil depths in a thinned control plot and a thinned plot subjected to long-term drought starting in May 2009. The color intensity in each cell corresponds to the percentage hyphal coverage at a particular soil depth in each image. Three minirhizotrons were installed in each plot ($n = 3$).

summer of 2010 and onwards to the next summer (Fig. 8.8, left panel). The drought plot had almost no hyphae in the upper 16 cm of the soil, but in deeper soil layers (below 47 cm) this plot had more hyphae than the control plot, with a steady increase in hyphal coverage over time (Fig. 8.8, right panel).

8.4 Discussion

The drought plot had considerably drier soil than the control plot throughout the experimental period (Fig. 8.3). Thus, the installed plastic roof successfully intercepted precipitation and created severe drought conditions, especially during the second growing season in 2010 and in the upper 30 cm of the soil (Fig. 8.3). Mean daily vapor pressure deficit (VPD) levels at our experimental site only intermittently reached high levels and peaked at about 1800 Pa, relative to the normal range of 800–950 Pa (Fig. 8.2). When VPD increases so does tree transpiration and the demand on fine roots to draw more water. Therefore, high VPD values indicate lower water availability for the fine roots (due to the low soil hydraulic conductivity) and possible water stress and fine root death. In the control plots that received normal precipitation the occasional spikes in VPD levels were soon quenched by precipitation (Fig. 8.2) and the soil experienced only short spells of low water potentials (Fig. 8.3). In the drought plot that received almost no precipitation, however, the combination of high VPD levels and low soil water potentials probably exposed the fine roots to severe water stress (Figs. 8.2 and 8.3). Drought intensified from May to October 2010 in the drought plot (Fig. 8.3) and this could also be seen by the much lower accumulated sap flow in the drought plot relative to the control plot (100 vs. 230 mm between May 17 and October 1; Gebauer et al. 2015).

8.4.1 Fine Root Growth and Biomass

There are different opinions in the scientific literature on how drought affects fine root growth; some studies report that drought leads to an increased investment in root growth, whereas other studies find the opposite effect (reviewed by Brunner et al. 2015). In our study we show that different full-sib families reacted differently to drought stress, indicating that there is a strong genetic influence on how Norway spruce trees react to drought stress. For example, in family 29 the number of root tips was not affected by drought, whereas in the other families drought dramatically reduced root tip numbers (Fig. 8.5). Overall, fine roots and associated fungal hyphae shifted their distribution into deeper soil layers in response to drought. The tendency to avoid drought by growing into deeper and moister soil layers is a manifestation of root plasticity and may result from root association with mycorrhizal hyphae which also function as water providers (e.g. Brownlee et al. 1983; Ekblad et al. 2013; Muhsin and Zwiazek 2002; Plamboeck et al. 2007).

Soil core estimates of total fine root biomass in the thinned control plot was about 50% higher in October 2010 than in May 2010, and at both time points most of the root biomass was found in the upper 10 cm of the soil (Fig. 8.6). In May 2010, the thinned drought plot had higher total fine root biomass than the thinned control plot at all soil depths (Fig. 8.6). This may be due to increased fine root growth in the drought plot as a first response to drought, probably to compensate for decreased water availability (Eldhuset et al. 2013). In October 2010, total fine root biomass in the drought plot was reduced to less than half of that in the control plot. The low and uniform distribution of fine roots in the upper 30 cm of soil in the drought plot (Fig. 8.6) was probably due to the near complete depletion of water in upper soil layers (Fig. 8.3), shifting fine root growth to deeper and more humid soil layers (see also Konôpka and Lukáč 2013; Lyr and Hoffmann 1967).

The results from our long term study of fine root dynamics after drought may partly explain the conflicting results in the literature on how fine roots respond to drought, where some authors report an increase in fine roots after drought while others have found the opposite trend. It seems that short term drought stress may stimulate root growth and lead to an increase in fine roots (Fig. 8.6, upper panel), but prolonged drought may decrease root growth (see Fig. 8.6, lower panel).

Thus, the duration and intensity of the drought period seems to be crucial for fine root dynamics and tree survival.

Minirhizotrons (Fig. 8.4) and nylon meshes (Fig. 8.7) provided much lower estimates of fine root growth in the drought plot than the soil cores did (Fig. 8.6). There may be two main explanations for this discrepancy. First, minirhizotrons and nylon meshes provide data on root growth dynamics from the time of their installation onwards, whereas soil cores describe the status *in situ* at the moment of extraction and include both older and newer fine roots. Therefore, the more dynamic measures (minirhizotrons, nylon meshes) emphasize current growth trends and tend to have higher temporal resolution than an accumulative measure such as soil cores. Second, although minirhizotrons, nylon meshes and soil cores were placed near each other they are different methods that provide results at different spatial scales. In addition, the different methods may reflect the inherent spatial diversity of fine root distribution in the soil. However, despite these sources of variability we show that results on root (and hyphal) growth from minirhizotrons, nylon meshes and soil cores complement each other. All three methods revealed reduced growth in response to prolonged drought and a tendency for roots and hyphae to shift to deeper soil layers.

8.4.2 Hyphal Growth and Biomass

The vertical distribution of hyphal biomass in our control plots agreed with previous studies, with the highest biomass occurring in the upper soil layers (Ekelund et al. 2001; Fig. 8.7). We expected drought to have a negative impact on hyphal biomass and to alter its vertical distribution in the soil. Indeed, the drought plot had less hyphal biomass in the soil compared to the controls and most of the biomass was found in deeper soil layers, with almost no hyphae in the upper 5 to 15 cm (Figs. 8.7 and 8.8). A similar vertical shift in hyphal biomass following drought was described by Ekblad et al. (2013). Such shifts in the vertical distribution of fungal hyphae may reflect the strategy of trees to obtain water from deeper soil layers, scavenging for water resources, in times of drought stress.

Our estimates of hyphal biomass in different soil layers based on results from nylon meshes were supported by our visual assessment of hyphal biomass from minirhizotron images. The two methods complement each other since nylon meshes and minirhizotrons monitored the same trees and the monitoring period for the two methods overlapped (Fig. 8.1). Although hyphal coverage in the minirhizotron images was estimated subjectively the method was consistent between treatments and time points since the assessment always was done by the same person. The use of two complementary methods that revealed similar patterns of hyphal biomass between soil layers strengthens the validity of our hyphal biomass estimates. Our estimates of hyphal biomass at different soil depths also agreed with our estimates of fine root biomass using nylon meshes and soil cores. For both hyphae and fine roots biomass was greatest in the upper soil layers in the control plots and decreased with depth, whereas in the drought plot biomass was greatest in deeper soil layers.

We do not know the species identity of the hyphae we quantified, but the fact that hyphae proliferated in deeper and more nutrient poor soil layers in the drought plot suggests that they were predominantly mycorrhizal hyphae that obtain carbohydrates from the roots and in return provide water and nutrients from the soil (Ekblad et al. 2013). The high hyphal biomass in the upper soil layers of the control plots probably consisted of mycorrhizal and saprotrophic fungi that should be able to find sufficient water and nutrients in this environment. Most mycorrhizal and saprotrophic hyphae would probably have died off in the upper soil layers of the drought plot, since even saprotrophic fungi need water to metabolize organic matter. We cannot fully explain why hyphal biomass was higher in the thinned than in the un-thinned control plot (Fig. 8.7), but perhaps thinning increased fine root growth and deposition of organic matter and that this in turn increased the abundance of both mycorrhizal and saprotrophic fungi. Resource competition between fungi may also have been less intense in the thinned than in the un-thinned control plot, permitting higher fungal abundance in the thinned plot. Buée et al. (2005) showed that thinning of a *Fagus sylvatica* forest resulted in significantly higher ectomycorrhizal diversity, and we might speculate that higher species diversity also may lead to higher total hyphal biomass.

8.5 Conclusions

Using different methodological approaches, we have demonstrated that the root system of Norway spruce shows great plasticity in its response to extended periods of drought. The main strategy of Norway spruce when exposed to extended drought seems to be to shift fine root growth to deeper soil layers, where moisture may be retained. This vertical shift in fine root growth is accompanied by a similar shift in the abundance of mycorrhizal fungi. The vertical redistribution of fine root and hyphal biomass may occur quite rapidly, from one growing season to the next. This drought avoidance strategy shows how Norway spruce and other forest trees may respond to future long term droughts induced by global warming and has implications for carbon storage and the carbon cycle. Drought responses at the individual tree level are likely to affect carbon storage at the whole forest level, since more carbon may be stored in forest soils if more carbon is deposited in deeper soil layers, where decomposition rates are slower.

Acknowledgements We thank Jaromíra Dreslerová for excellent technical assistance. This work was funded by Iceland, Liechtenstein and Norway through the EEA Financial Mechanism, the Norwegian Financial Mechanism (grant no. A/CZ0046/2/0009), Mendel University in Brno (grant IGA 73/2013), the EEA project FRAMEADAPT EHP-CZ02-OV-1-044-01-2014, and the project "Indicators of Tree Vitality" (Reg. No. CZ.1.07/2.3.00/20.0265), co-financed by the European Social Fund and the Czech Republic. We also acknowledge contribution by COST Actions FP1106 "STReESS" and FP1305 "BioLink".

References

Aber, J. D., Melillo, J. M., Nadelhoffer, K. J., McClaugherty, C. A., & Pastor, J. (1985). Fine root turnover in forest ecosystems in relation to quantity and form of nitrogen vailability: a comparison of two methods. *Oecologia, 66*, 317–321.

Allen, C. D., Macalady, A. K., Chenchouni, H., Bachelet, D., McDowell, N., et al. (2010). A global overview of drought and heat-induced tree mortality reveals emerging climate change risks for forests. *Forest Ecology and Management, 259*, 660–684.

Andreassen, K., Solberg, S., Tveito, O. E., & Lystad, S. L. (2006). Regional differences in climatic responses of Norway spruce (*Picea abies* L. Karst) growth in Norway. *Forest Ecology and Management, 222*, 211–221.

Borken, W., Savage, K., Davidson, E. A., & Trumbore, S. E. (2006). Effects of experimental drought on soil respiration and radiocarbon efflux from a temperate forest soil. *Global Change Biology, 12*, 177–193.

Bréda, N., Huc, R., Granier, A., & Dreyer, E. (2006). Temperate forest trees and stands under severe drought: A review of ecophysiological responses, adaptation processes and long-term consequences. *Annals of Forest Science, 63*, 625–644.

Brownlee, C., Duddridge, J. A., Malibari, A., & Read, D. J. (1983). The structure and function of mycelial systems of ectomycorrhizal roots with special reference to their role in forming inter-plant connections and providing pathways for assimilate and water transport. *Plant and Soil, 71*, 433–443.

Brunner, I., Herzog, C., Dawes, M. A., Arend, M., & Sperisen, C. (2015). How tree roots respond to drought. *Frontiers in Plant Science, 6*, 547.

Buée, M., Vairelles, D., & Garbaye, J. (2005). Year-round monitoring of diversity and potential metabolic activity of the ectomycorrhizal community in a beech (*Fagus sylvatica*) forest subjected to two thinning regimes. *Mycorrhiza, 15*, 235–245.

Čermák, J., Deml, M., & Penka, M. (1973). A new method of sap flow rate determination in trees. *Biologia Plantarum (Prague), 15*, 171–178.

Čermák, J., Kučera, J., & Nadezhdina, N. (2004). Sap flow measurements with some thermodynamic methods, flow integration within trees and scaling up from sample trees to entire forest stands. *Trees, 18*, 529–546.

Ditmarová, L., Kurjak, D., Palmroth, S., Kmeť, J., & Střelcová, K. (2010). Physiological responses of Norway spruce (*Picea abies*) seedlings to drought stress. *Tree Physiology, 30*, 205–213.

Ekblad, A., Wallander, H., Godbold, D. L., Cruz, C., Johnson, D., et al. (2013). The production and turnover of extramatrical mycelium of ectomycorrhizal fungi in forest soils: Role in carbon cycling. *Plant and Soil, 366*, 1–27.

Ekelund, F., Rønn, R., & Christensen, S. (2001). Distribution with depth of protozoa, bacteria and fungi in soil profiles from three Danish forest sites. *Soil Biology and Biochemistry, 33*, 475–481.

Eldhuset, T. D., Nagy, N. E., Volařík, D., Børja, I., Gebauer, R., et al. (2013). Drought affects tracheid structure, dehydrin expression, and above- and belowground growth in 5-year-old Norway spruce. *Plant and Soil, 366*, 305–320.

Gaul, D., Hertel, D., Borken, W., Matzner, E., & Leuschner, C. (2008). Effects of experimental drought on the fine root system of mature Norway spruce. *Forest Ecology and Management, 256*, 1151–1159.

Gebauer, R., Volařík, D., Urban, J., Børja, I., Nagy, N. E., et al. (2011). Effect of thinning on anatomical adaptations of Norway spruce needles. *Tree Physiology, 31*, 1103–1113.

Gebauer, R., Volařík, D., Urban, J., Børja, I., Nagy, N. E., et al. (2012). Effects of different light conditions on the xylem structure of Norway spruce needles. *Trees - Structure and Function, 26*, 1079–1089.

Gebauer, R., Volařík, D., Urban, J., Børja, I., Nagy, N. E., et al. (2015). Effects of prolonged drought on the anatomy of sun and shade needles in young Norway spruce trees. *Ecology and Evolution, 5*, 4989–4998.

Hendrick, R. L., & Pregitzer, K. S. (1993). The dynamics of fine-root length, biomass, and nitrogen-content in 2 northern hardwood ecosystems. *Canadian Journal of Forest Research, 23*, 2507–2520.

Hirano, Y., Noguchi, K., Ohashi, M., Hishi, T., Makita, N., et al. (2009). A new method for placing and lifting root meshes for estimating fine root production in forest ecosystems. *Plant Root, 3*, 26–31.

IPCC. (2013). IPCC fifth assessment report. *Weather, 68*, 310–310.

Jackson, R. B., Mooney, H. A., & Schulze, E. D. (1997). A global budget for fine root biomass, surface area, and nutrient contents. *Proceedings of the National Academy of Sciences, USA, 94*, 7362–7366.

Joslin, J. D., Wolfe, M. H., & Hanson, P. J. (2000). Effects of altered water regimes on forest root systems. *New Phytologist, 147*, 117–129.

Konôpka, B., & Lukáč, M. (2013). Moderate drought alters biomass and depth distribution of fine roots in Norway spruce. *Forest Pathology, 43*, 115–123.

Leuschner, C., Backes, K., Hertel, D., Schipka, F., Schmitt, U., et al. (2001). Drought responses at leaf, stem and fine root levels of competitive *Fagus sylvatica* L. and *Quercus petraea* (Matt.) Liebl. trees in dry and wet years. *Forest Ecology and Management, 149*, 33–46.

Leuschner, C., Hertel, D., Schmid, I., Koch, O., Muhs, A., & Holscher, D. (2004). Stand fine root biomass and fine root morphology in old-growth beech forests as a function of precipitation and soil fertility. *Plant and Soil, 258*, 43–56.

Lukáč, M., & Godbold, D. L. (2010). Fine root biomass and turnover in southern taiga estimated by root inclusion nets. *Plant and Soil, 331*, 505–513.

Lyr, H., & Hoffmann, G. (1967). Growth rate and growth periodicity of tree roots. *International Review of Forest Research, 2*, 181–206.

McDowell, N. G., & Sevanto, S. (2010). The mechanisms of carbon starvation: How, when, or does it even occur at all? *New Phytologist, 186*, 264–266.

McDowell, N., Pockman, W. T., Allen, C. D., Breshears, D. D., Cobb, N., et al. (2008). Mechanisms of plant survival and mortality during drought: Why do some plants survive while others succumb to drought? *New Phytologist, 178*, 719–739.

Meier, I. C., & Leuschner, C. (2008). Belowground drought response of European beech: Fine root biomass and carbon partitioning in 14 mature stands across a precipitation gradient. *Global Change Biology, 14*, 2081–2095.

Moser, B., Kipfer, T., Richter, S., Egli, S., & Wohlgemuth, T. (2015). Drought resistance of *Pinus sylvestris* seedlings conferred by plastic root architecture rather than ectomycorrhizal colonisation. *Annals of Forest Science, 72*, 301–309.

Muhsin, T. M., & Zwiazek, J. J. (2002). Ectomycorrhizas increase apoplastic water transport and root hydraulic conductivity in Ulmus americana seedlings. *New Phytologist, 153*, 153–158.

Plamboeck, A. H., Dawson, T. E., Egerton-Warburton, L. M., North, M., Bruns, T. D., & Querejeta, J. I. (2007). Water transfer via ectomycorrhizal fungal hyphae to conifer seedlings. *Mycorrhiza, 17*, 439–447.

Pregitzer, K. S., Hendrick, R. L., & Fogel, R. (1993). The demography of fine roots in response to patches of water and nitrogen. *New Phytologist, 125*, 575–580.

Roudier, P., Andersson, J. C. M., Donnelly, C., Feyen, L., Greuell, W., & Ludwig, F. (2016). Projections of future floods and hydrological droughts in Europe under a +2 degrees C global warming. *Climatic Change, 135*, 341–355.

Santantonio, D., & Hermann, R. K. (1985). Standing crop, production, and turnover of fine roots on dry, moderate, and wet sites of mature douglas-fir in western Oregon. *Annales des Sciences Forestières, 42*, 113–142.

Schall, P., Lödige, C., Beck, M., & Ammer, C. (2012). Biomass allocation to roots and shoots is more sensitive to shade and drought in European beech than in Norway spruce seedlings. *Forest Ecology and Management, 266*, 246–253.

Sevanto, S., McDowell, N. G., Dickman, L. T., Pangle, R., & Pockman, W. T. (2014). How do trees die? A test of the hydraulic failure and carbon starvation hypotheses. *Plant Cell and Environment, 37*, 153–161.

Sohn, J. A., Kohler, M., Gessler, A., & Bauhus, J. (2012). Interactions of thinning and stem height on the drought response of radial stem growth and isotopic composition of Norway spruce (*Picea abies*). *Tree Physiology, 32*, 1199–1213.

Solberg, S. (2004). Summer drought: A driver for crown condition and mortality of Norway spruce in Norway. *Forest Pathology, 34*, 93–104.

Steele, S. J., Gower, S. T., Vogel, J. G., & Norman, J. M. (1997). Root mass, net primary production and turnover in aspen, jack pine and black spruce forests in Saskatchewan and Manitoba, Canada. *Tree Physiology, 17*, 577–587.

Vogt, K. A., Vogt, D. J., Palmiotto, P. A., Boon, P., Ohara, J., & Asbjornsen, H. (1996). Review of root dynamics in forest ecosystems grouped by climate, climatic forest type and species. *Plant and Soil, 187*, 159–219.

Chapter 9
Ectomycorrhizal Diversity in Beech Dominated Stands in Central Europe

Christoph Rosinger, Hans Sandén, and Douglas L. Godbold

Abstract Mycorrhizal fungi are essential for ecosystem functioning. They are the most important pathway of plant-derived C into the soil and facilitate most of the N and P uptake. In boreal and temperate forest ecosystems, the subgroup of ectomycorrhizas is of special significance, and ectomycorrhizal symbiosis is obligatory for a number of tree species. These fungal assemblages can be highly diverse, which is often contrasting with the relatively low number of tree species. This is also true for small monoculture stands, where several studies showed dozens of different ectomycorrhizal species. The elemental mechanisms that drive ectomycorrhizal community composition and structure, especially on larger scales, are still subject of debate. In particular, the role and function of most of the rare species remain largely unknown. In this chapter, we investigate the structural traits of ectomycorrhizal communities of European beech (*Fagus sylvatica*), one of the most important tree species in temperate deciduous forest ecosystems. We therefore compiled data from ten different beech dominated sites across a European gradient to seek for ubiquitously valid statements regarding the species composition and community structure of ectomycorrhizae. A total of 205 ectomycorrhizal species were detected, of which 45% and 35% could be identified to the species and genus level, respectively. The majority of them are site-specific and found in very low abundances. A few species with narrow host range such as *Lactarius subdulcis*, *Laccaria amethystina* or *Lactarius pallidus* show low abundances on many sites. Three species, namely *Hebeloma sinapizans*, *Lactarius salmonicolor* and *Elaphomyces aculeatus*, appeared to be very dominant on one respective site. *Cenococcum geophilum* and *Russula ochroleuca*, both considered multi-host species, occur at all or rather several sites with relatively high abundances. Species ranking-abundance curves revealed a universal distribution pattern, with few dominant ectomycorrhizal species and a long tail of rare species. This shape is regardless of stand age and management practice, though no statistically significant correlations with species richness and the latter mentioned parameters could be found. By looking at several community structures, we were able to add another dimension of rarity and found most of

C. Rosinger (✉) • H. Sandén • D.L. Godbold
Institute of Forest Ecology, Department of Forest- and Soil Sciences, University of Natural Resources and Life Sciences, Peter-Jordan-Straße 82, 1190 Vienna, Austria
e-mail: christoph.rosinger@boku.ac.at

M. Lukac et al. (eds.), *Soil Biological Communities and Ecosystem Resilience*, Sustainability in Plant and Crop Protection, DOI 10.1007/978-3-319-63336-7_9

the species not only being rare at a particular site, but also rare in terms of single appearance across several analysed stands. However, a vast part of ectomycorrhizal fungal species still remains unidentified. This circumstance further impedes our understanding concerning the occurrence of rare species. Yet, understanding their functional behaviour is of particular significance, as they could pose a key factor in ecosystem functioning, especially under changing climatic conditions.

Keywords Community structure • Diversity • Ectomycorrhiza • European beech • *Fagus sylvatica*

9.1 Introduction

One of the most prominent mutualistic interactions in terrestrial ecosystems is the symbiosis of fungi with plant roots, also known as mycorrhiza. Mycorrhizal fungi are only a rather small group within the kingdom of fungi, however they play a critical role in nutrient cycling and ecosystem functioning (Smith and Read 2008). Mycorrhizal fungi improve plant growth and survival through a mutualistic relationship in which photoassimilates are exchanged for increased access to nutrients (Kernaghan 2005).

There are seven forms of mycorrhizal symbiosis covering most of the plant kingdom. Many economically and ecologically important trees in boreal and temperate ecosystems as well as some tropical tree species form ectomycorrhizas. Molina et al. (1992) estimated that between 5000–6000 fungal species, mainly within the groups Basidiomycota and Ascomycota, form ectomycorrhizal (EM) symbioses. However, estimates of the number of species are uncertain, and increased taxonomic efforts will further raise this number in the near future.

Common diversity theories do not seem to be appropriate to describe the community structure and diversity of ectomycorrhizal fungi. EM diversity is highest in temperate and boreal biomes, subtending with the latitudinal gradient of species richness. In temperate forest ecosystems, the richness and diversity of EM fungi further contrast with the relatively low number of tree species. This is also true for small monoculture stands, where several studies showed dozens of different ectomycorrhizal species (Kjøller 2006; Pena et al. 2010). Even single roots can be infested by several different species (pers. obs.). Despite the fact that we can estimate ectomycorrhizal species numbers, the mechanisms that drive ectomycorrhizal diversity, composition as well as community structure, particularly on a global biome scale, remain largely unexplained.

In this book chapter, we examine structural traits of ectomycorrhizal communities on a European scale. For this purpose, we have chosen the tree species *Fagus sylvatica*, one of the economically and ecologically most important forest tree species in temperate mid-European deciduous forests (Ellenberg and Leuschner 1996; Pritsch et al. 2009). Like all *Fagaceae*, it forms only EM symbioses, which are considered as ecologically obligatory for this species (Druebert et al. 2009), and the fine root tips are normally fully mycorrhizal (Taylor et al. 2000; Rumberger et al. 2004; Buée et al. 2005; Grebenc and Kraigher 2007). Beech is considered to exhibit extremely high ectomycorrhizal species diversity (Reich et al. 2009). Natural diversity levels were found to be around 25–30 EM fungal species for a mature beech forest (Grebenc and Kraigher 2007; Kraigher et al. 2007; Di Marino et al. 2009). However, intensive sampling over a whole vegetation period doubled the number of recognized EM species to sixty (Buée et al. 2005). Beech secures EM fungal species richness and is therefore ecologically important as a warrantor of EM fungal diversity (Lang et al. 2011).

9.2 Ectomycorrhizal Community Structure of Fagus Sylvatica Dominated Stands Across Europe

The objective of this study was to look for patterns in species composition and community structure of ectomycorrhizas in beech (dominated) stands. To investigate the structural traits of EM communities, we carried out a literature study. Prerequisites were comprehensive datasets with relative abundance values for each ectomycorrhizal species and analysis of EM root tips by molecular techniques (not by morphotyping or sporocarp counts). Furthermore, we concentrated on mature and pure or rather beech dominated stands that were not exposed to any experimental disturbance, e.g. the artificial application of a double ozone concentration as evaluated in one study. In studies including experimental disturbances only the untreated control plots were used for the analysis.

Pot experiments or studies on seedlings were also excluded for the community structure analysis. Six papers with species abundance data for ectomycorrhizal diversity in beech stands on ten different sites were considered appropriate and used for further analysis (Table 9.1). The sites show a broad spatial range across Central Europe, including unmanaged as well as managed stands of various intensities. Stand age varied between 60 and 200 years.

In all ten sites combined a total of 205 ectomycorrhizal species were detected, of which 92 (45% of total) and 72 (35%) could be identified to the species and genus level, respectively, and 41 (20%) species remain unidentified. The latter two categories of EM species may show some species overlap between the studied sites.

By analysing the relative abundance of the ectomycorrhizal species present, a prevalent pattern emerged (Fig. 9.1). The majority of the species are site specific - that is, low in site frequency - and found in very low abundances. These species are located in the lower-left part of the figure. A few species with narrow host range such as *Lactarius subdulcis, Laccaria amethystina* or *Lactarius pallidus* had low abundances on several sites. Three species, namely *Hebeloma sinapizans, Lactarius salmonicolor* and *Elaphomyces aculeatus*, were very dominant on one site. *Cenococcum geophilum* and *Russula ochroleuca*, both of them considered multi-

Table 9.1 Study sites with additional information used for the literature study

References	Location	Coordinates	Altitude	Stand age	Objective of the study
Grebenc and Kraigher (2007)	S-Germany	48°25′08″N, 11°39′41″E	485	60–70	Analysed double ambient ozone concentration
Kraigher et al. (2007)	Central Slovenia	46°7′33″N, 15°3′41″E	700–800	80–140	Effects of pollution (N, S) from power plant
		46°7′33″N, 15°3′41″E	700–800	80–140	
Di Marino et al. (2009)	N-Italy	46°10′8″N, 10°52′59″E	–	110–120	Analysed coppice and stand features
Grebenc et al. (2009)	SW Slovenia	45°39′36″N, 15°0′36″E	860–890	80–120	Studied gap-opening in natural and managed forests
		45°39′15″N, 15°1′40″E	860–890	Varying; up to 200	
	Denmark	55°51′N, 12°28′E	80–90	75	
		55°31′N, 11°54′E	80–90	Natural regeneration back in 1755	
Rineau and Garbaye (2010)	NE France	48°00′00″N, 6°29′28″E	570	60	Liming experiment
Lang et al. (2011)	C-Germany	51°05‘28″N, 10°31‘24″E	350	100	Analysed host-preferences and differential contributions

host species, occur at all or rather most sites with relatively high abundances. *C. geophilum* was by far the most abundant EM species across all sites. This species occurred on all plots, and it was found to be the most frequent species in half of the sites. *C. geophilum* is one of the most frequently encountered EM species in nature and exhibits an extremely wide host and habitat range (Cairney and Chambers 2013), which makes it one of the best studied EM species. Its high pioneer capability allows it to invade newly formed soils on which EM hosts become established (Trappe 1962). It has a high adaptation potential to extreme site conditions (Haselwandter and Read 1980).

Along with *C. geophilum*, EM species within the genus *Lactarius* were found to be dominant root colonizers on *F. sylvatica*. They are one of the largest known ectomycorrhizal genera within Basidiomycota and play a significant role as late-stage colonizers. Most of them exhibit a high degree of host specificity (Cairney and Chambers 2013), and *Lactarius subdulcis*, *L. pallidus* and *L. salmonicolor* are reported to be restricted to beech. In addition to these *Lactarius* species, *L. blennius*, *L. vellereus* and *L. camphoratus* were found on five, two and two sites, respectively,

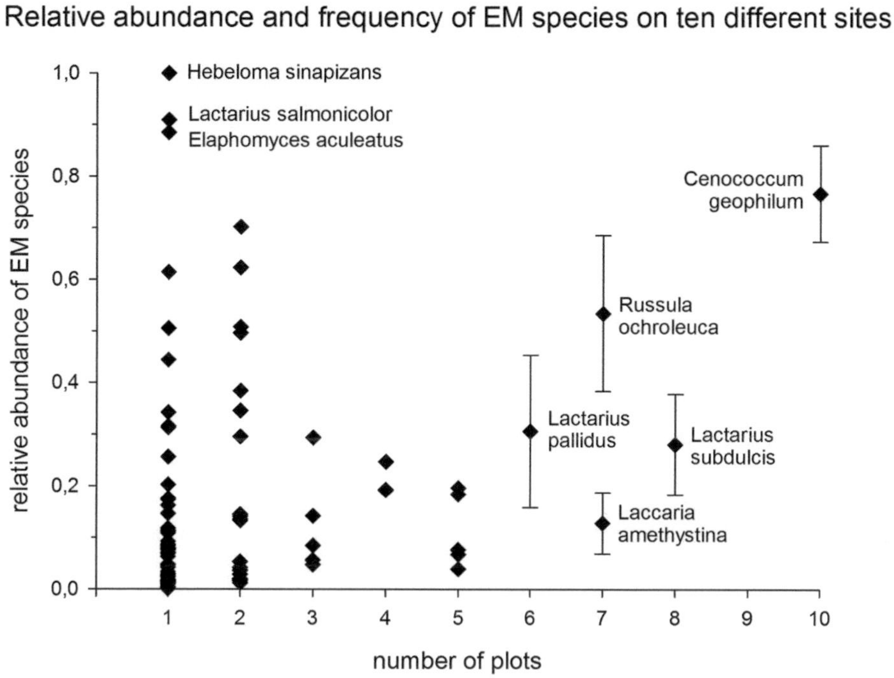

Fig. 9.1 Frequency and relative abundance of ectomycorrhizal fungal species on ten *Fagus sylvatica* sites; each symbol represents one species; relative abundance of a species is referred to as its abundance value in relation to the upmost abundant species on each site, respectively; e.g. *Lactarius pallidus* was found on 6 out of 10 sites with a mean relative abundance of 0.3; due to space limitation, *error bars* are solely shown for the 5 most frequent species; error bars indicate SE

making this genera one of the most dominant ectomycorrhizas on beech. However, their abundance shows quite a wide range.

The second-most dominant genus is *Russula*, with *R. ochroleuca* being the most prominent representative of this group. Five species of *Russula* (*R. ochroleuca*, *R. fellea*, *R. illota*, *R. mairei*, *R. cyanoxantha*) were found on at least four different sites. For *Russula*, little information on ecological traits is available. This is due to high host specificity and the difficulty of cultivating *Russula* in vitro (Cairney and Chambers 2013). It is known that increased deposition of N and acidification of forest soils decreases the abundance of certain *Lactarius* and *Russula* species (e.g. Jansen 1990). Some species also tend to be intolerant of drought stress (Smith and Read 2008). We still lack a comprehensive understanding of structural and functional traits of especially rare, single-host species. Further, their behaviour might be quite different in comparison to dominant multi-host species such as *C. geophilum*, which is known to be very competitive by means of inhibition at distance via mycotoxins (Baar and Stanton 2000; Koide et al. 2005).

Rarity can be interpreted in several different ways. At first, a species might be considered rare when the relative abundance on all sites is low. On all sites, the least frequently encountered 60% of all detected EM species colonized around 12% of *F. sylvatica* root tips. The ten most abundant species occupied more than 30% of all ectomycorrhizal tips. However, rarity can also be described as being restricted to one or a few sites. With regard to this aspect, 171 of 205 EM species or 83% were found solely on one site, while only ten species were found on five or more sites. Both aspects of scarcity illustrate the significance of rare ectomycorrhizal species in these forest ecosystems.

9.3 Diversity Measures by Means of Species Ranking-Abundance Curves and Simpson's Index

Several possibilities exist to describe diversity patterns. Usually, diversity is described by means of richness meaning the number of species, and evenness, which is a measure of the relative abundance of the species to each other. Combining these two traits into species ranking-abundance curves allows for description and comparison of structural traits among different studies.

In ranking-abundance curves, species are plotted in sequence from the most to the least abundant along the horizontal axis. Species abundance are typically displayed in a log10 format, so that species whose abundances might span several orders of magnitude can be easily illustrated on the same graph (Magurran 2004). Old as well as recent studies based on molecular identification techniques show a common pattern. A few ectomycorrhizal species show dominant root colonization, while there is a long list of species which occur rarely. This shape also seems to hold true for the ectomycorrhizal communities that we tested (Fig. 9.2), regardless of stand age or management practice. The shape of these curves can provide useful information about the community structure. Differences in evenness for example can be easily detected by considering the slope; a less steep slope indicates more even distribution and thus, higher evenness. The highest species richness seems to found in beech forests ranging from 80–140 years of stand age. However, no statistically significant correlation between stand age and species richness could be found ($n = 10$, p 0.204) (Fig. 9.2). Unmanaged stands do not exhibit higher species diversity compared to managed stands. And although most species were found in a 100-year old beech forest that has been unmanaged for the last four decades, another unmanaged beech forest in Denmark (diamond symbols in Fig. 9.2), subject to natural regeneration since 1755, shows surprisingly low species richness. In a nutshell, the curve patterns do not offer ubiquitously valid statements with regard to stand age or management practice.

Another way to describe or rather compare diversity levels is by means of diversity indices. The Simpson's index is one of the most meaningful and robust measure of diversity available, as it captures the variance of the species abundance distribution (Magurran 2004).

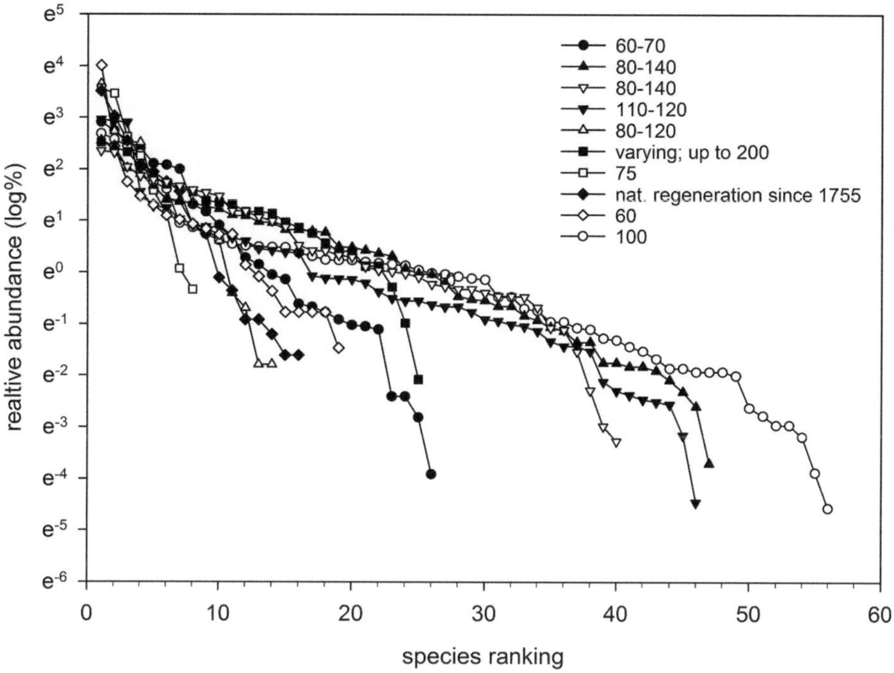

Fig. 9.2 Ectomycorrhizal species ranking-abundance curves for ten *Fagus sylvatica* stands in Europe; species are ranked from the most to the least abundant; symbols represent sequence of studies in Table 9.1

$$D = \sum_{i-1}^{S} p_i^2$$

Reciprocal (1/D) and complement (1-D) forms of the index are the most widely used forms. Due to the fact that severe variance problems can arise from this reciprocal form, a log natural transformation was recommended by Rosenzweig (1995), leading to easier interpretation as it reflects the underlying diversity and is independent of sample size. This independence of sampling size makes this index suitable for comparisons of our studies, as sampling volume and strategy differ greatly or are not clearly declared. Furthermore, a separate measure of evenness for the Simpson's index can be calculated (Smith and Wilson 1996).

The Simpson indices values found in our studies range between .673 and .966 (Table 9.2), and statistical analysis revealed no significant correlation between stand age and the calculated Simpson's Indices (n=10, p 0.178). Though, the highest Simpson's index was not evaluated for the site with the highest number of species (Lang et al. 2011), as this site exhibits a low Simpson's evenness measure, emphasizing the role of evenness in this particular diversity measure.

Our attempt to calculate the Simpson's indices revealed an overall problem, namely the availability of comprehensive meta information on sampling effort,

Table 9.2 Simpson's diversity index, complementary form and evenness measure, with information on root tip counts and species richness from the study sites

Reference		Simpson's index (D)	Simpson's index (1-D)	Simpson's evenness (1/D)/S	Root tip counts	Number of species
Grebenc and Kraigher (2007)		0.108	0.892	0.355	20500	26
Kraigher et al. (2007)	Polluted	0.037	0.963	0.575	11944	47
	Unpolluted	0.034	0.966	0.739	15552	40
Di Marino et al. (2009)		0.114	0.886	0.190	45946	46
Grebenc et al. (2009)	Managed	0.209	0.791	0.341	1818.5[a]	14
	Unmanaged	0.067	0.933	0.598	1818.5[a]	25
	Managed	0.256	0.744	0.488	1818.5[a]	8
	Unmanaged	0.189	0.811	0.331	1818.5[a]	16
Rineau and Garbaye (2010)		0.327	0.673	0.161	739	19
Lang et al. (2011)		0.068	0.932	0.261	53322	56

[a]Root tip counts calculated from a total number given

strategy or species accumulation curves indicating whether a species asymptote has been reached or not. In many cases, specific root tip counts are not declared specifically for each plot, site etc., which impedes meta-analyses and comparability between sites.

9.4 Pandora's Box: The Significance of Rare EM Species in Diversity and Ecosystem Functioning

Mycorrhizal fungi are immensely important in the forest ecosystem. They facilitate about 75% of annual P and 80% of annual N (Simard et al. 2003; Hobbie and Hobbie 2006). They act as the most important pathway of plant-derived C into the soil (Godbold et al. 2006). It has been shown that a few dominant EM species contribute remarkably to total EM root colonization (Taylor et al. 2000; Smith et al. 2007; Cox et al. 2010; Pickles et al. 2010).

It is suggested that rare mycorrhizal fungi might be an important pool of functional specialists for the adaptation of forest ecosystems to changing environmental conditions. This concept is based on the insurance hypothesis of biodiversity, where high biodiversity insures ecosystems against declines in their functioning caused by environmental fluctuations (Yachi and Loreau 1999). Pena et al. (2010) reported a loss of 50 fungal taxa from a total of 89 species, contributing 7.8% to root colonization, due to the effect of girdling. Furthermore, they found noticeable changes in the relative abundance of certain species due to the effect of girdling. As a consequence, rare EM species have a possible function as warrantors of important functional traits, leading to a stabilization of ecosystem functioning and services

(Johnson et al. 2005). Therefore, understanding the significance of rare species in dynamic ecosystem processes is essential.

Assuming that ectomycorrhizal species are specialized in certain beneficial functions for the tree, the concept of functional complementarity could give us another useful explanation for the coexistence of rare species (Buée et al. 2005). The distinction into different exploration types as an expression of functional niche adaptation might be one example (Agerer 2001). Enzyme profiling could be another means to search for functional traits. Many ectomycorrhizal types have been shown to exhibit specialization in their enzymatic profile, however enzyme activities were found to vary to a substantial degree among different samples of the same ectomycorrhizal fungi just taken few meters apart (Courty et al. 2005), indicating high plasticity.

Buée et al. (2005) suggest an emphasis on analysing functional activities on single EM root tips through (i) transcriptome analysis, (ii) isotope determinations, or other functions of root activity, such as (iii) metal complexing, protease and laccase activity or the release of protons and organic acids, with special emphasis on rare species. Recent studies extend the functional importance of EM fungi by depicting their participation in the regulation of plant hormone signalling, defences (Plett et al. 2014; Pozo et al. 2015), hyphosphere priming through the release of labile C compounds into the soil (Jansa et al. 2013) and the transfer of plant allelopathic compounds through mycelial networks (Johnson and Gilbert 2015). It is difficult to postulate general statements about ectomycorrhizal populations, and especially rare species could feature functions and activities of potential importance. Therefore, examining the functional differences among species is important to understand their role in ecosystem functioning (Koide et al. 2007).

In our opinion, better availability of meta-information on study site characteristics, sampling technique and effort, basic soil and climatic parameters and so on could be helpful for detailed and meaningful meta-analyses on structural traits of ectomycorrhizal communities.

Meta-analyses are still rare and progress is often hampered by a paucity of metadata on above-mentioned parameters which limits larger scale ecological analyses (Lilleskov and Parrent 2007). Diversity studies on different species and ecosystems especially emphasise the importance of the sampling effort (sample size, sample volume) on the interpretation of biodiversity-ecosystem function relationships (Wardle 1999), a factor often neglected. To address this problem, species accumulation curves can help clarify whether most of the EM species were detected by the sampling effort. It is known that spatial and temporal dynamics of EM communities enhance diversity data (Jonsson et al. 2001; Erland and Taylor 2002), and the usual way of sampling (which is often around bud break or leave fall) might drastically underestimate some morphotypes (Buée et al. 2005). Gardes and Bruns (1996) suggest intensive sampling over a minimal period of three years to obtain insight in EM species composition and community structure.

If it is unknown if asymptotic levels of EM species were reached within the analysed studies, then this impedes detailed comparisons between the studies for rare species (Taylor 2002). For the sites used in our analysis a regression analysis con-

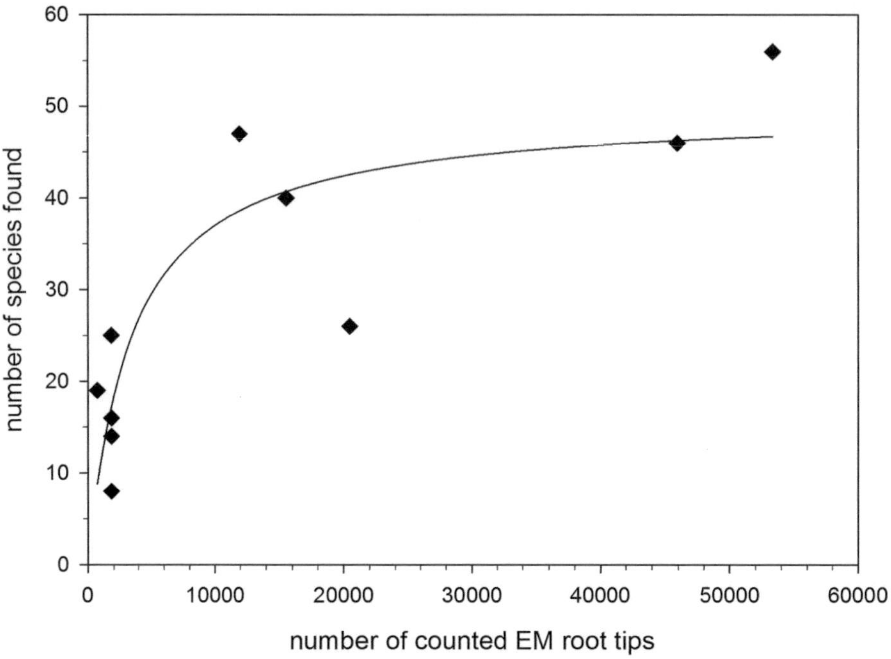

Fig. 9.3 Species accumulation curve with number of counted root tips as measure for sampling effort; diamonds represent sampling effort of our ten analysed sites

firms that the number of counted (vital) root tips has a highly significant effect on the number of ectomycorrhizal species detected (Fig. 9.3, hyperbolic fit, n = 10, p 0.0021). Not addressing this point could drastically underestimate species richness, and especially rare species might be neglected. Therefore, sampling effort and strategy must be considered when studies are analysed and compared with each other (Nara 2008). Figure 9.3 appears to give us a rough estimate of how many root tip counts would be necessary to approximately reach an asymptotic level of detected species. Evaluating around 20,000–30,000 vital root tips might be appropriate with respect to our ten study sites. Thus, certain factors such as diverse stands in terms of tree age, micro-sites etc. must be taken into consideration. The clustered nature of ectomycorrhizal species (Horton and Bruns 2001) further complicates the attempt of realistically assessing species richness. Therefore, pretesting with a random sampling design can be a meaningful measure to estimate the required amount of samples.

9.5 Conclusions

Ectomycorrhizal symbioses are of great significance and fulfil several important functions in temperate and boreal forests. However, their high richness and diversity often contrasts with the low number of tree species (Buée et al. 2007). Several dozens of different EM species on a stand scale can be detected. Yet, most of the stands are dominated by a few species, and a rather long tale of rare species colonizes a small portion of the root tips. This structural trait of a typical harp-shaped ranking-abundance curve seems valid for EM on a number of tree species across several forest biome/ecosystems (Jonsson et al. 1999; Grogan et al. 2000; Tedersoo et al. 2006). Our study revealed a similar pattern in beech stands across managed and unmanaged stands of different age in Europe. By looking at several EM community structures, we were able to add another dimension of rarity and found most of the species not only being rare at a particular site, but also rare in terms of single appearance across several analysed stands. The opposite is true for dominant species, where a few species such as *Cenococcum geophilum*, *Russula ochroleuca* and *Lactarius subdulcis* were found quite frequently on most of the sites. While much is known about these highly frequent species, the role and function of these so-called rare species remain largely unknown. Although it is likely that many EM fungal taxa fulfil largely similar ecological functions (Allen et al. 1995), communities of ectomycorrhizal fungi are still likely to retain a vast amount of functional heterogeneity, considering their high taxonomic diversity (Cairney 1999). Therefore, while possibly just a small number of species is essential under constant environmental conditions, these rare species could be important for maintaining ecosystem functioning under changing climatic conditions (Loreau et al. 2001). It is still an open question which species are of particular relevance on certain processes in certain ecosystems. More knowledge is required to comprehensively understand the role of EM for ecosystem functioning.

There is an ongoing debate how ectomycorrhizal communities can be enqueued into recent ecological theories. Certain diversity paradigms such as the high EM diversity in pure, even-aged forest stands (Dahlberg 2001; Horton and Bruns 2001) or the entity and function of rare ectomycorrhizal species remain subject of debate. Several stochastic (Hubbell 2001), deterministic (Vandermeer 1972; Yachi and Loreau 1999; Chagnon et al. 2012) as well as combined approaches (Tilman 2004) have been used to describe ecological networks. They all showed their legitimacy for certain communities, but failed to reach an overarching consensus when their approach was applied to (ecto)mycorrhizal taxa (Southworth et al. 2005). Studies on EM fungi indicate that several different types of interactions may be taking place (Pickles et al. 2012). In particular, explaining the abundance of rare taxa remains difficult (Buscot 2015), yet they could be a key factor for understanding mechanisms and missing links in ecosystem functioning.

Acknowledgements This work was supported by a Marie Curie grant GPF333996 LINKTOFUN to DG, and by the Ministry of Education, Youth and Sports of CR within the National Sustainability Program NPU I, grant No. LO1415.

References

Agerer, R. (2001). Exploration types of ectomycorrhizae. *Mycorrhiza, 11*, 107–114.

Allen, E. B., Allen, M. F., Helm, D. J., Trappe, J. M., Molina, R., & Rincon, E. (1995). Patterns and regulation of mycorrhizal plant and fungal diversity. *Plant and Soil, 170*, 47–62.

Baar, J., & Stanton, N. (2000). Ectomycorrhizal fungi challenged by saprotrophic basidiomycetes and soil microfungi under different ammonium regimes in vitro. *Mycological research, 104*, 691–697.

Buée, M., Vairelles, D., & Garbaye, J. (2005). Year-round monitoring of diversity and potential metabolic activity of the ectomycorrhizal community in a beech (*Fagus sylvatica*) forest subjected to two thinning regimes. *Mycorrhiza, 15*, 235–245.

Buée, M., Courty, P., Mignot, D., & Garbaye, J. (2007). Soil niche effect on species diversity and catabolic activities in an ectomycorrhizal fungal community. *Soil Biology and Biochemistry, 39*, 1947–1955.

Buscot, F. (2015). Implication of evolution and diversity in arbuscular and ectomycorrhizal symbioses. *Journal of Plant Physiology, 172*, 55–61.

Cairney, J. (1999). Intraspecific physiological variation: implications for understanding functional diversity in ectomycorrhizal fungi. *Mycorrhiza, 9*, 125–135.

Cairney, J. W., & Chambers, S. M. (2013). *Ectomycorrhizal fungi: Key genera in profile*. Berlin: Springer.

Chagnon, P. L., Bradley, R. L., & Klironomos, J. N. (2012). Using ecological network theory to evaluate the causes and consequences of arbuscular mycorrhizal community structure. *New Phytologist, 194*, 307–312.

Courty, P. E., Pritsch, K., Schloter, M., Hartmann, A., & Garbaye, J. (2005). Activity profiling of ectomycorrhiza communities in two forest soils using multiple enzymatic tests. *New Phytologist, 167*, 309–319.

Cox, F., Barsoum, N., Lilleskov, E. A., & Bidartondo, M. I. (2010). Nitrogen availability is a primary determinant of conifer mycorrhizas across complex environmental gradients. *Ecology Letters, 13*, 1103–1113.

Dahlberg, A. (2001). Community ecology of ectomycorrhizal fungi: An advancing interdisciplinary field. *New Phytologist, 150*, 555–562.

Di Marino, E., Montecchio, L., Scattolin, L., Abs, C., & Agerer, R. (2009). The ectomycorrhizal community structure in European beech forests differing in coppice shoot age and stand features. *Journal of Forestry, 107*, 250–259.

Druebert, C., Lang, C., Valtanen, K., & Polle, A. (2009). Beech carbon productivity as driver of ectomycorrhizal abundance and diversity. *Plant, Cell & Environment, 32*, 992–1003.

Ellenberg, H., & Leuschner, C. (1996). *Vegetation mitteleuropas mit den Alpen*. Stuttgart: Ulmer.

Erland, S., & Taylor, A. F. (2002). Diversity of ecto-mycorrhizal fungal communities in relation to the abiotic environment. In *Mycorrhizal ecology* (pp. 163–200). Dordrecht: Springer.

Gardes, M., & Bruns, T. (1996). Community structure of ectomycorrhizal fungi in a *Pinus muricata* forest: Above-and below-ground views. *Canadian Journal of Botany, 74*, 1572–1583.

Godbold, D. L., Hoosbeek, M. R., Lukac, M., Cotrufo, M. F., Janssens, I. A., et al. (2006). Mycorrhizal hyphal turnover as a dominant process for carbon input into soil organic matter. *Plant and Soil, 281*, 15–24.

Grebenc, T., & Kraigher, H. (2007). Types of ectomycorrhiza of mature beech and spruce at ozone-fumigated and control forest plots. *Environmental Monitoring and Assessment, 128*, 47–59.

Grebenc, T., Christensen, M., Vilhar, U., Căter, M., Martín, M. P., et al. (2009). Response of ectomycorrhizal community structure to gap opening in natural and managed temperate beech-dominated forests. *Canadian Journal of Forest Research, 39*, 1375–1386.
Grogan, P., Baar, J., & Bruns, T. (2000). Below-ground ectomycorrhizal community structure in a recently burned bishop pine forest. *Journal of Ecology, 88*, 1051–1062.
Haselwandter, K., & Read, D. (1980). Fungal associations of roots of dominant and sub-dominant plants in high-alpine vegetation systems with special reference to mycorrhiza. *Oecologia, 45*, 57–62.
Hobbie, J. E., & Hobbie, E. A. (2006). 15N in symbiotic fungi and plants estimates nitrogen and carbon flux rates in Arctic tundra. *Ecology, 87*, 816–822.
Horton, T. R., & Bruns, T. D. (2001). The molecular revolution in ectomycorrhizal ecology, peeking into the black box. *Molecular Ecology, 10*, 1855–1871.
Hubbell, S. P. (2001). *The unified neutral theory of biodiversity and biogeography (MPB-32)*. Princeton University Press.
Jansa, J., Bukovská, P., & Gryndler, M. (2013). Mycorrhizal hyphae as ecological niche for highly specialized hypersymbionts–or just soil free-riders. *Frontiers in Plant Science, 4*, 134.
Jansen, A. E. (1990). Conservation of fungi in Europe. *Mycologist, 4*, 83–85.
Johnson, D., & Gilbert, L. (2015). Interplant signalling through hyphal networks. *New Phytologist, 205*, 1448–1453.
Johnson, D., Ijdo, M., Genney, D. R., Anderson, I. C., & Alexander, I. J. (2005). How do plants regulate the function, community structure, and diversity of mycorrhizal fungi? *Journal of Experimental Botany, 56*, 1751–1760.
Jonsson, T., Kokalj, S., Finlay, R., & Erland, S. (1999). Ectomycorrhizal community structure in a limed spruce forest. *Mycological Research, 103*, 501–508.
Jonsson, L. M., Nilsson, M. C., Wardle, D. A., & Zackrisson, O. (2001). Context dependent effects of ectomycorrhizal species richness on tree seedling productivity. *Oikos, 93*, 353–364.
Kernaghan, G. (2005). Mycorrhizal diversity, cause and effect? *Pedobiologia, 49*, 511–520.
Kjøller, R. (2006). Disproportionate abundance between ectomycorrhizal root tips and their associated mycelia. *FEMS Microbiology Ecology, 58*, 214–224.
Koide, R. T., Xu, B., Sharda, J., Lekberg, Y., & Ostiguy, N. (2005). Evidence of species interactions within an ectomycorrhizal fungal community. *New Phytologist, 165*, 305–316.
Koide, R. T., Courty, P. E., & Garbaye, J. (2007). Research perspectives on functional diversity in ectomycorrhizal fungi. *New Phytologist, 174*, 240–243.
Kraigher, H., Petkovšek, S. A. S., Grebenc, T., & Simončič, P. (2007). Types of ectomycorrhiza as pollution stress indicators: Case studies in Slovenia. *Environmental Monitoring and Assessment, 128*, 31–45.
Lang, C., Seven, J., & Polle, A. (2011). Host preferences and differential contributions of deciduous tree species shape mycorrhizal species richness in a mixed Central European forest. *Mycorrhiza, 21*, 297–308.
Lilleskov, E. A., & Parrent, J. L. (2007). Can we develop general predictive models of mycorrhizal fungal community–environment relationships? *New Phytologist, 174*, 250–256.
Loreau, M., Naeem, S., Inchausti, P., Bengtsson, J., Grime, J., et al. (2001). Biodiversity and ecosystem functioning: current knowledge and future challenges. *Science, 294*, 804–808.
Magurran, A. (2004). *Measuring biological diversity*. Oxford: Blackwells.
Molina, R., Massicotte, H., & Trappe, J. M. (1992). Specificity phenomena in mycorrhizal symbioses, community-ecological consequences and practical implications. In M. F. Allen (Ed.), *Mycorrhizal functioning: An integrative plant-fungal process* (pp. 357–423). London: Chapman and Hall.
Nara, K. (2008). Community developmental patterns and ecological functions of ectomycorrhizal fungi, implications from primary succession. In *Mycorrhiza* (pp. 581–599). Dordrecht: Springer.
Pena, R., Offermann, C., Simon, J., Naumann, P. S., Geßler, A., et al. (2010). Girdling affects ectomycorrhizal fungal (EMF) diversity and reveals functional differences in EMF community composition in a beech forest. *Applied and Environmental Microbiology, 76*, 1831–1841.

Pickles, B. J., Genney, D. R., Potts, J. M., Lennon, J. J., Anderson, I. C., & Alexander, I. J. (2010). Spatial and temporal ecology of Scots pine ectomycorrhizas. *New Phytologist, 186*, 755–768.
Pickles, B. J., Genney, D. R., Anderson, I. C., & Alexander, I. J. (2012). Spatial analysis of ectomycorrhizal fungi reveals that root tip communities are structured by competitive interactions. *Molecular Ecology, 21*, 5110–5123.
Plett, J. M., Daguerre, Y., Wittulsky, S., Vayssières, A., Deveau, A., Melton, S. J., et al. (2014). Effector MiSSP7 of the mutualistic fungus *Laccaria bicolor* stabilizes the *Populus* JAZ6 protein and represses jasmonic acid (JA) responsive genes. *Proceedings of the National Academy of Sciences, 111*, 8299–8304.
Pozo, M. J., López-Ráez, J. A., Azcón-Aguilar, C., & García-Garrido, J. M. (2015). Phytohormones as integrators of environmental signals in the regulation of mycorrhizal symbioses. *New Phytologist, 205*, 1431–1436.
Pritsch, K., Esperschuetz, J., Haesler, F., Raidl, S., Winkler, B., & Schloter, M. (2009). Structure and activities of ectomycorrhizal and microbial communities in the rhizosphere of *Fagus sylvatica* under ozone and pathogen stress in a lysimeter study. *Plant and Soil, 323*, 97–109.
Reich, M., Kohler, A., Martin, F., & Buee, M. (2009). Development and validation of an oligonucleotide microarray to characterise ectomycorrhizal fungal communities. *BMC Microbiology, 9*, 1.
Rineau, F., & Garbaye, J. (2010). Effects of liming on potential oxalate secretion and iron chelation of beech ectomycorrhizal root tips. *Microbial Ecology, 60*, 331–339.
Rosenzweig, M. L. (1995). *Species diversity in space and time*. Cambridge: Cambridge University Press.
Rumberger, M. D., Münzenberger, B., Bens, O., Ehrig, F., Lentzsch, P., & Hüttl, R. F. (2004). Changes in diversity and storage function of ectomycorrhiza and soil organoprofile dynamics after introduction of beech into Scots pine forests. *Plant and Soil, 264*, 111–126.
Simard, S.W., Jones, M.D., Durall, D.M. (2003). Carbon and nutrient fluxes within and between mycorrhizal plants. In *Mycorrhizal ecology* (pp. 33–74). Dordrecht: Springer.
Smith, S., & Read, D. (2008). *Mycorrhizal symbiosis, 3rd*. San Diego: Academic.
Smith, B., & Wilson, J. B. (1996). A consumer's guide to evenness indices. *Oikos*, 70–82.
Smith, M. E., Douhan, G. W., & Rizzo, D. M. (2007). Ectomycorrhizal community structure in a xeric *Quercus* woodland based on rDNA sequence analysis of sporocarps and pooled roots. *New Phytologist, 174*, 847–863.
Southworth, D., He, X. H., Swenson, W., Bledsoe, C., & Horwath, W. (2005). Application of network theory to potential mycorrhizal networks. *Mycorrhiza, 15*, 589–595.
Taylor, A. F. (2002). Fungal diversity in ectomycorrhizal communities: sampling effort and species detection. In *Diversity and integration in Mycorrhizas* (pp. 19–28). Dordrecht: Springer.
Taylor, A., Martin, F., & Read, D. (2000). Fungal diversity in ectomycorrhizal communities of Norway spruce [*Picea abies* (L.) Karst.] and beech (*Fagus sylvatica* L.) along north-south transects in Europe. In *Carbon and nitrogen cycling in European forest ecosystems* (pp. 343–365). Dordrecht: Springer.
Tedersoo, L., Suvi, T., Larsson, E., & Kõljalg, U. (2006). Diversity and community structure of ectomycorrhizal fungi in a wooded meadow. *Mycological research, 110*, 734–748.
Tilman, D. (2004). Niche tradeoffs, neutrality, and community structure: A stochastic theory of resource competition, invasion, and community assembly. *Proceedings of the National Academy of Sciences, USA, 101*, 10854–10861.
Trappe, J. M. (1962). *Cenococcum graniforme* – Its distribution, ecology, mycorrhiza formation, and inherent variation. PhD dissertation, University of Washington, Seattle, Washington, USA.
Vandermeer, J. H. (1972). Niche theory. *Annual Review of Ecology and Systematics*, 107–132.
Wardle, D. A. (1999). Is "sampling effect" a problem for experiments investigating biodiversity-ecosystem function relationships? *Oikos*, 403–407.
Yachi, S., & Loreau, M. (1999). Biodiversity and ecosystem productivity in a fluctuating environment: the insurance hypothesis. *Proceedings of the National Academy of Sciences, USA, 96*, 1463–1468.

Chapter 10
Arbuscular Mycorrhizal Fungal Communities Pushed Over the Edge – Lessons from Extreme Ecosystems

Irena Maček

Abstract The diversity and structure of soil microbial communities are crucial elements in understanding the ecological impacts of rapidly changing environments. One important group of soil microbes is the ubiquitous plant symbiotic arbuscular mycorrhizal (AM) fungi. Their diverse communities are shaped by complex interactions of their abiotic and biotic environments. Locally extreme ecosystems have proven to be useful for natural long-term experiments in the ecology and evolution of AM fungi, giving an insight into much-needed processes of adaptation and acclimation of natural communities to abiotic stress. For example, data from natural CO_2 springs (mofettes) show that when exposed to extreme long-term stress (soil hypoxia and elevated soil CO_2 concentrations) specific and temporary stable AM fungal communities form with a high abundance of specialised, stress-tolerant taxa. Moreover, in both natural– and human-impacted ecosystems there are several such cases. This chapter covers a wide range of extremes (abiotic stresses) in the pedosphere, from high to low temperatures, drought and floods, hypoxia, salinity, and soil pollution. An overview of several specific stressed environments where AM fungal community ecology has been studied is presented. In some of these cases, locally extreme environments have already been used and could further serve as a powerful tool to study slow ecological and evolutionary processes that normally require long-term observations and experiments to study them.

Keywords Abiotic stress • Arbuscular mycorrhizal fungi • Soil biodiversity • Community ecology • Extreme ecosystems • Global change • Glomeromycota • Microbial ecology • Mofettes

I. Maček (✉)
Biotechnical Faculty, University of Ljubljana, Jamnikarjeva 101, 1000 Ljubljana, Slovenia

Faculty of Mathematics, Natural Sciences and Information Technologies (FAMNIT), University of Primorska, Glagoljaška 8, 6000 Koper, Slovenia
e-mail: irena.macek@bf.uni-lj.si

M. Lukac et al. (eds.), *Soil Biological Communities and Ecosystem Resilience*, Sustainability in Plant and Crop Protection, DOI 10.1007/978-3-319-63336-7_10

10.1 Introduction

Understanding mechanisms regulating the diversity and structure of soil microbial communities is urgently required for predicting the ecological impacts of rapidly changing environments. Arbuscular mycorrhizal (AM) fungal communities are diverse and near ubiquitous, symbiotically interacting with most terrestrial (Fitter and Moyersoen 1996; Smith and Read 2008), and some aquatic (e.g. Baar et al. 2011; Kohout et al. 2012; Sudová et al. 2015; Moora et al. 2016), plants. What structures their communities in natural ecosystems remains a matter of debate; however, it is clear that a complex combination of abiotic and biotic factors are determinants of AM fungal community composition. The direct effect of abiotic stress on AM fungal communities is less well understood than the impact it may have on their host plants. Recently, two hypotheses have been proposed predicting how AM fungi may respond to abiotic stress (Millar and Bennett 2016). The first, the stress exclusion hypothesis, predicts that AM fungal abundance and diversity will decrease with persistent abiotic stress, mostly by means of competitive exclusion of less tolerant taxa to a particular or combination of stress factors. The second, the mycorrhizal stress adaptation hypothesis, predicts that AM fungi will evolve in response to abiotic stress to maintain their fitness. While it is still not clear which are the main mechanisms behind AM fungal stress response, the authors of the study conclude that abiotic stress can have effects on AM fungi independent of the effects on the host plant (Millar and Bennett 2016). Moreover, there have been some studies in which the direct effects of abiotic factors on AM fungal communities have been well documented. Data from natural CO_2 springs (mofettes), for example, show AM fungal community response to long-term soil hypoxia. Mofettes are a valuable ecosystem in which studies in ecology and evolution are possible with the system serving as a long-term natural experiment (Maček et al. 2011, 2016a; Šibanc et al. 2014). The latter can be used to gain much-needed insight into the adaptation of natural communities and their ecological networks to the changing environment, including climate change.

However, apart from soil hypoxia (for a review see Maček 2017), there are many other stress factors that may even be more common in natural– and agro-ecosystems and still need more attention in the context of studies of soil microbial ecology. Here, AM fungi present just one but nevertheless functionally important and diverse group of organisms. The abiotic stress can be defined as a shift in any non-living factor within the environment away from the optimal condition or away from the condition to which most organisms in that environment have become adapted (Millar and Bennett 2016). Since natural ecosystems are often noisy and therefore challenging for research, one way of eliminating at least some of the variability is a study of locally extreme environments (for a review see Maček et al. 2016a). In extreme environments, selective pressures are sometimes long-term and relatively constant, providing ideal conditions for examining the genetic adaptation of biological communities to specific conditions (Maček et al. 2011, 2016a). Abiotic stresses, however, include a range of different factors, such as, for example, high and low temperatures, drought and floods, hypoxia, salinity, and different types of soil pollution. In this chapter, an overview of some specific environments where AM fungal diversity and community ecology has been studied will be presented. As a well-documented example, mofettes or natural CO_2 springs, serving as a model natural ecosystem for soil hypoxia research, are presented as a case study (Maček et al. 2016a; see Sect. 10.3.2).

10.2 Arbuscular Mycorrhiza

About 80% of all vascular plant species form arbuscular mycorrhiza, which is an underground symbiotic association between plants and AM fungi (Smith and Read 2008) (Figs. 10.1 and 10.2). This is a functionally important group of soil fungi involved in many terrestrial ecosystem processes (Fitter 2005; Rosendahl 2008). The symbiosis is ancient, over 450 million years old, and was significant in enabling the colonisation of land by plants (Redecker et al. 2002). AM fungi were accompanying plants in their transition from water to land from the very beginning and have been evolving in a range of diverse terrestrial and aquatic ecosystems. They are important players in key soil functions such as biogeochemical cycling of essential macronutrients and minerals, and maintenance of soil structure. For plants, there are several benefits of being mycorrhizal; AM fungi facilitate plant mineral nutrient uptake from the soil, affect plant water relations, and pathogen and pollutant resistance and tolerance (Smith and Read 2008). Indeed, as the most exposed function, mycorrhizal fungi provide significant amounts of N and P to their host plants in

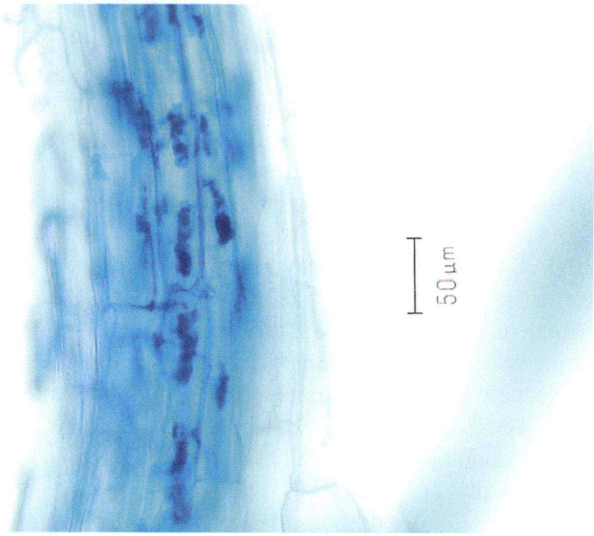

Fig. 10.1 AM fungi – root colonisation with abundant arbuscules in the root cortex of a C4 grass *Setaria pumila*

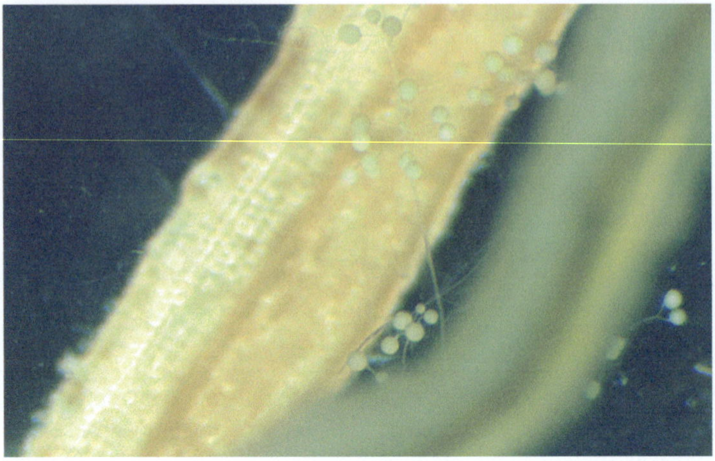

Fig. 10.2 Spores and extra-radical hyphae of AM fungus *Rhizophagus irregularis* around transformed *Medicago truncatula* roots in *in vitro* culture

natural ecosystems, especially those with reduced soil nutrient availability (see Fitter 2005; van der Heijden et al. 2015).

Therefore, AM fungi are common in many stressed environments and have been shown to increase plant survival and vitality in such ecosystems. They acquire all their carbon from the host plants and have central roles (e.g. nutrient cycling) in many habitats. Several indicators exist that the benefits provided to plants by AM fungi will become even more important due to increased abiotic stresses caused by climate change (e.g. Hanson Welzin 2000).

Understanding AM fungal ecology and identification of the main predictors of their community-level processes applies to a wide range of habitats. By delivering to the plant a range of benefits, they have a profound effect on plant community dynamics and diversity (Fitter 2005; van der Heijden et al. 2015). Recent molecular studies have shown, that communities of AM fungi in nature are more diverse than originally thought on the basis of spore morphology. AM fungal spores still serve as the main taxonomic characteristic since AM fungi cannot be identified morphologically in roots (Merryweather and Fitter 1998) (Figs. 10.1 and 10.2). It is known, however, that soil spore counts do not reflect the AM fungal diversity in plant roots and are not necessarily correlated with physiologically active AM fungal taxa forming mycorrhiza (e.g. Clapp et al. 1995; Renker et al. 2005). Since the morphological features of the structures of AM fungi in plant roots only allow low levels of identification, molecular approaches are needed for a more detailed description of their communities. The majority of the ecological studies on AM fungi are constructed on DNA-based techniques that have been developed to quantify AM fungi in field-collected soil and plant roots since the 1990s (e.g. Helgason et al. 1998).

Although over 250 morpho-species of the ph. Glomeromycota have been described (Walker and Trappe 1993; Schüßler 2008), molecular data show that significantly more AM fungal taxa exist, however those are currently known solely by their environmental sequences (e.g. Helgason et al. 2002; Öpik et al. 2013, 2014). With improved DNA-based identification methods, such as next generation sequencing (NGS), our ability to study soil microbial diversity has started to increase (e.g. Roesch et al. 2007; Öpik et al. 2009; Schloss 2009; Lemos et al. 2011; Dumbrell et al. 2011), allowing the characterisation of important mechanisms structuring natural AM fungal communities and tracking their seasonal dynamics (e.g. Dumbrell et al. 2011). The heterogeneous and dynamic nature of soil ecosystems, however, still makes it challenging to study the effect of the soil environment on natural microbial communities *in situ*.

10.3 AM Fungal Diversity in Extreme Ecosystems

Extreme environments have been defined as having one or more environmental parameters showing values permanently close to the lower or upper limits known for life (ESF – European Science Foundation report 2007). Therefore, for non-adapted organisms, an extreme environment is simultaneously a highly stressful one. The diversity of extreme environments is vast and can include several abiotic factors, ranging from physical (e.g. extremely high and extremely low temperatures), resource availability (e.g. limited water and O_2), to different types of pollution.

AM fungal diversity has been studied in several extreme ecosystems (Fig. 10.3); however, the use of molecular methods in the characterisation of their communities,

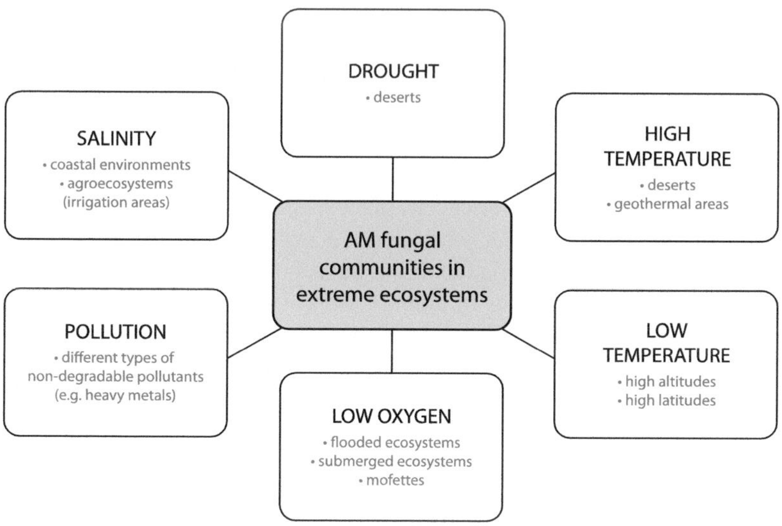

Fig. 10.3 AM fungal diversity (communities) and abiotic stress in selected extreme ecosystems that show potential for serving as long-term experimental sites in ecology and evolution

along with detailed description of relevant abiotic and biotic factors impacting their community structure, or even their temporal stability, is still limited. In most cases, AM fungal communities have not been sampled to saturation, and more intensive sampling might result in detecting additional taxa and would allow for a more realistic description of the patterns in the community ecology of this group of organisms. For example, the next generation sequencing (NGS) approaches (e.g. Dumbrell et al. 2011, 2016) may reveal other new taxa to be present, in particular, the rare ones, which are usually found only with the use of a higher-throughput methodology. The latter allows more intense sampling and is increasing the number of the analysed sequences (Dumbrell et al. 2016). Specifically, there is a need to target some less studied habitats, including aquatic and extreme environments (Fig. 10.3), where a combination of different approaches and expertise is needed. In particular, uniting species– (taxonomy) and community–oriented (ecology) approaches would be a major advantage in studying AM fungal community ecology and global diversity patterns (Öpik and Davison 2016).

10.3.1 Temperature and Drought

Both extremes, very high and very low temperatures can be problematic for life, since temperature stress disrupts metabolic processes in living organisms (e.g. protein stability, enzymatic reactions, changes in membrane fluidity and electron flow). Along with challenging temperatures, limited water availability is accompanying both extremes. Cell dehydration is not only common in dry and hot places, like deserts, but also occurs during freezing stress caused by extracellular ice crystal formation.

The community composition of AM fungi in northern latitudes remains poorly investigated (Francini et al. 2014; Öpik and Davison 2016), while there have been some reports of AM fungi in Swiss alpine sites (e.g. Oehl et al. 2012; Oehl and Körner 2014). One of the reasons may also be a limited number of projects involving AM fungal researchers that have access to those sites. In addition, arctic and alpine tundra are dominated by ericaceus plant species which host diverse networks of root-associated fungi forming other types of symbiosis in which AM fungi are not common (Toju et al. 2016). Indeed, in a recent study using high-throughput sequencing, fungal communities in alpine tundra were shown to be dominated by the taxa from the ascomycete order Helotiales and basidiomycete order Sebacinales, which are known to embrace ectomycorrhizal fungal lineages allied to ericoid mycorrhizal ones (Tedersoo et al. 2009, 2014). Despite the deep sequencing of root-associated fungal community compositions in the study of Toju et al. (2016), only a very small number of AM fungal sequences were found, which shows that arbuscular mycorrhiza is rare in those habitats and that diversity of AM fungi is low, as also seen in other studies (Francini et al. 2014; Varga et al. 2015). This is in accordance with the general knowledge that AM symbiosis usually dominates in milder and warmer abiotic environments (Read 1991). Nevertheless, other vegetation types and plant hosts that are known to form a symbiosis with AM fungi should also be more thoroughly examined for AM fungal diversity, both in alpine and arctic areas.

Indeed, the few reports on AM fungal diversity in arctic and alpine areas show that there also is a high number of taxa that are adapted to extreme conditions in the indigenous AM fungal communities (e.g. Francini et al. 2014). A similar observation was also found in an alpine site in Switzerland (Dom summit, at 4,505 m asl) known as the coldest place where angiosperm plant life is found, where five different species of AM fungi were found colonising *Saxifraga oppositifolia* (Oehl and Körner 2014). In those conditions the capability of the fungi to form long-surviving spores and retaining their viability after long-term cold conditions with freezing temperatures is vital (Varga et al. 2015). In addition, root colonisation from other AM fungal structures (e.g. vesicles) is also possible, and in such conditions some adaptive traits of fungi may be present, either colonising new roots from the existent structures in the roots or the ability for germination under cold conditions just above freezing temperatures, as observed for *Archaeospora trappei* from Dom summit, a species that is not known to form vesicles (Oehl and Körner 2014).

On the other end of temperature extremes, there have been several reports on the diversity of AM fungi in deserts, for example from the Arabian Peninsula (Al-Yahya'ei et al. 2010; Symanczik et al. 2014a), and Eritrea in East Africa (Harikumar et al. 2015), with several descriptions of new species based on spore morphology and in some cases also molecular identification (e.g. Symanczik et al. 2014b). Furthermore, geothermal vents are another specific high-temperature environment where AM fungal communities were studied. This was done in the geothermal and non-thermal grasslands in the Yellowstone National Park (USA), where plants experience rooting zone temperatures of 45 degrees C or more (Appoloni et al. 2008; Lekberg et al. 2011), and in geothermal areas of Iceland (Appoloni et al. 2008). Five of the seven operational taxonomic units (OTUs) detected in Iceland were also found in Yellowstone National Park. A subset of three OTUs was determined to be associated with geothermal conditions in the field sites analysed, while the AM fungal communities in geothermal soils include both unique OTUs and generalist fungi that occur across a broad range of environmental conditions (Appoloni et al. 2008). With the exception of an apparent generalist fungal type closely related to *Glomus intraradices*[1], AM fungal community composition in the Yellowstone National Park was highly correlated with soil pH and pH-driven changes in soil chemistry, and the large differences in soil temperature and differences in plant community composition were only secondary factor affecting the AM fungal community (Lekberg et al. 2011). Soil pH has also been confirmed in other studies as the main abiotic driver of AM fungal community composition (e.g. Dumbrell et al. 2010).

[1]The names from the original papers have been used in this Chapter, as some authors are following the nomenclature preceding the major modifications published by Schüßler and Walker (2010), and Oehl et al. (2011). See also Öpik et al. (2013) for comparison between names.

10.3.2 Anaerobic Stress

Hypoxia (low O_2 concentration) or even anoxia (lack of O_2) in plant rhizosphere are common phenomena that can be consequence of flooding, submergence, soil compaction, or are a specific characteristic of some extreme ecosystems (e.g. due to geological CO_2 release in natural CO_2 springs (mofettes), after displacing soil O_2 with CO_2 or sometimes water) (Maček et al. 2016a; Maček 2017). Plants and soil fungi are known to be obligate aerobes and are sensitive to O_2 deficiencies in their environment since they need a sufficient amount of this gas to support their aerobic metabolism. Large areas of land are flooded each year and climate change models projections show the extreme events resulting in increased areas of flooding to become even more frequent and severe in the future (Hirabayashi et al. 2013). Therefore, anaerobic soil conditions will be an even more common phenomenon both in natural and agroecosystems.

Most of the studies of AM fungal diversity in submerged ecosystems (in the context of permanent plant root and aerial system flooding) come from aquatic macrophyte vegetation from the oligotrophic and ultraoligotrophic lakes of northern Europe (e.g. Baar et al. 2011; Sudová et al. 2015; Moora et al. 2016, for review see also Maček 2017). Only very recently, and empowered by the newly developed molecular tools, researchers have looked further into AM fungal community composition in these specific ecosystems. Diverse AM fungal communities have been found in the roots of an aquatic macrophyte *Littorella uniflora*, with several AM fungal taxa present, including the taxa from the genera *Glomus, Acaulospora*, and *Archaeospora* (Baar et al. 2011; Kohout et al. 2012), and a significant share of previously non-recorded sequences. The latter has been reported also from a recent study on AM fungal communities inhabiting the roots of submerged aquatic plant *Lobelia dortmanna* (Moora et al. 2016). A new AM fungal species, *Rhizoglomus melanum*, was also recently described and isolated from the rhizosphere of aquatic macrophytes from the freshwater lake Avsjøen in Norway (Sudová et al. 2015). Those aquatic plants have characteristic aeration systems allowing rapid O_2 diffusion from the shoots to the roots and are known for high radial O_2 losses from their roots into the sediment (Smolders et al. 2002). The AM fungi appear to be dependent on the high O_2 concentrations in the roots and surrounding root zones of the aquatic plants (Wigand et al. 1998), and this appears to be a consistent and important component of AM fungal habitats in hypoxic soils.

Natural CO_2 springs, or mofettes, are another ecosystem with long-term soil hypoxia and relatively well-described communities of AM fungi (Maček et al. 2011, 2016a, see Fig. 10.4). These are tectonically or volcanically active sites with CO_2 gas vents, through which ambient temperature geological CO_2 reaches the surface (Vodnik et al. 2006, 2009; Maček et al. 2012, 2016a; Maček 2013). Here, O_2 in the soil atmosphere is largely displaced by CO_2. Thus far, only the research by Maček et al. (2011, 2016a), studying the impact of elevated CO_2 and soil hypoxia on the diversity of AM fungal communities, has focused on the diversity of AM fungi from these habitats. In this study, significant levels of AM fungal community turnover (beta diversity) between soil types and the numerical dominance of specific AM

Fig. 10.4 Mofette area (natural CO_2 springs) with geological CO_2 exhalations causing long-term soil hypoxia (Stavešinci mofette, Slovenia). Inhibition in vegetation growth due to elevated geological CO_2 and soil hypoxia can be seen centrally around the mofette (*white line*)

fungal taxa when exposed to soil hypoxia were found. This work strongly suggests that direct environmental selection acting on AM fungi is a major factor regulating AM fungal communities (Maček et al. 2011). Moreover, mofette research has allowed a further step of using locally extreme environments as natural experiments where important questions in ecology and evolution can be investigated (Maček et al. 2016a). Extreme, persistent, and directed abiotic pressures have been shown to result in a more stable system with highly specific and partially predictable composition of microbial communities, dominated by the adapted taxa consistently present in high abundance in those soils (e.g. Maček et al. 2016a). This has already been confirmed for several groups of mofette microbes: archaea and bacteria (Šibanc et al. 2014), AM fungi (Maček et al. 2011, 2016a), as well as some higher organisms, including soil fauna (Hohberg et al. 2015; Schulz and Potapov 2010), and plants (Maček et al. 2016a). For more details, mofettes have been throughly presented as a case study and as a natural long-term experimental site in ecology in a chapter of a recent thematic issue of *Advances in Ecological Research*, titled '*Large Scale Ecology: Model Systems to Global Perspectives*' (Maček et al. 2016a).

10.3.3 Salinity

Some plants can survive (salt-tolerant plants) or even thrive (halophytes) in high salt concentrations. Salinity stress has both osmotic (causing water deficits) and cytotoxic effects (accumulation of toxic ions in the cells) that affect both groups of organisms in mycorrhizal symbiosis, plants, and fungi. Along with the coastal areas where salinity is naturally present, it is estimated that 20% of all irrigated land is currently affected by salinity stress. The occurrence and community composition of AM fungi was investigated in several coastal areas where biota has been naturally exposed to long-term salt stress. Some example ecosystems across the globe would be: (i) a NaCl salt marsh site of the island Terschelling, Atlantic coast, the Netherlands, (ii) a K_2CO_3 marsh at Schreyahn, Northern Germany (Wilde et al. 2009), (iii) coastal vegetation on Okinawa island in Japan (Yamato et al. 2008, 2012), (iv) Saemangeum reclaimed land in South Korea (Krishnamoorthy et al. 2014), and marsh and sand dunes vegetation in Cabo de Gata Natural Park in Spain (Estrada et al. 2013). The overall biodiversity of AM fungi, based on molecular and morphological analyses, was reported to be relatively low at most of those sites, though a relatively high diversity of 30 AM fungal morphospecies from 13 genera, including some newly described species has also been reported for some areas (e.g. Estrada et al. 2013). As for other extreme environments, there is a need for more diversity and community ecology studies, including those using higher-throughput techniques to thoroughly characterise these ecosystems regarding their soil biological communities composition.

10.3.4 Pollution – Toxins (e.g. Heavy Metal Polluted Soil)

Soil contamination, caused by large amounts of pollutants, is becoming a major problem on a global scale (EEA 2007, The EU Environment, State and Outlook). In particular, heavy (toxic) metals are a major concern as they are non-degradable and persist in soil, thus causing permanent long-term stress to all biota present (e.g. Maček et al. 2016b). Heavy metals are known to cause severe toxicity and include zinc, copper, cobalt, nickel, mercury, lead, cadmium, silver and chromium.

By using molecular methods, AM fungal communities have also been described in sites polluted with heavy metals (e.g. Zarei et al. 2008, 2010; Hassan et al. 2011; Maček et al. 2016b). Among other abiotic factors toxic metal concentration in soils have been shown to have an important impact on the composition of AM fungal communities (Zarei et al. 2008, 2010; Hassan et al. 2011). Many reports indicate a reduction of AM fungal diversity in heavy metal polluted areas, based both on spore morphology (e.g. Griffioen 1994; Pawlowska et al. 1996; Leyval et al. 1997; del Val et al. 1999) and molecular data (e.g. Zarei et al. 2008, 2010; Hassan et al. 2011). A predominance of the taxa within the genus *Glomus*[1] has been reported in most of the studied areas with severe heavy metal disturbance (e.g. Whitfield et al. 2004; Vallino et al. 2006; Zarei et al. 2008; Sonjak et al. 2009; Hassan et al. 2011), as well as other human-impacted environments, such as agricultural sites, phosphate-contaminated sites (Daniell et al. 2001; Renker et al. 2005), and sites with fungicide treatments (Helgason et al. 2007). Zarei et al. (2008) analysed the diversity of AM fungal associated to *Veronica rechingeri* growing in the heavy metal-polluted soil of the Anguran Zn and Pb mining region in Iran. Three species could be separated morphologically, while phylogenetic analyses revealed seven different AM fungal MOTUs in plant roots, all within the genus *Glomus*[1]. Some MOTUs were only found at sites with the highest and lowest soil heavy metal contents and some in both, which is a pattern also observed in other studies (e.g. Zarei et al. 2010; Hassan et al. 2011). Thus, the patterns of new taxa identifications are also showing in extreme environments that entirely originate from the human-impacted pollution.

10.4 Conclusions

Extreme environments have for a long time been explored as ecosystems of particular interest. The ecology of extremophiles has long been a rich source of knowledge about the evolution and functions relevant to stress adaptation, of microbes from different phylogenetic groups (e.g. Gostinčar et al. 2010). Many initial studies have aimed to describe those systems from different aspects, including the diversity of abiotic factors and biological communities. AM fungi are diverse in many extreme environments, but those ecosystems have not yet been largely used to study general principles in AM fungal biology, ecology and evolution (Maček et al. 2016a).

Experimental manipulation of soils is challenging, and gives conflicting results, possibly because few studies have the long-term manipulation necessary to observe adaptive change (e.g. Jansa et al. 2003). This is also one of the reasons that in locally extreme environments some important fundamental, but still unanswered, questions in microbial ecology may be addressed, including those that involve the response of soil microbes to long-term disturbance or environmental change, and questions about the predictability of microbial community composition patterns (Maček et al. 2016a). Therefore, as a potential natural source for studying general principles in microbial ecology, extreme ecosystems have a wider importance to the scientific community that extends beyond stress ecology *sensu stricto* and descriptions of extreme ecosystems *per se*.

Importantly, questions on long-term (press) related changes in soil microbial communities are relevant to many human drivers of long-term nature, including climate change, pollution, nutrient input, land-use change and others. However, in particular questions about the stability against press perturbations have received relatively little attention so far. The introduction of NGS methods into ecology can largely change this (see Dumbrell et al. 2016). These tools are allowing us to obtain an exceptional amount of data in a relatively short time. This is a key condition to be fulfilled in order to understand the complex temporal and spatial patterns in soil microbial communities along with their environmental drivers and consequently increase our capacity to predict their response to human impacts and global change.

Acknowledgements Work supported by the Slovenian Research Agency (ARRS), projects J4-5526 and J4-7052, ARRS programme P4-0085, and Swiss National Science Foundation project SCOPES (Scientific Co-operation between Eastern Europe and Switzerland).

References

Al-Yahya'ei, M. N., Oehl, F., Vallino, M., Lumini, E., Redecker, D., Wiemken, A., & Bonfante, P. (2010). Unique arbuscular mycorrhizal fungal communities uncovered in date palm plantations and surrounding desert habitats of Southern Arabia. *Mycorrhiza, 21*, 195–209.

Appoloni, S., Lekberg, Y., Tercek, M. T., Zabinski, C. A., & Redecker, D. (2008). Molecular community analysis of arbuscular mycorrhizal fungi in roots of geothermal soils in Yellowstone National Park (USA). *Microbial Ecology, 56*, 649–659.

Baar, J., Paradi, I., Lucassen, E. C. H. E. T., Hudson-Edwards, K. A., Redecker, D., Roelofs, J. G. M., & Smolders, A. J. P. (2011). Molecular analysis of AMF diversity in aquatic macrophytes: A comparison of oligotrophic and utra-oligotrophic lakes. *Aquatic Botany, 94*, 53–61.

Clapp, J. P., Young, J. P. W., Merryweather, J. W., & Fitter, A. H. (1995). Diversity of fungal symbionts in arbuscular mycorrhizas from a natural community. *New Phytologist, 130*, 259–265.

Daniell, T. J., Husband, R., Fitter, A. H., & Young, J. P. W. (2001). Molecular diversity of arbuscular mycorrhizal fungi colonizing arable crops. *FEMS Microbiology Ecology, 36*, 203–209.

del Val, C., Barea, J. M., & Azon-Aguilar, C. (1999). Diversity of arbuscular mycorrhizal fungus populations in heavy-metal-contaminated soils. *Applied and Environmental Microbiology, 65*, 718–723.

Dumbrell, A. J., Nelson, M., Helgason, T., Dytham, C., Fitter, A. H., Nelson, H., Dytham, C., & Fitter, A. H. (2010). Relative roles of niche and neutral processes in structuring a soil microbial community. *The ISME Journal, 4*, 337–345.

Dumbrell, A. J., Ashton, P. D., Aziz, N., Feng, G., Nelson, M., Dytham, C., Fitter, A. H., & Helgason, T. (2011). Distinct seasonal assemblages of arbuscular mycorrhizal fungi revealed by massively parallel pyrosequencing. *New Phytologist, 190*, 794–804.

Dumbrell, A. J., Ferguson, R. M. W., & Clark, D. R. (2016). Microbial community analysis by single-amplicon high-throughput next generation sequencing: Data analysis – From raw output to ecology. In T. J. McGenity, K. N. Timmis, & B. Nogales (Eds.), *Hydrocarbon and lipid microbiology protocols, Springer protocols handbooks*. Heidelberg: Springer.

Estrada, B., Beltran-Hermoso, M., Palenzuela, J., Iwase, K., Ruiz-Lozano, J. M., Barea, J. M., & Oehl, F. (2013). Diversity of arbuscular mycorrhizal fungi in the rhizosphere of *Asteriscus maritimus* (L.) Less., a representative plant species in arid and saline Mediterranean ecosystems. *Journal of Arid Environments, 97*, 170–175.

European Environment Agency. (2007). *State of the environment No 1/2007 Chapter 2*. Office for official publications of the European communities.

European Science Foundation. (2007). *Annual report*. Available at http://www.esf.org/fileadmin/Public_documents/Publications/AnnualReport2007.pdf

Fitter, A. H. (2005). Darkness visible: Reflections on underground ecology. *Journal of Ecology, 93*, 231–243.

Fitter, A. H., & Moyersoen, B. (1996). Evolutionary trends in root–microbe symbioses. *Philosophical Transactions of the Royal Society B: Biological Sciences, 351*, 1367–1375.

Francini, G., Männistö, M., Alaoja, V., & Kytöviita, M. M. (2014). Arbuscular mycorrhizal fungal community divergence within a common host plant in two different soils in a subarctic Aeolian sand area. *Mycorrhiza, 24*, 539–550.

Gostinčar, C., Grube, M., de Hoog, S., Zalar, P., & Gunde-Cimerman, N. (2010). Extremotolerance in fungi: evolution on the edge. Mini review. *FEMS Microbiology Ecology, 71*, 2–11.

Griffioen, W. A. J. (1994). Characterization of a heavy metal-tolerant endomycorrhizal fungus from the surroundings of a zinc refinery. *Mycorrhiza, 4*, 197–200.
Hanson, P. J., & Weltzin, J. F. (2000). Drought disturbance from climate change: Response of United States forests. *Science of the Total Environment, 262*, 205–220.
Harikumar, V. S., Blaszkowski, J., Medhanie, G., Kanagaraj, M. K., & Samuel, V. D. (2015). Arbuscular mycorrhizal fungi colonizing the plant communities in Eritrea, Northeast Africa. *Applied Ecology and Environmental Research, 13*, 193–203.
Hassan, S. E. D., Boon, E., St-Arnaud, M., & Hijri, M. (2011). Molecular biodiversity of arbuscular mycorrhizal fungi in trace metal-polluted soils. *Molecular Ecology, 20*, 3469–3483.
Helgason, T., Daniell, T. J., Husband, R., Fitter, A. H., & Young, J. P. W. (1998). Ploughing up the wood-wide web? *Nature, 394*, 431.
Helgason, T., Merryweather, J. W., Denison, J., Wilson, P., Young, J. P. W., & Fitter, A. H. (2002). Selectivity and functional diversity in arbuscular mycorrhizas of co-occurring fungi and plants from a temperate deciduous woodland. *Journal of Ecology, 90*, 371–384.
Helgason, T., Merryweather, J. W., Young, J. P. W., & Fitter, A. H. (2007). Specificity and resilience in the arbuscular mycorrhizal fungi of a natural woodland community. *Journal of Ecology, 95*, 623–630.
Hirabayashi, Y., Mahendran, R., Koirala, S., Konoshima, L., Yamazaki, D., et al. (2013). Global flood risk under climate change. *Nature Climate Change, 3*, 816–821.
Hohberg, K., Schulz, H. J., Balkenhol, B., Pilz, M., Thomalla, A., et al. (2015). Soil faunal communities from mofette fields: Effects of high geogenic carbon dioxide concentration. *Soil Biology Biochemistry, 88*, 420–429.
Jansa, J., Mozafar, A., Kuhn, G., Anken, T., Ruh, R., et al. (2003). Soil tillage affects the community structure of mycorrhizal fungi in maize roots. *Ecological Applications, 13*, 1164–1176.
Kohout, P., Sýkorová, Z., Ctvrtlíková, M., Rydlová, J., Suda, J., et al. (2012). Surprising spectra of root-associated fungi in submerged aquatic plants. *FEMS Microbiology Ecology, 80*, 216–235.
Krishnamoorthy, R., Kim, K., Kim, C., & Sa, T. (2014). Changes of arbuscular mycorrhizal traits and community structure with respect to soil salinity in a coastal reclamation land. *Soil Biology Biochemistry, 72*, 1–10.
Lekberg, Y., Meadow, J., Rohr, J. R., Redecker, D., & Zabinski, C. A. (2011). Importance of dispersal and thermal environment for mycorrhizal communities: Lessons from Yellowstone National Park. *Ecology, 92*, 1292–1302.
Lemos, L. N., Fulthorpe, R. R., Triplett, E. W., & Roesch, L. F. W. (2011). Rethinking microbial diversity analysis in the high throughput sequencing era. *Journal of Microbiological Methods, 86*, 42–51.
Leyval, C., Turnau, K., & Haselwandter, K. (1997). Effect of heavy metal pollution on mycorrhizal colonization and function: Physiological, ecological and applied aspects. *Mycorrhiza, 7*, 139–153.
Maček, I. (2013). A decade of research in mofette areas has given us new insights into adaptation of soil microorganisms to abiotic stress. *Acta Agriculturae Slovenica, 101*, 209–217.
Maček, I. (2017). Arbuscular mycorrhizal fungi in hypoxic environments. In A. Varma et al. (Eds.), *Mycorrhiza – Function, diversity, state of art*. Cham: Springer. doi:10.1007/978-3-319-53064-2_16.
Maček, I., Dumbrell, A. J., Nelson, M., Fitter, A. H., Vodnik, D., & Helgason, T. (2011). Local adaptation to soil hypoxia determines the structure of an arbuscular mycorrhizal fungal community in roots from natural CO_2 springs. *Applied and Environmental Microbiology, 77*, 4770–4777.
Maček, I., Kastelec, D., & Vodnik, D. (2012). Root colonization with arbuscular mycorrhizal fungi and glomalin-related soil protein (GRSP) concentration in hypoxic soils from natural CO_2 springs. *Agricultural and Food Science, 21*, 62–71.
Maček, I., Vodnik, D., Pfanz, H., Low-Décarie, E., Dumbrell, A.J. (2016a). Locally extreme environments as natural long-term experiments in ecology. In: A. J. Dumbrell, R. Kordas, & G. Woodward (Eds), *Advances in ecological research, large scale ecology: Model systems to global perspectives* (Vol. 55). Elsevier Ltd.

Maček, I., Šibanc, N., Kavščšek, M., & Leštan, D. (2016b). Diversity of arbuscular mycorrhizal fungi in metal polluted and EDTA washed garden soils before and after soil revitalization with commercial and indigenous fungal inoculum. *Ecological Engineering, 95*, 330–339.

Merryweather, J. W., & Fitter, A. H. (1998). The arbuscular mycorrhizal fungi of *Hyacinthoides non-scripta* I. Diversity of fungal taxa. *New Phytologist, 138*, 117–129.

Millar, N., & Bennett, A. E. (2016). Stressed out symbiotes: Hypotheses for the influence of abiotic stress on arbuscular mycorrhizal fungi. *Oecologia*, doi:10.1007/s00442-016-3673-7.

Moora, M., Öpik, M., Davison, J., Jairus, T., Vasar, M., et al. (2016). AM fungal communities inhabiting the roots of submerged aquatic plant *Lobelia dortmanna* are diverse and include a high proportion of novel taxa. *Mycorrhiza, 26*, 735–745.

Oehl, F., & Körner, C. (2014). Multiple mycorrhization at the coldest place known for Angiosperm plant life. *Alpine Botany, 124*, 193–198.

Oehl, F., Sieverding, E., Palenzuela, J., Ineichen, K., & da Silva, G. A. (2011). Advances in Glomeromycota taxonomy and classification. *IMA Fungus, 2*, 191–199.

Oehl, F., Palenzuela, J., Sanchez-Castro, I., Kuss, P., Sieverding, E., & da Silva, G. A. (2012). *Acaulospora nivalis*, a new fungus in the Glomeromycetes, characteristic for high alpine and nival altitudes of the Swiss Alps. *Nova Hedwigia, 95*, 105–122.

Öpik, M., & Davison, J. (2016). Uniting species – And community-oriented approaches to understand arbuscular mycorrhizal fungal diversity. *Fungal Ecology*. doi:10.1016/j.funeco.2016.07.005.

Öpik, M., Metsis, M., Daniell, T. J., Zobel, M., & Moora, M. (2009). Large-scale parallel 454 sequencing reveals host ecological group specificity of arbuscular mycorrhizal fungi in a boreonemoral forest. *New Phytologist, 184*, 424–437.

Öpik, M., Zobel, M., Cantero, J. J., Davison, J., Facelli, J. M., et al. (2013). Global sampling of plant roots expands the described molecular diversity of arbuscular mycorrhizal fungi. *Mycorrhiza, 23*, 411–430.

Öpik, M., Davison, J., Moora, M., & Zobel, M. (2014). DNA-based detection and identification of Glomeromycota: The virtual taxonomy of environmental sequences. *The Botanical Review, 147*, 135–147.

Pawlowska, T. E., Blaszkowski, J., & Rühling, A. (1996). The mycorrhizal status of plants colonizing a calamine spoil mound in southern Poland. *Mycorrhiza, 6*, 499–505.

Read, D. J. (1991). Mycorrhizas in ecosystems. *Experientia, 47*, 376–391.

Redecker, D., Kodner, R., & Graham, L. E. (2002). *Palaeoglonius grayi* from the Ordovician. *Mycotaxon, 84*, 33–37.

Renker, C., Blanke, V., & Buscot, F. (2005). Diversity of arbuscular mycorrhizal fungi in grassland spontaneously developed on area polluted by a fertilizer plant. *Environmental Pollution, 135*, 255–266.

Roesch, L. F. W., Fulthorpe, R. R., Riva, A., Casella, G., Hadwin, A. K. M., et al. (2007). Pyrosequencing enumerates and contrasts soil microbial diversity. *The ISME Journal, 1*, 283–290.

Rosendahl, S. (2008). The first glance into the *Glomus* genome: An ancient asexual scandal with meiosis? *New Phytologist, 193*, 546–548.

Schloss, P. D. (2009). A high-throughput DNA sequence aligner for microbial ecology studies. *PloS One, 4*, e8230.

Schulz, H. J., & Potapov, M. B. (2010). A new species of *Folsomia* from mofette fields of the Northwest Czechia (Collembola, Isotomidae). *Zootaxa, 2553*, 60–64.

Schüßler, A. (2008). *Glomeromycota species list*. Available at: http://schuessler.userweb.mwn.de/amphylo/

Schüßler, A., & Walker, C. (2010). *The Glomeromycota: A species list with new families and new genera*. The Royal Botanic Garden Edinburgh (UK), The Royal Botanic Garden, Kew (UK), Botanische Staatssammlung Munich (DE), and Oregon State University, Gloucester (USA). Available online at http://www.amf-phylogeny.com

Šibanc, N., Dumbrell, A. J., Mandić-Mulec, I., & Maček, I. (2014). Impacts of naturally elevated soil CO_2 concentrations on communities of soil archaea and bacteria. *Soil Biology Biochemistry, 68*, 348–356.
Smith, S. E., & Read, D. J. (2008). *Mycorrhizal symbiosis* (3rd ed.787 pp). Academic: London.
Smolders, A. J. P., Lucassen, E., & Roelofs, J. G. M. (2002). The isoetid environment: biogeochemistry and threats. *Aquatic Botany, 73*, 325–350.
Sonjak, S., Beguiristain, T., Leyval, C., & Regvar, M. (2009). Temporal temperature gradient gel electrophoresis (TTGE) analysis of arbuscular mycorrhizal fungi associated with selected plants from saline and metal polluted environments. *Plant and Soil, 314*, 25–34.
Sudová, R., Sýkorová, Z., Rydlová, J., Čtvrtlíková, M., & Oehl, F. (2015). *Rhizoglomus melanum*, a new arbuscular mycorrhizal fungal species associated with submerged plants in freshwater lake Avsjøen in Norway. *Mycological Progress, 14*, 1–9.
Symanczik, S., Blaszkowski, J., Koegel, S., Boller, T., Wiemken, A., & Al-Yahya'ei, M. (2014a). Isolation and identification of desert habituated arbuscular mycorrhizal fungi newly reported from the Arabian Peninsula. *Journal of Arid Land, 6*, 488–497.
Symanczik, S., Blaszkowski, J., Chwat, G., Boller, T., Wiemken, A., & Al-Yahya'ei, M. N. (2014b). Three new species of arbuscular mycorrhizal fungi discovered at one location in a desert of Oman: *Diversispara omaniana*, *Septoglomus nakheelum* and *Rhizophagus arabicus*. *Mycologia, 106*, 243–259.
Tedersoo, L., Pärtel, K., Jairus, T., Gates, G., Põldmaa, K., & Tamm, H. (2009). Ascomycetes associated with ectomycorrhizas: molecular diversity and ecology with particular reference to the Helotiales. *Environmental Microbiology, 11*, 3166–3178.
Tedersoo, L., Bahram, M., Ryberg, M., Otsing, E., Kõljalg, U., & Abarenkov, K. (2014). Global biogeography of the ectomycorrhizal/sebacina lineage (Fungi, Sebacinales) as revealed from comparative phylogenetic analyses. *Molecular Ecology, 23*, 4168–4183.
Toju, H., Tanabe, A. S., & Ishii, H. S. (2016). Ericaceous plant-fungus network in a harsh alpine-subalpine environment. *Molecular Ecology, 25*, 3242–3257.
Vallino, M., Massa, N., Lumini, E., Bianciotto, V., Berta, G., & Bonfante, P. (2006). Assessment of arbuscular mycorrhizal fungal diversity in roots of *Solidago gigantea* growing in a polluted soil in Northern Italy. *Environmental Microbiology, 8*, 971–983.
van der Heijden, M. G. A., Martin, F. M., Selosse, M., & Sanders, I. R. (2015). Mycorrhizal ecology and evolution: The past, the present, and the future. *New Phytologist, 205*, 1406–1423.
Varga, S., Finozzi, C., Vestberg, M., & Kytöviita, M. (2015). Arctic arbuscular mycorrhizal spore community and viability after storage in cold conditions. *Mycorrhiza, 25*, 335–343.
Vodnik, D., Kastelec, D., Pfanz, H., Maček, I., & Turk, B. (2006). Small-scale spatial variation in soil CO_2 concentration in a natural carbon dioxide spring and some related plant responses. *Geoderma, 133*, 309–319.
Vodnik, D., Videmšek, U., Pintar, M., Maček, I., & Pfanz, H. (2009). The characteristics of soil CO_2 fluxes at a site with natural CO_2 enrichment. *Geoderma, 150*, 32–37.
Walker, C., & Trappe, J. M. (1993). Names and epithets in the Glomales and Endogonales. *Mycological Research, 97*, 339–344.
Whitfield, L., Richards, A. J., & Rimmer, D. L. (2004). Relationships between soil heavy metal concentration and mycorrhizal colonization in *Thymus polytrichus* in northern England. *Mycorrhiza, 14*, 55–62.
Wigand, C., Andersen, F. O., Christensen, K. K., Holmer, M., & Jensen, H. S. (1998). Endomycorrhizae of isoetids along a biogeochemical gradient. *Limnology and Oceanography, 43*, 508–515.
Wilde, P., Manal, A., Stodden, M., Sieverding, E., Hildebrandt, U., & Bothe, H. (2009). Biodiversity of arbuscular mycorrhizal fungi in roots and soils of two salt marshes. *Environmental Microbiology, 11*, 1548–1561.
Yamato, M., Ikeda, S., & Iwase, K. (2008). Community of arbuscular mycorrhizal fungi in a coastal vegetation on Okinawa island and effect of the isolated fungi on growth of sorghum under salt-treated conditions. *Mycorrhiza, 18*, 241–249.

Yamato, M., Yagame, T., Yoshimura, Y., & Iwase, K. (2012). Effect of environmental gradient in coastal vegetation on communities of arbuscular mycorrhizal fungi associated with *Ixeris repens* (Asteraceae). *Mycorrhiza, 22*, 623–630.

Zarei, M., König, S., Hempel, S., Nekouei, M. K., Savaghebi, G., & Buscot, F. (2008). Community structure of arbuscular mycorrhizal fungi associated to *Veronica rechingeri* at the Anguran zinc and lead mining region. *Environmental Pollution, 156*, 1277–1283.

Zarei, M., Hempel, S., Wubet, T., Schäfer, T., Savaghebi, G., et al. (2010). Molecular diversity of arbuscular mycorrhizal fungi in relation to soil chemical properties and heavy metal contamination. *Environmental Pollution, 158*, 2757–2765.

Chapter 11
An Intact Soil Core Bioassay for Cultivating Forest Ectomycorrhizal Fungal Communities

Peter G. Avis, Ina C. Meier, and Richard P. Phillips

Abstract Ectomycorrhizal and other root-associated fungi play a critical role in forests by influencing plant community structure and nutrient dynamics; however, given that most fungal taxa are difficult to cultivate, methods for cultivating natural fungal communities are needed to better understand the ecological consequences of species gains and losses. We grew loblolly pine (*Pinus taeda*) seedlings in intact, undisturbed soil cores collected from the Duke Forest (North Carolina, USA) and used molecular tools to measure how the fungal communities that colonized the seedlings resembled those reported in previous field-based investigations. Overall, we found substantial overlap between fungi, especially ectomycorrhizal fungi, recovered with our method and those from *in situ* studies, with over 70% of the sequences and nearly 50% of the OTU in common. This study shows that the intact core 'baiting' method greatly facilitates the study of natural fungal communities composited by ectomycorrhizal and other fungi, is a significant improvement over bioassays that homogenize soils, and can be used to experimentally study functioning of ecologically-relevant fungal communities of forest ecosystems.

Keywords Bioassay • Ectomycorrhiza • Fungal functions • Indigenous fungal community • Loblolly pine

P.G. Avis
Department of Biology, Indiana University Northwest, 3400 Broadway, Gary, IN 46408, USA

I.C. Meier (✉)
Department of Biology, Indiana University, 1001 East Third Street, Bloomington, IN 47405, USA

Present address: Plant Ecology, Albrecht von Haller Institute for Plant Sciences, University of Göttingen, Untere Karspüle 2, 37073 Göttingen, Germany
e-mail: imeier1@uni-goettingen.de

R.P. Phillips
Department of Biology, Indiana University, 1001 East Third Street, Bloomington, IN 47405, USA

M. Lukac et al. (eds.), *Soil Biological Communities and Ecosystem Resilience*, Sustainability in Plant and Crop Protection, DOI 10.1007/978-3-319-63336-7_11

11.1 Introduction

An elusive goal in the field of forest ecology is to understand how mycorrhizal fungal communities influence biogeochemical processes in ecosystems. For instance, ectomycorrhizal (EcM) fungal communities mediate critical biogeochemical processes such as soil acidification (Van Breemen et al. 2000), mineral weathering (Landeweert et al. 2001; Rosling et al. 2007), and nutrient release from organic matter decomposition (Talbot et al. 2013). However, our understanding of taxon-specific effects of EcM fungi and of EcM community effects on soil biogeochemistry has been limited by methodological challenges. For example, taxon specific effects are often investigated using inoculum approaches whereby 'plugs' of fungi grown on medium are added to soils containing plants. This approach, while useful for investigating fungal-specific responses to varying environmental conditions (e.g. Lilleskov and Bruns 2003), is inherently biased by which fungal taxa can actually grow well in artificial media. As such, there is much less information about fungi that do not grow well on such artificial media such as those referred to as 'late stage' fungi (e.g. Deacon and Fleming 1993) or 'mature ecosystem' fungi (Dickie et al. 2013), despite their importance to biogeochemical cycling in many ecosystems (e.g. in mature forests).

One possible reason that certain ectomycorrhizal fungal taxa are difficult to culture relates to the sensitivity of fungi to soil disturbance. Conventional soil processing generally involves the homogenization of soil (e.g. mixing and sieving), a process which destroys mycelial networks and likely favors disturbance-adapted species. In the case of EcM fungi, many taxa may only colonize roots from intact mycelial networks that would be broken if soils were homogenized. This is supported by previous studies that have reported that the commonly deployed practice of sieving and drying influences which mycorrhizal fungi colonize host plants (Taylor and Bruns 1999; Avis and Charvat 2005).

While field-based investigations have reported EcM community shifts owing to global change factors such as N deposition (Lilleskov et al. 2002), ozone (Andrew and Lilleskov 2009), and elevated CO_2 (Andrew and Lilleskov 2009), our ability to understand the consequences of these changes is hindered by our inability to culture and measure the traits of the most responsive fungal taxa (Dickie et al. 2012). In this study, we tested how well we could culture EcM fungi (and other soil fungi) on tree seedlings. We used an approach that minimized soil disturbance and hypothesized that this approach would culture and provide a large number of tree seedling-root associated fungi from a well-studied forest ecosystem, the Duke Forest, where numerous culture independent (i.e. molecular based) studies have widely characterized the fungal community with first-generation (O'Brien et al. 2005; Parrent et al. 2006; Parrent and Vilgalys 2007; Hersh et al. 2012) or next-generation DNA sequencing approaches in continent wide studies (Talbot et al. 2014; Glassman et al. 2015). Specifically, we tested how well this approach was able to recover the known fungi identified by the lone previous field study that targeted tree seedling roots (Parrent et al. 2006) as root associated fungi were our specific targets in a parallel study of ecological process (Meier et al. 2013). We also compared our result to the other three studies exclusive to Duke Forest although these studies did not specifically target root-associated fungi. We propose that this kind of method can be used in any field setting but especially in forests and can be useful in field sites that have fungal communities previously described by field-based molecular ecological approaches.

11.2 Material and Methods

11.2.1 The Method Overview

Intact soil cores were collected from a field site of interest with a tool that minimized disturbance to the core (a modification of Brundrett et al. 2003). Ideal tools included a wide-diameter soil corer, a 'cup cutter' – the tool used by greenkeepers for making holes on golf course greens – or even a shovel if great care were taken to keep the core intact. Cores were returned to a controlled growth facility like a greenhouse or growth chamber as soon as possible. Surface sterilized seeds or seedlings of relevant plants were sown on the top of the cores and then grown for an appropriate length of time. Subsequent analysis of the fungi for a particular study then examined these cores with the desired post-culturing methodology that combined fungal identity (e.g. from the molecular analysis of mycorrhizal roots) with process (e.g. from the measurement of root exudates).

11.2.2 Study Site and Seedling Inoculation

We used this method at the Duke Forest free-air carbon enrichment (FACE) experiment, North Carolina, USA, in October 2009. At this site, a loblolly pine (*Pinus taeda* L.) plantation was established in 1983. Pines occupy greater 80% of the basal area and nearly all pine roots are colonized by EcM fungi or non-mycorrhizal fungal associates (Parrent et al. 2006). To date, six published studies have conducted molecular-based methods to describe the fungal communities in this specific study site or very nearby in Duke Forest (O'Brien et al. 2005; Parrent et al. 2006; Parrent and Vilgalys 2007; Hersh et al. 2012; Talbot et al. 2014; Glassman et al. 2015). Because of the extensive amount of research this site has received, it served as a good location to conduct this study as it provided an extensively documented community of fungi against which we could compare our results.

While the specific focus of this investigation was to use this method to culture and then compare the fungi we found in our study to those of previous studies, the collection within the Duke FACE experiment allowed additional comparisons to FACE study variables.

We collected soil cores (5.1 cm in diameter) from the top 15 cm of soil at random locations in each FACE treatment at the site (ambient CO_2, elevated CO_2, N fertilization, and elevated CO_2 with N fertilization). Nine or 10 replicate cores were collected from each treatment (37 cores total). Soil cores were kept intact, not homogenized, and as undisturbed as possible in Plexiglas tubes, sealed in plastic bags, and transported on ice to the growth facilities. In this case, these were in Bloomington, Indiana, USA. Five loblolly pine seeds (surface-sterilized in 1% bleach solution and cold-stratified at 4°C for three weeks) from a first generation orchard mix from the North Carolina piedmont (NCSU seed lot #SOM8) were sown into each intact core. Soil cores were set up in a randomized block design in a climate chamber. A 14 h diurnal photoperiod was maintained in each chamber with cool-white fluorescent lamps (~120 μmol m^{-2} s^{-1}). The chamber temperature and relative humidity averaged 23/17°C day/night and 75%, respectively. All cores were irrigated when needed, and rotated within the growth chamber. After four weeks, pine seedlings were moved to a greenhouse until the end of this experiment in February 2010 (14 weeks after pine seedling germination). The photoperiod in the greenhouse was maintained at 14 h with high-pressure sodium lights (approximately 500 μmol PAR m^{-2} s^{-1}).

11.2.3 Ectomycorrhizal Colonization and Molecular Identification of Root-Associated Fungi

We quantified the degree of EcM colonization on 50 root tips per seedling by examining the size, color and morphology of fungal structures (mantle type and extra-radical hyphae) found on root surfaces (Agerer 1991; Goodman et al. 1996; *cf.* Meier et al. 2013). We randomly selected 16 healthy and mature EcM root tips per plant from the 37 plants representing all FACE treatments at the site. Root tips were collected using a dissecting microscope and by dividing the total root system into eight even sections, then randomly picking the first two EcM tips observed from each section. Each tip was placed by itself into a well of a 96-well plate for DNA analysis. DNA was extracted from each sample with REDExtract-N-Amp Plant PCR Kit (Sigma, St. Louis, USA). We used the primers ITS1-F and ITS4 (Gardes and Bruns 1993) to amplify the ribosomal internal transcribed spacers (ITS1 and ITS2) and 5.8S ribosomal RNA gene by PCR following Avis et al. (2008). Successful PCR reactions were excised and purified by GELase Agarose Gel-Digesting Preparation (Epicentre Biotechnologies, Madison, USA), then cycle sequenced using Big Dye version 3.1 and screened on a 3730 DNA Analyzer (Applied Biosystems, Carlsbad, USA) at the Pritzker Laboratory for Molecular Evolution and Systematics at the Field Museum of Natural History in Chicago, Illinois, USA. Sequence length and quality were then examined using Sequencher 3.0 (GeneCodes, Ann Arbor, USA).

Sequences were compared to those in GenBank by BLAST analysis to check for matches (>96% BLAST ID) to fungi found in the four previous studies conducted exclusively at Duke Forest (O'Brien et al. 2005; Parrent et al. 2006; Parrent and Vilgalys 2007; Hersh et al. 2012).

Fungal identification based on these sequences was made in two ways. First, the BLAST comparisons conducted to find matches to previously identified Duke Forest fungi were used to identify a sequence to genus or family in most cases. This first approach was called 'Inclusive OTU' as the operational taxonomic units (OTU) were generally inclusive for a given lineage – this could be considered monophyletic groupings at the genus, family, or larger taxonomic level. Because this approach could use nearly all of the sequences generated (even short, sometimes poor quality sequences) but was conservative, it lumped closely related but different taxa (e.g. species of the same genus). To determine how many additional taxa were in the Inclusive OTU we also used a more restrictive and exclusive approach (basically a phylogenetic species estimate) on a subset of sequences, which excluded shorter, poorer quality sequences. For this second approach, we used ClustalW2 (http://www.ebi.ac.uk/Tools/msa/clustalw2/) to align sequences within the same genus (or family if a genus level distinction was not made) and then conducted phylogenetic analyses on these alignments using maximum likelihood as by the RAxML-HPC2 7.2.8 software package (Stamatakis 2006; Stamatakis et al. 2008) via the CIPRES portal (www.phylo.org/sub_sections/ portal/). OTU were determined as terminal clades with >70 fast bootstrap support – this level selected since we needed to be more thorough than a simple BLAST search but wanted to maintain a relatively efficient and conservative estimate. This approach resulted in more exclusive groups and was referred to as 'Exclusive OTU'. The trophic status or nutritional mode for each OTU (e.g. ectomycorrhizal, pathogenic and saprobic) was assigned based on previous studies (Parrent and Vilgalys 2007; Hersh et al. 2012) and known phylogenetic relationships.

11.2.4 Data Analyses

Diversity indices were calculated with PAST 3.11 (PAleontological STatistics; Hammer and Harper 2001). Species diversity was expressed as (i) the number of taxa, (ii) the Shannon index (H'; entropy), where $H' = -\sum p_i \ln p_i$ and p_i is the relative number of sequences of taxon *i*, and (iii) the dominance (D), where $D = \sum(p_i)^2$. Equitability was calculated as a measure of the evenness with which sequences were divided among the taxa present. Taxonomic diversity and taxonomic distinctness was analyzed according to Clarke and Warwick (1998), using taxonomic division, class, order and family as group information. A species-effort curve of fungal taxa richness was estimated by individual rarefaction (algorithm by Krebs 1989).

11.3 Results

We examined 608 root tips collected from 37 seedlings in this study. From these root tips, 438 sequences were generated with the remainder (170 tips) failing to produce sequences. Sequence length (~200 to >700 bp) and quality varied but most were above 400 bp and of high quality. We identified 61 'Exclusive OTUs' (out of 319 sequences) and 28 'Inclusive OTUs' (out of 438 sequences; Tables 11.1 and 11.2, respectively) associated with the roots, including EcM, pathogenic and saprobic taxa; most fungi were Basidiomycota and Ascomycota (Fig. 11.1a). We found EcM fungi on the roots of all seedlings but pathogens or saprotrophs on less than half (Table 11.3). Seedlings had on average just over three OTUs per root system. We were not successful in identifying all of the root tips as some did not produce sequences as mentioned above; but of the root tips that were successfully sequenced, on average 55% and 73% of the root tips per seedling were identified to an Inclusive OTU or Exclusive OTU, respectively. Based on rarefaction, our sampling effort had not plateaued yet (Fig. 11.2) and it is likely that more taxa would be found with more extensive sampling.

Table 11.1 Exclusive Operational Taxonomic Units (Exclusive OTU): fungal identification based on an approach which excludes shorter, poorer quality sequences. Shown are phyla, GenBank accession numbers, trophic status, number of seedlings on which a sequence occurred, and frequency of sequences of the identified fungal species colonizing loblolly pine (*Pinus taeda* L.) seedlings grown in soil from different origins of the Duke FACE experiment

Exclusive OTU	Family / Class	Phyl-um	GenBank accession No.	Trophic type	Seedling freq.	Sequence freq.	Prev. found	Blast seq. similarity [%]	Top Duke Match
Atheliaceae_1	Atheliaceae	b	JQ616814	E	14	45	Yes[3]	99	EF619844
Thelephoraceae_2	Theleporaceae	b	JQ616805	E	8	71	Yes[1,2]	99	EF619784
Wilcoxina_2	Pyronemataceae	a	JQ616785	E	8	21	No	92	EF619914
Wilcoxina_1	Pyronemataceae	a	KC109137	E	6	13	Yes[3]	100	EF619913
Lactarius_1	Russulaceae	b	KC109126	E	4	13	No	88	EF619712
Cenococcum_2	Gloniaceae	a	JQ616778	E	4	7	Yes[3]	97	EF619645
Atheliaceae_5	Atheliaceae	b	KC109119	E	3	5	No	83	EF619637
Atheliaceae_7	Atheliaceae	b	KC109121	E	3	4	Yes[1,2]	99	EF619633
Russula_3	Russulaceae	b	KC109132	E	3	4	Yes[1,3]	99	EF619749
Sebacinaceae_2	Sebacinaceae	b	JQ616803	E	3	4	Yes[3]	99	EF619763
Atheliaceae_4	Atheliaceae	b	KC109118	E	2	11	Yes[1]	97	EF619637
Wilcoxina_3	Pyronemataceae	a	KC109138	E	2	4	No	92	EF619914
Atheliaceae_2	Atheliaceae	b	KC109116	E	2	2	Yes[3]	98	EF619845
Cenococcum_3	Gloniaceae	a	KC109123	E	2	2	No	91	EF619648
Russula_2	Russulaceae	b	JQ616801	E	2	2	No	94	EF619755
Russula_4	Russulaceae	b	KC109133	E	2	2	No	94	EF619749
Thelephoraceae_1	Theleporaceae	b	JQ616804	E	2	2	No	79	EF619781
Tomentella_6	Theleporaceae	b	KC109136	E	1	9	Yes[3]	99	EF619818
Tomentella_1	Theleporaceae	b	JQ616808	E	1	8	Yes[3]	99	EF619828
Atheliaceae_6	Atheliaceae	b	KC109120	E	1	7	No	93	EF619637
Amanita_1	Amanitaceae	b	JQ616790	E	1	6	Yes[3]	100	EF619627
Thelephoraceae_3	Theleporaceae	b	JQ616806	E	1	6	Yes[3]	98	EF619788
Clavulinaceae_1	Clavulinaceae	b	JQ616791	E	1	3	Yes[3]	99	EF619643
Inocybe_1	Inocybaceae	b	JQ616795	E	1	3	Yes[3]	98	EF619709
Russula_5	Russulaceae	b	KC109134	E	1	3	No	82	EF619755
Tomentella_2	Theleporaceae	b	JQ616809	E	1	3	No	93	EF619820
Laccaria_1	Hydnangiaceae	b	JQ616796	E	1	2	No	78	EF619842

(continued)

Table 11.1 (continued)

Exclusive OTU	Family / Class	Phyl-um	GenBank accession No.	Trophic type	Seedling freq.	Sequence freq.	Prev. found	Blast seq. similarity [%]	Top Duke Match
Tomentella_3	Theleporaceae	b	JQ616810	E	1	2	Yes[3]	99	EF619820
Tomentella_4	Theleporaceae	b	JQ616811	E	1	2	No	81	EF619832
Atheliaceae_3	Atheliaceae	b	KC109117	E	1	1	No	91	EF619845
Cenococcum_1	Gloniaceae	a	KC109122	E	1	1	Yes[3]	99	EF619645
Cenococcum_4	Gloniaceae	a	KC109124	E	1	1	Yes[3]	99	EF619645
Hydnellum_1	Bankeraceae	b	JQ616794	E	1	1	No	79	EF619782
Inocybe_2	Inocybaceae	b	KC109125	E	1	1	No	88	EF619708
Pezizaceae_1	Pezizaceae	a	JQ616784	E	1	1	No	82	EF619728
Piloderma_2	Atheliaceae	b	JQ616799	E	1	1	Yes[3]	99	EF619740
Rhizopogon_1	Rhizopgonaceae	b	JQ616800	E	1	1	No	88	EF619768
Russula_1	Russulaceae	b	KC109131	E	1	1	Yes[1]	99	EF619752
Sebacinaceae_1	Sebacinaceae	b	JQ616802	E	1	1	Yes[3]	98	EF619757
Sebacinaceae_3	Sebacinaceae	b	KC109135	E	1	1	Yes[3]	99	EF619765
Thelephoraceae_4	Theleporaceae	b	JQ616807	E	1	1	No	86	EF619812
Tomentella_5	Theleporaceae	b	JQ616812	E	1	1	Yes[1]	97	EF619784
Wilcoxina_4	Pyronemataceae	a	KC109139	E	1	1	No	90	EF619914
Wilcoxina_5	Pyronemataceae	a	KC109140	E	1	1	No	91	EF619914
Cylindrocarpon_1	Nectriaceae	a	JQ616786	P	5	7	Yes[4]	99	GQ996085
Neonectria_1	Nectriaceae	a	JQ616787	P	2	3	No	95	GQ996084
Phialocephala_2	Leotiomycetes	a	JQ616782	P	2	2	Yes[3]	99	EF619734
Phialocephala_3	Leotiomycetes	a	KC109130	P	2	2	No	81	GQ996117
Cochliobolus_1	Pleosparaceae	a	JQ616777	P	1	1	No	90	EF619897
Ophiostoma_1	Ophiostomataceae	a	JQ616788	P	1	1	No	81	GQ996106
Phialocephala_1	Leotiomycetes	a	KC109129	P	1	1	No	93	GQ996117
Tricholomataceae_1	Tricholomataceae	b	JQ616813	S	4	4	No	77	EF619842
Oidiodendron_1	Myxotrichaceae	a	KC109127	S	3	4	No	94	GQ9966105
Ceratobasidiaceae_1	Ceratobasidiaceae	b	JQ616792	S	3	3	Yes[3]	99	EF619650
Oidiodendron_2	Myxotrichaceae	a	JQ616779	S	3	3	Yes[4]	99	GQ9966105
Mortierella	Mortierellaceae	z	JQ616817	S	2	2	Yes[3]	100	EF619716
Helotiales_1	Leotiomycete	a	JQ616780	S	1	1	No	96	EF619698
Oidiodendron_3	Myxotrichaceae	a	KC109128	S	1	1	No	96	GQ9966105
Trichoderma_1	Hypocreaceae	a	JQ616789	S	1	1	No	94	EF619840
Uncultured Chytrid	Chytridiomycota	c	JQ616816	S	1	1	Yes[3]	100	EF619659
Ascomycota_1	Leotiomycete	a	JQ616776	U	1	1	No	96	EF619864

Abbreviations: *a* Ascomycota, *b* Basidiomycota, *c* Chytridiomycota, *z* Zygomycota; *E* ectomycorrhizal fungus, *S* saprotrophic fungus, *P* putative pathogen fungus, *U* unknown trophic status
[1]O'Brien et al. (2005); [2] Parrent et al. (2006); [3] Parrent and Vilgalys (2007); [4] Hersh et al. (2012)

Even though not taken at the exact same location in the Duke Forest and in the same year, over 70% of the sequences and nearly 50% of the OTUs were identical to those found in previous Duke Forest studies (Table 11.1). Most of the overlap in fungi was due to the recovery of EcM fungi found in the previous studies (Fig. 11.1c). In addition, we report the presence of certain species of Hydnellum, Inocybe, and Russula for the first time in this site. Most of the pathogens we found were also previously found at Duke Forest, including the very abundant Cylindrocarpon (Hersh et al. 2012).

In contrast, most of the saprotroph fungi encountered in our study were not previously found in earlier Duke Forest studies with the exception of some members of

Table 11.2 Inclusive Operational Taxonomic Units (Inclusive OTU): approach which conservatively determines inclusion. Shown are trophic types, number of seedlings on which a sequence occurred, and frequency of sequences of the identified fungal species colonizing loblolly pine (Pinus taeda L.) seedlings grown in soil from different origins of the Duke FACE experiment

Inclusive OTU	Trophic type	Seedling frequency	Sequence frequency	% root tips	Previously found
Atheliaceae	E	21	111	25.3	Yes[3]
Thelephoraceae	E	17	132	30.1	Yes[1,2]
Wilcoxina	E	17	55	12.6	Yes[3]
Russula	E	10	18	4.1	Yes[1,3]
Cenococcum	E	10	15	3.4	Yes[3]
Sebacinaceae	E	5	7	1.6	Yes[3]
Lactarius	E	4	24	5.5	No
Inocybe	E	2	4	0.9	Yes[3]
Amanita	E	1	6	1.4	Yes[3]
Clavulinaceae	E	1	3	0.7	Yes[3]
Laccaria	E	1	3	0.7	No
Hydnellum	E	1	1	0.2	No
Pezizales	E	1	1	0.2	No
Piloderma	E	1	1	0.2	Yes[3]
Rhizopogon	E	1	1	0.2	No
Oidiodendron	S	6	10	2.3	No
Tricholomataceae	S	4	18	4.1	No
Ceratobasidiaceae	S	3	4	0.9	No
Helotiales	S	2	1	0.2	No
Mortierella	S	2	1	0.2	No
Chytridiomycota	S	1	1	0.2	Yes[3]
Trichoderma	S	1	1	0.2	No
Cylindrocarpon	P	5	7	1.6	Yes[4]
Phialocephala	P	5	5	1.1	No
Neonectria	P	3	4	0.9	No
Ophiostoma	P	1	1	0.2	No
Cochliobolus	P	1	1	0.2	No
Ascomycota	U	1	1	0.2	No

Abbreviations: *E* ectomycorrhizal fungus, *S* saprotrophic fungus, *P* putative pathogen fungus, *U* unknown trophic status

[1]O'Brien et al. (2005); [2] Parrent et al. (2006); [3] Parrent and Vilgalys (2007); [4]Hersh et al. (2012)

the order Helotiales. Fungal species abundance (as numbers of sequences) and frequency (as numbers of seedlings on which a taxon was found) varied. Relatively few fungi were common and most were rare (Tables 11.1 and 11.2). Frequent taxa were found on many seedlings but not necessarily abundant by sequence count on those seedlings (e.g. Atheliaceae_1), or were found on many root tips but on fewer seedlings (e.g. Thelephoraceae_2).

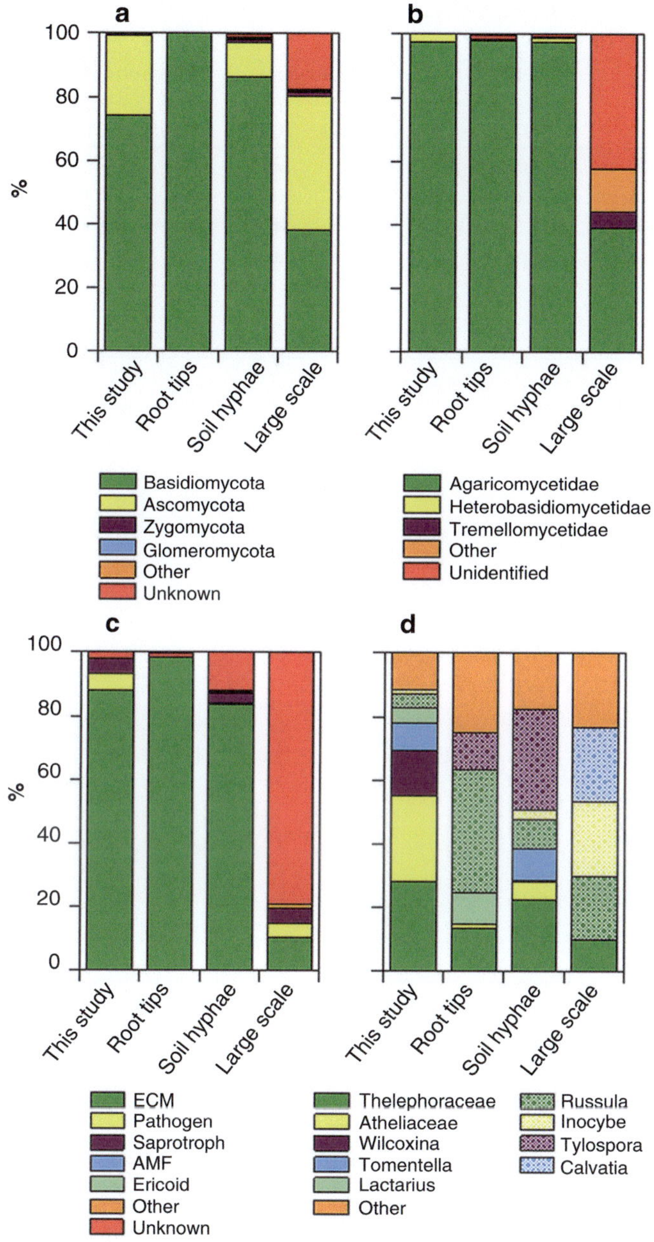

Fig. 11.1 (**a**) Proportional distribution of different divisions, (**b**) subclasses of the Basidiomycetes, (**c**) nutrition types, and (**d**) ectomycorrhizal species in the investigated fungal community. Compared are three in situ methods ('root tips' based on Parrent et al. (2006), 'soil hyphae' based on Parrent and Vilgalys (2007), and 'large scale' based on O'Brien et al. (2005)) with the soil baiting method of the current study, all performed at the Duke Forest FACE experiment.

Table 11.3 Number of Inclusive Operational Taxonomic Units (OTU) and Exclusive Operational Taxonomic Units by trophic group

	# ECM		# Pathogens		# Saprotrophs		# Unknown	
	Incl. OTU	Excl. OTU	Incl. OTU	Excl. OTU	Incl. OTU	Excl. OTU	Incl. OTU	Excl. OTU
Sum (all seedlings)	96	92	25	25	8	8	2	2
Mean per seedl.	2.6	2.5	0.7	0.7	0.2	0.2	0.1	0.1
Maximum per seedl.	6	5	4	4	2	2	1	1
Colonization rate [%]	100	100	41	41	19	19	5	5

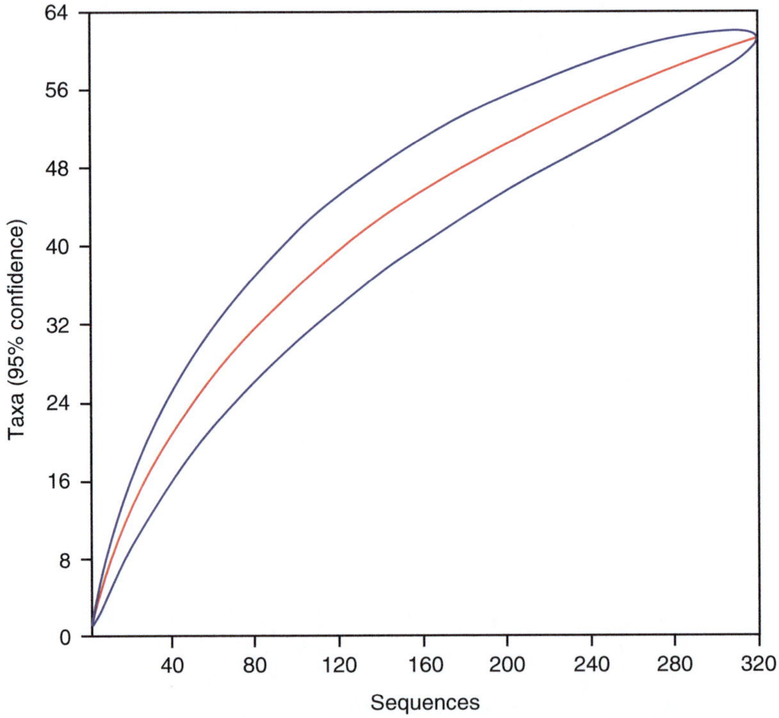

Fig. 11.2 Species-effort curve of fungal taxa richness (with 95 % confidence interval).

Because of the bioinformatic challenges presented in attempting to consolidate the results of all of the previous Duke studies into one single database (and thus generate a 'master list' of fungi for Duke Forest to compare against), we took a simpler approach and compared our study to each of the four studies that exclusively examined Duke Forest by using a set of diversity metrics (Table 11.4).

We recovered fewer taxa (with the exception of the Hersh et al. (2012) study), the lowest evenness and Shannon diversity, but the highest dominance, taxonomic

Table 11.4 Diversity indices for fungal communities investigated by different methodological means.

Study type	Soil baiting	Root tips (*in situ*)	Soil hyphae (*in situ*)	Large scale (*in situ*)	Pathogens (*in situ*)
Study	This study Excl. OTU	Parrent et al. (2006)	Parrent and Vilgalys (2007)	O'Brien et al. (2005)	Hersh et al. (2012)
Number of taxa	61	72	294	412	36
Number of sequences	319	401	2137	862	493
Taxa : sequences [%]	19	18	14	48	7
Shannon diversity index	3.2	3.9	4.3	5.6	3.3
Evenness	0.41	0.69	0.76	0.64	0.92
Dominance	0.08	0.03	0.04	0.01	0.05
Taxonomic diversity	2.5	0.98	1.9	2.2	1.6
Taxonomic distinctness	2.7	1.0	2.0	2.2	1.6

diversity and taxonomic distinctness when compared to these four Duke Forest studies. Our taxa recovered per sequence effort (e.g. 61/319 for exclusive OTU = 19%) was very similar to the two studies that focused on EcM fungi (Parrent et al. 2006; Parrent and Vilgalys 2007). When comparing taxonomic level (phyla, Fig. 11.1a; subclass Fig. 11.1b) and nutritional type (Fig. 11.1c), similar patterns were found between our study and the two EcM-focused studies as well (Parrent et al. 2006; Parrent and Vilgalys 2007). However, patterns differed substantially from the large-scale soil fungi sequencing study of O'Brien et al. (2005). In a comparison to the two recently published next-generation studies (Talbot et al. 2014; Glassman et al. 2015), we recovered two of the top five genera of the spore bank study (Glassman et al. 2015) and 11 of the top 20 fungal families found in mature forest soil examined by Talbot et al. (2014).

11.4 Discussion

The intact core method we used recovered a substantial number of EcM fungi that had previously been found at Duke Forest. Additionally, the most abundant taxa found in our study corresponded to the same abundant taxa as those found by Parrent et al. (2006) and Parrent and Vilgalys (2007) (i.e., the Thelephoraceae, the Atheliaceae, and Russula; Table 11.2). Many of the EcM fungi we found (e.g. Russula) are considered 'late-stage' or 'mature ecosystem' (Deacon and Fleming 1993; Dickie et al. 2013), are hard to culture by other commonly used culture methods and were not found in expansive resistant propagule studies (Glassman et al. 2015).

We also regularly found Cylindrocarpon/Neonectria, the pathogen determined to have the most impact on pine seedlings at this site (Hersh et al. 2012). The community structure of the fungal communities we sampled was similar to the structure measured in the field studies. For instance, metrics like taxonomic groups (e.g. Fig. 11.1), rank abundance patterns (Tables 11.1 and 11.2), mean number of taxa per root system examined, and total number of taxa were comparable to field studies. The EcM fungal taxa included some that were found on many seedlings but were relatively rare on those seedlings; while other taxa were found on many root tips but only on a few seedlings. Although we were not successful in sequencing all root tips, our coverage (72% success) was higher than the 52% success of Parrent et al. (2006) and comparable to the recovery found by other field studies (e.g. 65% in Branco and Ree 2010). Given these results, we believe this method is a very useful way to capture ecologically-relevant EcM fungal communities and culture them on tree seedling roots.

In our primary comparison to the Duke Forest root tip study, Parrent et al. (2006) found 69 basidiomycete EcM fungi OTU and 3 putative saprobes whereas we recovered 34 basidiomycete EcM fungi OTU, 10 ascomycete EcM fungi OTU, and seven saprobes. Thus, the species richness in terms of basidiomycete EcM fungi of our study was about 50% of the field based root tip study. However, the number of fungi we were able to culture with this method (and then provide for testing of ecological process in a parallel study; Meier et al. 2013) was much greater than what other soil bioassay studies typically culture. In soil bioassay studies that not only disturbed the soil in some way (mixing, sieving, drying, etc.) but also for which a comparable field study existed that estimated the diversity of fungi on roots of trees *in situ*, the number recovered was on average about 25% of the known field fungi (Table 11.5). Remarkably, this is highly consistent with the findings of two recent next-generation DNA sequencing studies conducted at Duke Forest, where Glassman et al. (2015) found 29 OTU in the spore bank as measured by a bioassay that dried and homogenized soils while Talbot et al. (2014) found 110 OTU in mature forest soil. Our method thus is a substantial improvement on these typical bioassay studies.

In comparison to the two exclusive Duke Forest studies focused on fungi represented in total soil (O'Brien et al. 2005; Parrent and Vilgalys 2007), our study found fewer fungi. This is not surprising, however, given the difference in the target tissues and DNA sources of these two other studies.

In a large scale sequencing study of environmental DNA extracted from soil cores, O'Brien et al. (2005) found 412 total fungal OTUs that included many fungi in each major trophic group (i.e., 179 basidiomycete and 197 ascomycete EcM fungi). In a study of hyphae colonizing ingrowth bags, Parrent and Vilgalys (2007) found 134 EcM fungal taxa, 17 saprotrophs, seven arbuscular mycorrhizal fungi, six pathogens, a group of 27 pathogen and/or saprobe taxa and 98 taxa that could not be determined to any trophic group. We presume that if we had followed the procedures of these studies (i.e. had we targeted soil DNA and/or soil hyphae instead of only fungi colonizing root tips) our numbers would have been higher. Therefore, the absolute difference is not a concern. What is important to note, though, is that we had comparable OTU/species recovery per sequencing effort (Table 11.4), suggesting

Table 11.5 Bioassay studies examined for comparison to this study

Vegetation	Soil treatment	Species richness in field study [n]	Species richness in bioassay [% of total in field study]	References
Pinus taeda	Intact	72	49	This study
Pinus muricata	Sieved	48	44	Peay et al. (2009)
Pinus muricata	Mixed	48	17	Baar et al. (1999)
Pinus muricata	Dried and sieved	37	11	Taylor and Bruns (1999)
Fir-pine forest	Sieved	101	23	Izzo et al. (2006a, b)
Fir-pine forest	Sieved, other soil added	101	17	Izzo et al. (2006a, b)
Oak with chestnut	Dried, other soil added	71	20	Dulmer et al. (2014)
Scrub-oak serpentine	Homogenized	79	34	Branco and Ree (2010)
Oak savanna	*In situ* seedling	72	54	Dickie and Reich (2005)
Oak savanna	Dried and sieved	72	7	Avis and Charvat (2005)
Lowland heathlands	Homogenized	22	33	Collier and Bidartondo (2009)

that our method could allow for the detection of even more diversity with more exhaustive detection methodology like next-generation sequencing if we were so inclined. Since our goal was not only to culture ecologically-relevant fungal communities on seedlings but also to test the roles of these fungi in a relatively natural context, complete enumeration of all fungal taxa contributing DNA to a soil core was of less interest.

Another difference of our study was that we found many Wilcoxina and Cenococcum which were not abundant taxa in the previous Duke Forest studies (Fig. 11.1d, Table 11.2). These taxa contributed to the higher taxonomic diversity and distinctness found in our study (Table 11.4). Both taxa are considered early successional taxa (Dickie et al. 2013) and are often found in disturbed bioassays (Avis and Charvat 2005; Izzo et al. 2006a, 2006b; Peay et al. 2009; Glassman et al. 2015) and with regular natural soil disturbance (Dickie and Reich 2005). This indicates that our method still selects, to some extent, for taxa that are not common in the field – a typical problem of other types of bioassay studies.

Consistent with the other Duke Forest studies, we found no effects of CO_2 and N on fungal richness or the frequency at which we found individual fungal OTU in these treatments. This is the same result found in the field by Parrent et al. (2006) and Parrent and Vilgalys (2007) where the primary responses to these treatments were by individual taxa and only four (of 72 total) out of the most abundant taxa responded to N-fertilization. Although we did not see any significant shifts in taxon response, the trend of more frequent Russula in fertilized plots was consistent with

Russula G in Parrent et al. (2006) and the behavior of nitrophilic Russula taxa (Avis 2012). The reasons we may not have seen responses to N, especially given how often N is a major factor that structures soil fungal communities (e.g. Lilleskov et al. 2011), may be a seasonal effect or that sample sizes were too small.

Only a few other studies of which we are aware have used an intact core approach (Avis and Charvat 2005; Dickie et al. 2012). In the former, intact soil cores recovered about three times as many fungi as disturbed soil cores. The latter study did not analyze the fungal community but the authors contended that the method simulated a natural enough context for their purposes.

One of the benefits of the intact core approach is the ability to culture hard to culture fungi. In an extensive bioassay study in lowland heathlands, Collier and Bidartondo (2009) cautioned that bioassays are problematic in general because they do not adequately recover field dominant species like many members of the Russulaceae. It is unfortunate they did not consider using an intact core bioassay in their study as it would have ideally tested the reasons why Russulaceae are recalcitrant. The success of the intact core approach in culturing these challenging fungi and the results of our study suggest that the main limitation is intact (enough) mycelium from which these fungi colonize roots (Avis and Charvat 2005), an inability to colonize from spores and less to do with age of host (Collier and Bidartondo 2009).

Some studies have used an alternative approach and employed 'field seedlings' (i.e. seedlings grown *in situ* and then removed after a period of growth for analysis). The results of these studies have tended to be like the intact core method and have recovered a greater percentage of fungi than disturbed bioassays (Dickie and Reich 2005; Ashkannejhad and Horton 2006; Nunez et al. 2009). We do not know of any direct comparison of field seedlings vs. intact core (and that should be a future study). However, the challenges in using field seedlings in subsequent ecological process analyses seem greater than working with intact cores grown in the greenhouse because of the problems of logistics (i.e. just one trip to the field to collect the samples for the intact method) and recovering the seedlings intact (e.g. Dickie and Reich 2005; Dickie et al. 2012).

Like for all bioassay methods, though, numerous caveats should be considered with the intact core method. One limitation was indicated by the lower evenness and higher dominance of taxa with our method versus the other Duke Forest studies (Table 11.4). The combination of these metrics suggests that some taxa are overrepresented in the cultures with respect to what was found in the field. This is due to the relative speed at which taxa spread (i.e. colonize roots) within the cultures. If a fungus spreads quicker than others within the limited space and time of the culture, our method would sample it more. This is not an uncommon issue in culturing studies (e.g. Avis and Charvat 2005) and will be an inherent issue of any culturing approach. Therefore, a crucial consideration is the time at which cultures are harvested and assayed. In this study, we took into consideration the time for fungal communities to develop and the time required for root exudate assays (Meier et al. 2013). Therefore, we grew cultures for 14 weeks, which was based on the development of mycelium (sample seedlings were examined and nearly all root tips examined were colonized by EcM fungi) and what appeared to be optimal growth for the parallel study examining root exudation.

Additional caveats remained, all of which are inherent to any bioassay. These included the altered microenvironments of the soil core when extracted from the field and moved to the greenhouse. Despite the care taken to keep the soil structure intact, the act of taking the core severs a portion of extensive fungal mycelia (that are larger than the dimensions of the core) and may change the biochemistry of the fungal material resident in the cores compared to intact fungi. Nonetheless, our approach appears to keep enough mycelium intact to allow for the growth and development of recalcitrant and likely mycelial-network demanding fungi like the Russulaceae. How similar the amount of disturbance our method imposes in comparison to natural soil disturbance events (e.g. grazing of roots and hyphae and burrowing by animals) remains to be determined.

Another caveat is that the seeding, subsequent germination and initial growth may select for fast growing fungal taxa or those taxa with carbon reserve able to wait out the formation of appropriate feeder roots to colonize. Unlike the disturbed soil bioassays, the intact method does not stack the deck in favor of these rapid growers since the mycelial demanding taxa have at least minimal access to some undisturbed soil. Moreover, our rank abundance patterns are very similar to field results and relevant field studies (e.g. Parrent et al. 2006), further indicating that the method can generate reasonable and useful proxies of patterns of fungal communities in the field.

This method could be modified in a variety of ways to suit the specific needs of a study intent on linking fungal identity to ecosystem process. We chose to grow pine seedlings since they were the dominant tree in the forest under study and of most interest to the parallel study of root exudation (Meier et al. 2013). However, matching the primary plant host in a habitat of interest to use in these cultures may be more challenging if there are multiple dominant hosts. In such cases, experimental designs could add more plant hosts per culture and/or test if different hosts in the same intact cores recovered the same fungal community.

11.5 Conclusions

Based on these results and the use of this method in simultaneously conducted parallel study that examined the relationship between fungal nutritional type and root exudates (Meier et al. 2013), we believe this is an important, valuable and easy-to-use method for fungal ecology. We encourage all those interested to incorporate such methods when looking to determine fungal community function in biogeochemical processes.

Acknowledgements The authors would like to thank Robert Nettles (Brookhaven National Laboratory) and the staff of Duke Forest for technical assistance and operation of the Duke Forest FACTS-1 site. We acknowledge the financial support by the Office of Science (BER), U.S. Department of Energy, grant no. DE-FG02-95ER62083. This project was supported by Agriculture and Food Research Initiative Competitive, grant no. 2008-35107-04500 from the USDA National Institute of Food and Agriculture (NIFA). The project was also supported by the Indiana University Northwest Department of Biology. Thanks to R. Healy for help in identifying pezizalean fungi.

References

Agerer, R. (1991). Characterization of ectomycorrhiza. *Methods in Microbiology, 23*, 25–73.

Andrew, C., & Lilleskov, E. A. (2009). Productivity and community structure of ectomycorrhizal fungal sporocarps under increased atmospheric CO_2 and O_3. *Ecology Letters, 12*, 813–822.

Ashkannejhad, S., & Horton, T. R. (2006). Ectomycorrhizal ecology under primary succession on coastal sand dunes, interactions involving *Pinus contorta*, suilloid fungi and deer. *New Phytologist, 169*, 345–354.

Avis, P. G. (2012). Ectomycorrhizal iconoclasts: The ITS rDNA diversity and nitrophilic tendencies of foetid *Russula*. *Mycologia, 104*, 998–1007.

Avis, P. G., & Charvat, I. (2005). The response of ectomycorrhizal fungal inoculum to long-term increases in nitrogen supply. *Mycologia, 97*, 329–337.

Avis, P. G., Mueller, G. M., & Lussenhop, J. (2008). Ectomycorrhizal fungal communities in two North American oak forests respond to nitrogen addition. *New Phytologist, 179*, 472–483.

Baar, J., Horton, T. R., Kretzer, A. M., & Bruns, T. D. (1999). Mycorrhizal colonization of *Pinus muricata* from resistant propagules after a stand-replacing wildfire. *New Phytologist, 143*, 409–418.

Branco, S., & Ree, R. H. (2010). Serpentine soils do not limit mycorrhizal fungal diversity. *PLoS ONE, 5*, e11757. doi:10.1371/journal.pone.0011757.

Brundrett, M. C., Scade, A., Barry, A. L., Dixon, K. W., & Sivasithamparam, K. (2003). Development of *in situ* and *ex situ* seed baiting techniques to detect mycorrhizal fungi from terrestrial orchid habitats. *Mycological Research, 107*, 1210–1220.

Clarke, K. R., & Warwick, R. M. (1998). A taxonomic distinctness index and its statistical properties. *Journal of Applied Ecology, 35*, 523–531.

Collier, F. A., & Bidartondo, M. I. (2009). Waiting for fungi: The ectomycorrhizal invasion of lowland heathlands. *Journal of Ecology, 97*, 950–963.

Deacon, J. W., & Fleming, L. V. (1993). Interactions of ectomycorrhizal fungi. In *Mycorrhizal functioning – An integrative plant-fungal process* (pp. 249–300). New York: Chapman and Hall.

Dickie, I. A., & Reich, P. B. (2005). Ectomycorrhizal fungal communities at forest edges. *Journal of Ecology, 93*, 244–255.

Dickie, I. A., Davis, M., & Carswell, F. E. (2012). Quantification of mycorrhizal limitation in beech spread. *New Zealand Journal of Ecology, 36*, 210–215.

Dickie, I. A., Martinez-Garcia, L. B., Koele, N., Grelet, G.-A., Tylianakis, J. M., Peltzer, D. A., & Richardson, S. J. (2013). Mycorrhizas and mycorrhizal fungal communities throughout ecosystem development. *Plant and Soil, 367*, 11–39.

Dulmer, K. M., Leduc, S. M., & Horton, T. R. (2014). Ectomycorrhizal inoculum potential of northeastern U.S. forest soils for American chestnut: results from field and laboratory bioassays. *Mycorrhiza, 24*, 65–74.

Gardes, M., & Bruns, T. D. (1993). ITS primers with enhanced specificity for basidiomycetes – Application to the identification of mycorrhizae and rusts. *Molecular Ecology, 2*, 113–118.

Glassman, S. I., Peay, K. G., Talbot, J. M., Smith, D. P., Chung, J. A., Taylor, J. W., Vilgalys, R., & Bruns, T. D. (2015). A continental view of pine-associated ectomycorrhizal fungal spore banks: A quiescent functional guild with a strong biogeographic pattern. *New Phytologist, 205*, 1619–1631.

Goodman, D. M., Durall, D. M., Trofymow, J. A., & Berch, S. M. (1996). *A manual of concise descriptions of North American ectomycorrhizae*. Victoria: Mycologue Publications.

Hammer, Ø., & Harper, D. A. T. (2001). PAST: Paleontological statistics software package for education and data analysis. *Palaeontologia Electronia, 4*, 1–9.

Hersh, M. H., Vilgalys, R., & Clark, J. S. (2012). Evaluating the impacts of multiple generalist fungal pathogens on temperate tree seedling survival. *Ecology, 93*, 511–520.

Izzo, A., Canright, M., & Bruns, T. D. (2006a). The effects of heat treatments on ectomycorrhizal resistant propagules and their ability to colonize bioassay seedlings. *Mycological Research, 110*, 196–202.

Izzo, A., Nguyen, D. T., & Bruns, T. D. (2006b). Spatial structure and richness of ectomycorrhizal fungi colonizing bioassay seedlings from resistant propagules in a Sierra Nevada forest: Comparisons using two hosts that exhibit different seedling establishment patterns. *Mycologia, 98*, 374–383.

Krebs, C. J. (1989). *Ecological methodology* (624 pp). New York: Harper and Row.

Landeweert, R., Hoffland, E., Finlay, R. D., Kuyper, T. W., & van Breemen, N. (2001). Linking plants to rocks: Ectomycorrhizal fungi mobilize nutrients from minerals. *Trends in Ecology and Evolution, 16*, 248–254.

Lilleskov, E. A., & Bruns, T. D. (2003). Root colonization dynamics of two ectomycorrhizal fungi of contrasting life history strategies are mediated by addition of organic nutrient patches. *New Phytologist, 159*, 141–151.

Lilleskov, E. A., Fahey, T. J., Horton, T. R., & Lovett, G. M. (2002). Belowground ectomycorrhizal fungal community change over a nitrogen deposition gradient in Alaska. *Ecology, 83*, 104–115.

Lilleskov, E. A., Hobbie, E. A., & Horton, T. R. (2011). Conservation of ectomycorrhizal fungi: Exploring the linkages between functional and taxonomic responses to anthropogenic N deposition. *Fungal Ecology, 4*, 174–183.

Meier, I. C., Avis, P. G., & Phillips, R. P. (2013). Fungal communities influence root exudation rates in pine seedlings. *FEMS Microbiology Ecology, 83*, 585–595.

Nunez, M. A., Horton, T. R., & Simberloff, D. (2009). Lack of belowground mutualisms hinders Pinaceae invasions. *Ecology, 90*, 2353–2359.

O'Brien, H. E., Parrent, J. L., Jackson, J. A., Moncalvo, J. M., & Vilgalys, R. (2005). Fungal community analysis by large-scale sequencing of environmental samples. *Applied and Environmental Microbiology, 71*, 5544–5550.

Parrent, J. L., & Vilgalys, R. (2007). Biomass and compositional responses of ectomycorrhizal fungal hyphae to elevated CO_2 and nitrogen fertilization. *New Phytologist, 176*, 164–174.

Parrent, J. L., Morris, W. F., & Vilgalys, R. (2006). CO_2-enrichment and nutrient availability alter ectomycorrhizal fungal communities. *Ecology, 87*, 2278–2287.

Peay, K. G., Garbelotto, M., & Bruns, T. D. (2009). Spore heat resistance plays an important role in disturbance-mediated assemblage shift of ectomycorrhizal fungi colonizing *Pinus muricata* seedlings. *Journal of Ecology, 97*, 537–547.

Rosling, A., Suttle, K. B., Johansson, E., van Hees, P. A. W., & Banfield, J. F. (2007). Phosphorous availability influences the dissolution of apatite by soil fungi. *Geobiology, 5*, 265–280.

Stamatakis, A. (2006). RAxML-VI-HPC: maximum likelihood-based phylogenetic analyses with thousands of taxa and mixed models. *Bioinformatics, 22*, 2688–2690.

Stamatakis, A., Hoover, P., & Rougemont, J. (2008). A rapid bootstrap algorithm for the RAxML web servers. *Systematic Biology, 57*, 758–771.

Talbot, J. M., Bruns, T. D., Smith, D. P., Branco, S., Glassman, S. I., Erlandson, S., Vilgalys, R., & Peay, K. G. (2013). Independent roles of ectomycorrhizal and saprotrophic communities in soil organic matter decomposition. *Soil Biology and Biochemistry, 57*, 282–291.

Talbot, J. M., Bruns, T. D., Taylor, J. W., Smith, D. P., Branco, S., Glassman, S. I., Erlandson, S., Vilgalys, R., Liao, H. L., Smith, M. E., & Peay, K. G. (2014). Endemism and functional convergence across the North American soil mycobiome. *Proceedings of the National Academy of Sciences, USA, 111*, 6341–6346.

Taylor, D. L., & Bruns, T. D. (1999). Community structure of ectomycorrhizal fungi in a *Pinus muricata* forest: Minimal overlap between the mature forest and resistant propagule communities. *Molecular Ecology, 8*, 1837–1850.

Van Breemen, N., Finlay, R., Lundström, U., Jongmans, A. G., Giesler, R., & Olsson, M. (2000). Mycorrhizal weathering: A true case of mineral plant nutrition? *Biogeochemistry, 49*, 53–67.

Chapter 12
Potential Role of Beneficial Soil Microorganisms in Plant Tolerance to Abiotic Stress Factors

Raffaella Balestrini, Walter Chitarra, Vasileios Fotopoulos, and Michelina Ruocco

Abstract Abiotic stress conditions such as drought, salinity and extreme temperatures are an increasing problem in agriculture. Research efforts are aimed to develop strategies to make agriculture more resilient and to mitigate the stress effects on crop production. In this context, the use of root-associated microbial communities able to improve plant tolerance is attracting increasing attention. In this chapter, we will offer an overview of the researches on the use of soil beneficial microorganisms, focusing mainly on mycorrhizal fungi and biocontrol agents such as *Trichoderma* species, to improve plant tolerance to different abiotic stresses (e.g. water stress, salinity, extreme temperatures).

Keywords Plant tolerance • Abiotic stresses • Drought • Symbiotic fungi • Arbuscular mycorrhizal fungi • *Trichoderma*

12.1 Introduction

It is well known that environmental stresses such as drought, salinity and cold can limit plant growth and development dramatically reducing the yields of important crops (Foley et al. 2011). Drought and land degradation following the salinization of soils are increasing considerably worldwide and many food crops are growing in suboptimal climatic conditions in different parts of the world (Lesk et al. 2016).

R. Balestrini (✉) • W. Chitarra • M. Ruocco
Institute for Sustainable Plant Protection, CNR, Torino, Italy
e-mail: raffaella.balestrini@ipsp.cnr.it

V. Fotopoulos
Department of Agricultural Sciences, Biotechnology and Food Science, Cyprus University of Technology, Limassol, Cyprus

M. Lukac et al. (eds.), *Soil Biological Communities and Ecosystem Resilience*, Sustainability in Plant and Crop Protection, DOI 10.1007/978-3-319-63336-7_12

The expected climate changes could further worsen this scenario Sheffield et al. (2012). Recent evidence shows that plants respond to combination of stresses by regulating the expression of an array of defense-related genes, which differs from their single-stress responses and is related to the exact environmental conditions encountered (Atkinson and Urwin 2012). Plant physiology studies show that plants might adapt to abiotic stresses by reprogramming their patterns of gene expression, and a large number of plant stress-responsive genes have been recently identified by transcriptomic and proteomic analyses (Hirayama and Shinozaki 2010; Soda et al. 2015). During the last decades, a variety of strategies have been deployed to improve stress tolerance in crops, including traditional selection methods and genetic engineering (Wang et al. 2003; Fleury et al. 2010). Research efforts are aimed to develop strategies to make agriculture more resilient and to mitigate the stress effects on crop yield, e.g. through the selection of crop varieties adapted to drought or by improving soil management and irrigation techniques (Campbell 2012), while plant priming with chemical compounds against stress factors is also attracting increasing attention (Savvides et al. 2016). Designing new strategies to improve abiotic stress tolerance in important crops will have positive impact on the plant productivity and on the expansion of arable areas. As an alternative strategy, the use of root-associated microbial communities able to improve plant tolerance is attracting increasing attention (Coleman-Derr and Tringe 2014). Only recently, researchers widened their point of view considering plants as part of a diverse and dynamic phytobiome, where beneficial and detrimental organisms (from microbes to vertebrates) interact with each other and with the environment. Given that environmental conditions are continuously altered, plants benefit from their phenotypic plasticity, further maximize through interactions with beneficial organisms (Dicke 2016). Plants live in closely related association with soil microbes, and the presence of the complex root-associated soil microbial community (rizospheric microbiota) is crucial for plant performance and health (Schlaeppi and Bulgarelli 2014; Balestrini et al. 2015). The root-associated microbiome is nowadays considered as a key factor for managing crop production.

In this chapter, we will offer an overview of the researches on the use of soil beneficial microorganisms, focusing mainly on mycorrhizal fungi and biocontrol agents such as *Trichoderma* species, to improve plant tolerance to different abiotic stresses (water stress, salinity, extreme temperatures).

12.2 Soil Microorganisms Influence Plant Tolerance to Abiotic Stresses

Although drought, salinity and extreme temperatures are an increasing problem in agriculture, the contribution of the root-associated microbiome in plant adaptation to abiotic stresses has not been extensively studied yet (Rolli et al. 2015). Abiotic-stress tolerance is a complex trait and, although several efforts on this topic have been recently developed by researchers, the physiological and molecular mechanisms underlying increased stress tolerance in the presence of microorganisms still remain to be elucidated. Additionally, the composition of the plant-associated microbiome is continuously modified by the occurrence of specific stresses (biotic and/or abiotic), including pathogen attack and either water or nutrient shortage (Lareen et al. 2016). On the other hand, these plant-associated microorganisms (bacteria and fungi) help the plant to react and in some cases overcome the resulting constraint. Some plant growth-promoting bacteria (PGPB; or plant growth-promoting rhizobacteria, PGPR) appear to be useful in conferring tolerance to drought as well as other abiotic and biotic threats (Timmusk and Wagner 1999; Mayak et al. 2004; Rojas-Tapias et al. 2012; Timmusk et al. 2013; Rolli et al. 2015; Staudinger et al. 2016). A process called rhizobacterial-induced drought endurance and resilience (RIDER), which includes physiological and biochemical changes, has been proposed to be involved in the mitigation of the impact of drought on plants by PGPR (Kaushal and Wani 2016). Multiple mechanisms are involved including changes in phytohormonal levels, antioxidant defence, osmolytes (e.g., proline) and polyamines, production of heat-shock proteins (HSPs), dehydrins and volatile organic compounds (VOCs). Beneficial microorganisms are able to produce and induce the formation of plant drought-induced volatiles as signals for developing priming and systemic responses for themselves and neighbouring plants (Cho et al. 2008). Timmusk et al. (2014) demonstrated the effect of bacterial priming on wheat drought stress tolerance improvement, resulting in up to 78% greater plant biomass and five-fold higher survival under severe drought. Interestingly, these authors monitored the emissions of seven stress-related volatiles from bacterially-primed drought-stressed seedlings, suggesting that monitoring VOC emissions could be a promising method to characterize the efficiency of different bacterial strains in priming for drought stress tolerance. Exopolysaccharides (EPS) produced and released by bacteria as component of biofilms can also play a role in the alleviation of water deficit effects, increasing soil aggregation and improving water retention around the roots (Sandhya et al. 2009). Nevertheless, bacterial priming towards improved crop plant performance shows to be a useful strategy to be explored for the alleviation of the impact of low temperature as well (Osman et al. 2013). Together with rhizobacteria, which can establish root-nodule symbiosis with legumes, the most important bio-fertilizer microorganisms are arbuscular mycorrhizal fungi (AM), since their best understood function in the symbiosis is to enhance the acquisition of plant mineral nutrients (such as phosphorous) in exchange for carbon compounds (Balestrini et al. 2015). While ectomycorrhizal (ECM) fungi dominate and play important roles in forest ecosystems, AM fungi form mutualistic symbioses with the roots of most plants of agronomic importance (van der Heijden et al. 2015). They are considered to be essential elements for plant nutrition as their hyphae can extend for many meters in the ground helping the plants to acquire water and mineral nutrients present in the soil (Bucher et al. 2014). In the last few years, several papers have reported the influence of AM symbiosis on plant response to abiotic stresses such as drought, salinity, and flooding (Lenoir et al. 2016). AM fungi are also known to alleviate heavy metal toxicity in the host plants and to tolerate high metal concentrations in the soil (Cornejo et al. 2013; Tamayo et al. 2014; Meier et al. 2015). Due to their several beneficial effects on terrestrial ecosystems, AM fungi are widely used in organic agriculture and plant nurseries to improve the growth of economically important species (Corradi and Bonfante 2012). In this context, the majority of studies on the impact of mycorrhizal fungi on plant performance under abiotic stress conditions have been focused on AM symbioses. In recent years, the availability of the mycorrhizal fungi genome sequences (both endo- and ectomycorrhizal fungi), and of the transcriptomics data obtained in parallel, allowed to obtain important information about the biology of these fungi and the mechanisms involved in the interactions with host plants (Martin et al. 2008, 2010; Tisserant et al. 2013; Kohler et al. 2015). The genome sequence of ECM fungus *Cenococcum geophylum* has been recently published (Peter et al. 2016), and changes in the expression of genes coding for aquaporins (AQPs), which channel water and/or small solutes across membranes (Maurel et al. 2015), in symbiosis have been revealed. It has been suggested that the high expression of two highly water permeable AQPs in functioning ectomycorrhizae may be provoked by plant water and/or nutrient demand during the interaction (Peter et al. 2016). Expression analyses also indicate that a fine-tuned regulation of the AQP genes occurs under drought conditions, although no significant effect of *C. geophilum* colonization on plant physiological parameters under drought was revealed. Interestingly, this high upregulation of AQP genes seems to be a feature of the *C. geophilum* symbiotic interaction when compared with other mycorrhizal fungi, although important functions of fungal AQPs in the ECM symbiosis have been shown for other ECM fungi (Dietz et al. 2011; Navarro-Ródenas et al. 2012).

In addition to mycorrhizal fungi, other important soil beneficial fungi are those belonging to the genus *Trichoderma*. Although these fungi are widely used in agriculture for their activity both as plant-growth promoting microorganisms and bio-fungicides, information on their involvement in plant tolerance to abiotic stresses is still poor.

12.2.1 AM Fungi Versus Drought

Among abiotic factors, drought represents one of the major constraints of crop yield and agricultural productivity, with a vast impact on food security issues (Boyer et al. 2013). For comparative physiological studies, drought would be described as independent from plant species/genotype characteristics, but related to the local environment. During drought events, water losses in a particular agro/ecosystem are greater than water inputs, leading to soil water deficit (SWD; Gilbert and Medina 2016). During the last decade, several studies have been aimed to verify the impact of this stress on plant physiology, trying to further elucidate the complex network of responses and to plan strategies focused on sustainable crop management (Osakabe et al. 2014; Daryanto et al. 2016). It is well known that water stress adversely impacts many aspects of plant production and physiology, involving morphological, biochemical and molecular traits that actively respond to limiting damages and to permit recovery after water restoration (Chitarra et al. 2014; Secchi et al. 2016). More specifically, water deficiency is primarily sensed from root apparatus, then a root-to-leaf signal cascade induces physiological and morphological changes, leading to a reduction in long distance water transport (by forming embolisms) and to stomatal closure and reduction in photosynthetic performances (Flexas et al. 2004; Chitarra et al. 2014; Salazar et al. 2015). Behind this, a complex biological and molecular network, involving phythormone regulation, metabolite accumulation and signal transduction at cellular level, is at the basis of these modifications (Osakabe et al. 2014). Among phytohormones, abscisic acid (ABA) plays a key role in plant responses to water stress by mediating regulatory events such as gene expression, signalling, ion and water transport (e.g., stomatal opening and AQP gene regulation). Furthermore, ABA influences the activities of several transcription factors that are orchestrated into molecular pathways enabling the plants to cope with stressors (González-Guzmán et al. 2014; Tombesi et al. 2015; McAdam et al. 2016). Drought-induced damage depends on the intensity and persistence of the water-shortened period and may or may not be recoverable (Gilbert and Medina 2016). Water deficit effects have been studied in several important crops such as maize (Kamara et al. 2003), rice (Lafitte et al. 2007), wheat (Rampino et al. 2006), tomato (Secchi et al. 2013) and grapevine (Perrone et al. 2012; Chitarra et al. 2014). Thanks to the constant development of *-omics* technologies, researchers are defining new ways to better understand plant responses to water stress (Dugas et al. 2011; Kakumanu et al. 2012; Iovieno et al. 2016). It has been also demonstrated that the responses to water deficit conditions can be diverse among different genotypes/accessions (Galmés et al. 2011, 2013). Furthermore, it has been predicted that plants will be exposed to more extreme and multiple abiotic stresses in the future, resulting in numerous studies currently being carried out to develop sustainable and adaptation techniques in agriculture (Chatzidaki and Ventura 2010). In this context, AM fungal colonization has shown noteworthy results in increasing tolerance to water stress (Ruiz-Lozano and Aroca 2010; Augé et al. 2015). The thin AM fungal hyphae can explore soil pores inaccessible to root hairs, thus reaching water sources not available to non-AM plants. Furthermore, hyphal water transport to the root under drought conditions was previously demonstrated (Khalvati et al. 2005). A potential water transport via AM fungus to the host plant has been also suggested on the basis of the gene expression profiles for two functionally characterized fungal AQP genes, thus supporting the existence of a direct AM fungus involvement in plant tolerance to drought (Li et al. 2013a, b). AM fungi, which facilitate the formation of water-stable aggregates, can also influence soil moisture retention properties (Augé et al. 2001). In addition to an increase in nutrient and water uptake, different mechanisms have been proposed to be involved in water stress alleviation in AM plants. Water use efficiency (WUE), i.e., the ratio between net assimilation (A_N) and transpiration rates (E), is used as key selection trait by breeders for the selection of plants with higher survival rate and productivity under stress events (Osakabe et al. 2014). Influence of AM symbiosis on stomatal conductance (g_s), which exerts a control in assimilation rates and water exchanges, has been largely reported (reviewed in Augé et al. 2015). Since the first studies regarding the activity and the expression of superoxide dismutase (SOD) in lettuce plants subjected to drought (Ruiz-Lozano et al. 1996, 2001; Ruiz-Lozano 2003), the impact of AM symbiosis on plant performance under water deficit has been largely studied, considering mainly plant performance and growth (in terms of WUE and biomass), osmolyte accumulation (i.e., proline), antioxidant enzyme activities (e.g., SOD), as well as AQPs expression (Porcel et al. 2004, 2006; Bárzana et al. 2012, 2014, 2015; Hazzoumi et al. 2015; Calvo-Polanco et al. 2014; Chitarra et al. 2016; Mo et al. 2016; Sánchez-Romera et al. 2016; Ruiz-Lozano et al. 2016, and references therein). However, the precise mechanisms involved in the improved tolerance by AM symbiosis still remain to be fully elucidated. Additionally, previous reports suggested that the plant responses on stress alleviation are environment- and symbiont-specific (Augé 2001; Augé et al. 2015). Focusing on more recent papers, following a multidisciplinary approach including eco-physiological, morphometric, biochemical and molecular methods, the positive effects of two AM fungi (*Funneliformis mosseae* and *Rhizophagus intraradices*) in increasing tolerance to water deficit in tomato plants were verified by Chitarra et al. (2016), although differences between the two fungi were observed. Under severe water stress, AM symbiosis affects plant physiology parameters: greater intrinsic WUE (iWUE, calculated as A_N/g_s ratio) values were observed in AM+ plants coupled with lower level of ABA compared with AM- plants. Furthermore, *R. intraradices*-colonized plants, besides the enhanced morphometric values, showed significantly higher stomatal density and thus A_N rates in both well-watered and water stress conditions. Mirroring the interest and the research efforts on this topic, Ruiz-Lozano and colleagues have recently reported a correlation between AM root colonization, strigolactone levels and drought severity, suggesting that plants under these unfavourable conditions might increase strigolactone production in order to promote symbiosis establishment to cope with the stress (Ruiz-Lozano et al. 2016).

12.2.2 AM Fungi Versus High Salinity and Extreme Temperature

Drought, salinity, extreme temperatures and oxidative stress are often interconnected, and may induce similar cellular damage (Krasensky and Jonak 2016). The progressive salinization of land is among the major environmental factors limiting plant growth and productivity, mainly in arid and semiarid regions. The development of salt-tolerant crops is an important challenge for breeders and farmers. Although the excess of salinity can negatively affect AM fungi due to osmotic and/or toxic effect of salt (Juniper and Abbott 2006), several studies showed an improved growth and performance of AM-colonized stressed plants in several species such as maize, clover, tomato, sweet basil, cucumber and lettuce (Porcel et al. 2012; Elhindi et al. 2016; Chandrasekaran et al. 2016). Interestingly, alleviation of salinity by AM fungi has been also reported for woody fruit crops such as olive and apple tree, and *Citrus* spp. (Porras-Soriano et al. 2009; Navarro et al. 2014; Yang et al. 2014).

Several processes have been proposed to be involved in the enhanced tolerance to salt stress by AM fungal colonization, including improved host plant nutrition, higher K+/Na+ ratio in plant tissue, production of osmoregulators, increase in photosynthesis ability and WUE, as well as water uptake facilitation in AM colonized plants. An increase in the activity of antioxidant enzymes in order to cope with the reactive oxygen species (ROS) generated by salinity has also been reported, as well as a regulation in the expression of genes involved in proline biosynthesis and genes coding for AQPs and late embryogenesis abundant (LEA) proteins (reviewed in Porcel et al. 2012).

Pedranzani et al. (2016) studied the impact of AM symbiosis in *Digitaria eriantha cv.* Sudafricana plants subjected to drought, cold and high salinity. A positive role of AM fungi to cope with stressful condition was suggested, by modulating jasmonate and antioxidant levels in stressed plants. An impact of AM symbiosis on low temperature has also been reported in cucumber and maize (Zhu et al. 2010a, b; Zhang et al. 2009). AM fungi can induce higher enzymatic activity and secondary metabolite content in plants leading to greater cold tolerance in cucumber plants (Chen et al. 2013). Interestingly, it has also been suggested that AM fungi may be significant drivers of plant response to increased soil temperature associated with global climate change (Bunn et al. 2009). Researchers' attention has also focused towards the identification of AM fungal species/isolates physiologically and genetically adapted to the stress conditions in a specific environment. This could be an important point for success in recovering saline areas both in natural environments and in agricultural lands affected by salinity (Estrada et al. 2013a, b).

12.2.3 **Trichoderma** *spp. Versus Abiotic Stresses in Plants*

Among plant beneficial microorganisms, *Trichoderma* spp. are some of the most studied and applied thanks to their multifactorial characteristics (Zeilinger et al. 2016). Long known for the mycoparasitism and for the production of cell wall degrading enzymes (CWDEs) (Bischof et al. 2016), they are now more and more characterized for the overall performance. Thanks to the systems biology technologies, it has been possible to deeply investigate the physiological changes that occur in plant interacting with *Trichoderma* spp. in the last ten years. In addition, new insights are emerging regarding the induction of plant resilience to abiotic stresses due to the interaction with these fungi (Alfano et al. 2007; Moran-Diez et al. 2012; Brotman et al. 2013; De Palma et al. 2016; Pandey et al. 2016). A few recent papers clearly demonstrated that fungi belonging to *Trichoderma* genus can alleviate plants from abiotic stresses (salinity, drought, extreme temperature).

Yildirim and collaborators (2006) found that squash plants treated with *T. harzianum* T22 or other beneficial microorganisms were more tolerant to salinity than untreated ones. On the same direction, inoculation with *T. hamatum* 382 actively induced expression of genes in tomato leaves with a putative function in the response to biotic and abiotic stresses, such as two salt-induced genes (TC122281 and TC127846) (Alfano et al. 2007). Arabidopsis and cucumber (*Cucumis sativus* L.) plants treated with *T. asperelloides* T203 prior salt stress imposition also showed significantly improved seed germination. In addition, *Trichoderma* T203 treatment affected the expression of several genes related to osmoprotection and general oxidative stress in roots of both plant species (Brotman et al. 2013). *T. asperellum* Q1 isolated from the rhizosphere of cucumber plants grown under greenhouse conditions significantly promoted seedling growth and alleviated the growth suppression induced by salt stress. This activity was related to its siderophores production and phosphate solubilization (Qi and Zhao 2013).

Iron (Fe) deficiency is a major plant nutritional disorder in many parts of the world, particularly in areas with saline soils. A high siderophore-producing strain of *T. asperellum* T6, was isolated from the rhizosphere of cucumber plants, and its application resulted in increased soil levels of Fe^2 and siderophores, as well as to an increase of Fe^2 and Fe^3-chelate reductase (FCR) activity in cucumber tissues. Cucumber plants treated with *T. asperellum* T6 had a growth promotion under salt stress compared with the untreated control plants (Zhao et al. 2014). Furthermore, *T. parareesei* T6 was able to prime defence responses in tomato plants against biotic and abiotic stresses by exerting beneficial effects in terms of seedling lateral root development, as well as by improved defence against *Botrytis cinerea* and growth promotion under salt stress in adult plants (Rubio et al. 2014).

It is interesting to note that Mastouri and collaborators (2010) demonstrated how applications of *T. harzianum* T22 on tomato seeds were more effective when plants were subjected to several stresses. In fact, *T. harzianum* T22-treated seeds germinated consistently faster and more uniformly than untreated seeds under osmotic and salt stress, or suboptimal temperatures. These authors suggested that abiotic stress tolerance, at least for osmotic stress, is alleviated through amelioration of damage caused by ROS, similar to what was previously observed in plants treated with endophytic root-colonising fungus *Piriformospora indica* (Baltruschat et al. 2008). *T. harzianum* T22 inoculated tomato plants resulted in enhanced redox state by higher activity of ascorbate and glutathione recycling enzymes and more tolerant to drought when compared with the non-inoculated control plants. It has been hypothesized (Mastouri et al. 2012) that enhanced tolerance of colonized plants to water deficit is at least partly due to higher capacity to scavenge ROS and recycle oxidized ascorbate and glutathione, a mechanism that is expected to enhance tolerance to both abiotic and biotic stresses. Interestingly, the colonization of *Theobroma cacao* (cacao) with *T. hamatum* DIS 219b isolate enhanced seedling growth, altered gene expression, and delayed the onset of the cacao drought response in leaves. Results showed that DIS 219b colonization delayed the drought-altered expression of all seven expressed sequence tags (ESTs) responsive to drought in leaves, while it had less influence on the expression pattern of the drought-responsive ESTs in roots. Drought induced an increase in the concentration of several amino acids in cacao leaves, while the DIS 219b colonization caused a decrease in aspartic acid and glutamic acid concentrations and an increase in alanine and γ-aminobutyric acid (GABA) concentrations. The fact that concentrations of several amino acids (i.e., GABA) were found directly responsive to DIS 219b colonization suggested the possibility that colonization may pre-adapt cacao to drought as a component of the altered signal transduction pathways observed in direct response to colonization (Bae et al. 2009).

Soil contamination by heavy metal and metalloids could also cause major problems to plant growth and development. Arsenic (As) is a ubiquitous metalloid extremely toxic for plant tissues, because a small amount of the inorganic forms, such as the arsenite (As^{III}) can accumulate through aquaporin channels and the arsenate (As^{V}) through the phosphate transporter system, affecting photosynthesis respiration, growth regulation and reproduction and, in the most severe conditions, leading to plant death (Tripathi et al. 2007). Soil microorganisms have been previously employed for As decontamination (Srivastava et al. 2011), and few studies have demonstrated the role of *Trichoderma* spp. in ameliorating As stress in plant. Tripathi and collaborators (2013) showed that chickpea (*Cicer arietinum* L.) plants inoculated with *T. reesei* BRI0716 were more tolerant to As-mediated stress under greenhouse conditions. Similarly, Caporale and collaborators (2014) demonstrated that two species of *Trichoderma* (*T. harzianum* T22 and *T. atroviride* P1) alleviated, at least in part, the phytotoxicity of As, essentially by decreasing its accumulation in the tissues and enhancing plant growth, P status and net photosynthesis rate.

Additionally, the selection of *Trichoderma* isolates from various agro-climatic zones have been reported, with the aim to select isolates tolerant to high temperatures and salinity, with an applicative potential in stressed soils (Poosapati et al. 2014).

12.3 Conclusions

In the frame of the development of more sustainable agricultural systems, with a positive impact on the environment as well as on food security and quality, soil biologists have recently highlighted the potential of the interactions between roots and soil microorganisms in improving plant tolerance to abiotic and nutrient stresses (Fig. 12.1). Results so far obtained, allow us to draw a more precise picture about plant and soil microbe interactions, underlying the importance of these microorganisms in applicable programs. Recent evidence shows that plants respond to multiple and combined stresses by activating a program of gene expression, which differs from their single-stress responses and is related to the exact environmental conditions encountered (Sewelam et al. 2014; Pandey et al. 2015). On the other hand, these soil microorganisms also will be exposed to the expected climate changes and extreme drought events that will cause changes in biodiversity and abundance as suggested by Millar and Bennet (2016).

For these reasons, single species or microbial consortia should be isolated and characterized from the stress-affected soils in order to select adapted strains/isolates that should be more efficient against specific stress conditions (Ortiz et al. 2015).

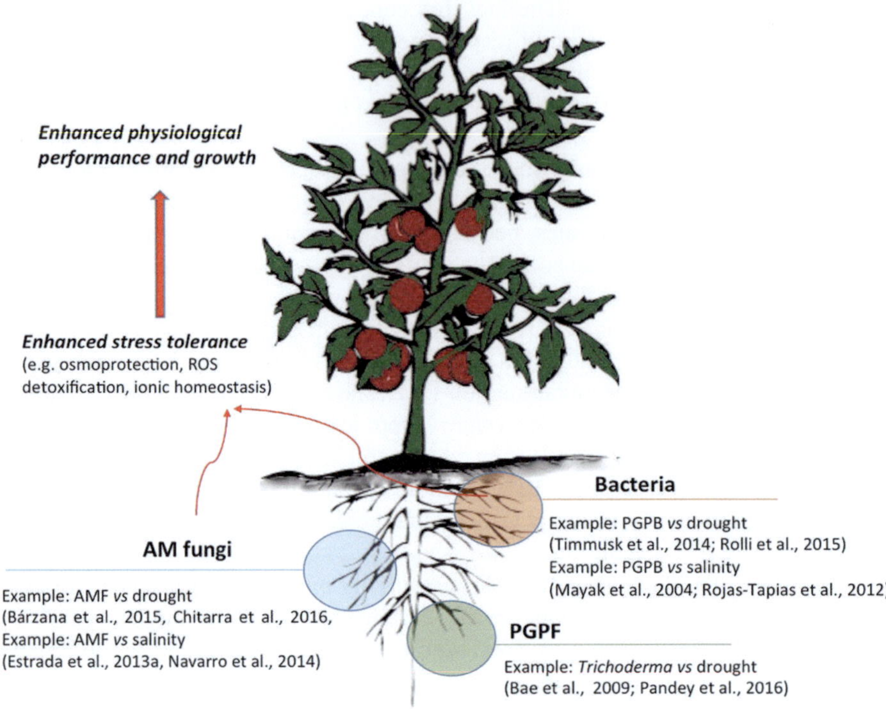

Fig. 12.1 Alleviation of abiotic stress effects by soil microorganisms. *PGPF* plant growth-promoting fungi

Understanding the physiological and molecular mechanisms involved in plant responses to multiple simultaneous stresses is therefore crucial for the development of broad-spectrum strategies applicable for the improvement of stress tolerance in crops.

The use of a high throughput phenotyping platforms (Singh et al. 2016) to quantify crop performance and tolerance traits of different genotypes may be very promising to verify the impact of these interactions under stress conditions.

Studying the mechanism at the base of such interactions will be fundamental to verify the role of soil microbial communities in the modulation of plant traits relevant to enhanced crop performances under stress conditions. In addition, this will allow evaluation of the crop capacity/efficiency to interact with these microbes under stress conditions, and to obtain benefits from them, as a trait to be considered in breeding programs.

Acknowledgements The work has in part been funded by the AQUA project (Progetto Premiale, CNR).

References

Alfano, G., Ivey, M. L. L., Cakir, C., Bos, J. I. B., Miller, S. A., Madden, L. V., et al. (2007). Systemic modulation of gene expression in tomato by *Trichoderma hamatum* 382. *Phytopathology, 97*, 429–437.

Atkinson, N. J., & Urwin, P. E. (2012). The interaction of plant biotic and abiotic stresses: from genes to the field. *Journal of Experimental Botany, 63*, 3523–3543.

Augé, R. M. (2001). Water relations, drought and vesicular-arbuscular mycorrhizal symbiosis. *Mycorrhiza, 11*, 3–42.

Augé, R. M., Stodola, A. J. W., Tims, J. E., & Saxton, A. M. (2001). Moisture retention properties of a mycorrhizal soil. *Plant and Soil, 230*, 87–97.

Augé, R. M., Toler, H. D., & Saxton, A. M. (2015). Arbuscular mycorrhizal symbiosis alters stomatal conductance of host plants more under drought than under amply watered conditions: a meta-analysis. *Mycorrhiza, 25*, 13–24.

Bae, H., Sicher, R. C., Kim, M. S., Kim, S.-H., Strem, M. D., Melnick, R. L., & Bailey, B. A. (2009). The beneficial endophyte *Trichoderma hamatum* isolate DIS 219b promotes growth and delays the onset of the drought response in *Theobroma cacao*. *Journal of Experimental Botany, 60*, 3279–3295.

Balestrini, R., Lumini, E., Borriello, R., Bianciotto, V. (2015) Plant-soil biota interactions. In Chapter 11: *Soil microbiology, ecology and biochemistry* (4th ed.). Elsevier: Academic.

Baltruschat, H., Fodor, J., Harrach, B. D., Niemczy, K. E., Barna, B., et al. (2008). Salt tolerance of barley induced by the root endophyte *Piriformospora indica* is associated with a strong increase in antioxidants. *New Phytologist, 180*, 501–510.

Bárzana, G., Aroca, R., Paz, J. A., Chaumont, F., Martinez-Ballesta, M. C., et al. (2012). Arbuscular mycorrhizal symbiosis increases relative apoplastic water flow in roots of the host plant under both well-watered and drought stress conditions. *Annals of Botany, 109*, 1009–1017.

Bárzana, G., Aroca, R., Bienert, P., Chaumont, F., & Ruiz-Lozano, J. M. (2014). New insights into the regulation of aquaporins by the arbuscular mycorrhizal symbiosis in maize plants under drought stress and possible implications for plant performance. *Molecular Plant-Microbe Interactions, 27*, 349–363.

Bárzana, G., Aroca, R., & Ruiz-Lozano, J. M. (2015). Localized and non-localized effects of arbuscular mycorrhizal symbiosis on accumulation of osmolytes and aquaporins and on antioxidant systems in maize plants subjected to total or partial root drying. *Plant, Cell & Environment, 38*, 1613–1627.

Bischof, R. H., Ramoni, J., & Seiboth, B. (2016). Cellulases and beyond: the first 70 years of the enzyme producer *Trichoderma reesei*. *Microbial Cell Factories, 15*, 106.

Boyer, J. S., Byrne, P., Cassman, K. G., Cooper, M., Delmer, D., Greene, T., et al. (2013). The U.S. drought of 2012 in perspective: A call to action. *Global Food Security, 2*, 139–143.

Brotman, Y., Landau, U., Cuadros-Inostroza, A., Takayuki, T., Fernie, A. R., et al. (2013). *Trichoderma*-plant root colonization: Escaping early plant defense responses and activation of the antioxidant machinery for saline stress tolerance. *PLoS Pathogens, 9*, e1003221.

Bucher, M., Hause, B., Krajinski, F., & Küster, H. (2014). Through the doors of perception to function in arbuscular mycorrhizal symbioses. *New Phytologist, 204*, 833–840.

Bunn, R., Lekberg, I., & Zabinski, C. (2009). Arbuscular mycorrhizal fungi ameliorate temperature stress in thermophilic plants. *Ecology, 90*, 1378–1388.

Calvo-Polanco, M., Molina, S., Zamarreno, A. M., Garcıa-Mina, J. M., & Aroca, R. (2014). The symbiosis with the arbuscular mycorrhizal fungus *Rhizophagus irregularis* drives root water transport in flooded tomato plants. *Plant and Cell Physiology, 55*, 1017–1029.

Campbell, B. (2012). The global imperative. Drought affects us all. In Perspectives: Legislating change. What should governments do to enhance sustainable agriculture and mitigate droughts? *Nature, 501*, Outlook Agriculture and Drought, s12–s14.

Caporale, A. G., Sommella, A., Lorito, M., Lombardi, N., Azam, S. M. G. G., Pigna, M., & Ruocco, M. (2014). *Trichoderma* spp. alleviate phytotoxicity in lettuce plants (*Lactuca sativa* L.) irrigated with arsenic-contaminated water. *Journal of Plant Physiology, 171*, 1378–1384.
Chandrasekaran, M., Kim, K., Krishnamoorthy, R., Walitang, D., Sundaram, S., et al. (2016). Mycorrhizal symbiotic efficiency on C3 and C4 plants under salinity stress – A meta-analysis. *Frontiers in Microbiology, 7*, 1246.
Chatzidaki, E., & Ventura, F. (2010). Adaptation to climate change and mitigation strategies in cultivated and natural environments. *A review. Italian Journal of Agrometeorology, 3*, 21–42.
Chen, S., Jin, W., Liu, A., Zhang, S., Liuc, D., et al. (2013). Arbuscular mycorrhizal fungi (AMF) increase growth and secondary metabolism in cucumber subjected to low temperature stress. *Scientia Horticulturae, 160*, 222–229.
Chitarra, W., Balestrini, R., Vitali, M., Pagliarani, C., Perrone, I., et al. (2014). Gene expression in vessel-associated cells upon xylem embolism repair in *Vitis vinifera* L. petioles. *Planta, 239*, 887–899.
Chitarra, W., Pagliarani, C., Maserti, B., Lumini, E., Siciliano, I., et al. (2016). Insights on the impact of arbuscular mycorrhizal symbiosis on tomato tolerance to water stress. *Plant Physiology, 171*, 1009–1023.
Cho, S. M., Kang, B. R., Han, S. H., Anderson, A. J., Park, J. Y., & Lee, Y. H. (2008). 2R,3R-Butanediol, a bacterial volatile produced by *Pseudomonas chlororaphis* O6, is involved in induction od systemic tolerance to drought in *Arabidopsis thaliana. Molecular Plant-Microbe Interactions, 21*, 1067–1075.
Coleman-Derr, D., & Tringe, S. G. (2014). Building the crops of tomorrow: advantages of symbiont-based approaches to improving abiotic stress tolerance. *Frontiers in Microbiology, 5*, 283.
Cornejo, P., Pérez-Tienda, J., Meier, S., Valderas, A., Borie, F., Azcón-Aguilar, C., et al. (2013). Copper compartmentalization in spores as a survival strategy of arbuscular mycorrhizal fungi in Cu-polluted environments. *Soil Biology and Biochemestry, 57*, 925–928.
Corradi, N., & Bonfante, P. (2012). The arbuscular mycorrhizal symbiosis: Origin and evolution of a beneficial plant infection. *PLoS Pathogens, 8*, e1002600.
Daryanto, S., Wang, L., Jacinthe, P.A. (2016). Global synthesis of drought effects on cereal, legume, tuber and root crops production: A review. *Agricultural Water Management* (in press). doi:10.1016/j.agwat.2016.04.022.
De Palma, M., D'Agostino, N., Proietti, S., Bertini, L., Lorito, M., et al. (2016). Suppression subtractive hybridization analysis provides new insights into the tomato (*Solanum lycopersicum* L.) response to the plant probiotic microorganism *Trichoderma longibrachiatum* MK1. *Journal of Plant Physiology, 190*, 79–94.
Dicke, M. (2016). Plant phenotypic plasticity in the phytobiome: a volatile issue. *Current Opinion in Plant Biology, 32*, 17–23.
Dietz, S., von Bülow, J., Beitz, E., & Nehls, U. (2011). The aquaporin gene family of the ectomycorrhizal fungus *Laccaria bicolor*: Lessons for symbiotic functions. *New Phytologist, 190*, 927–940.
Dugas, D. V., Monaco, M. K., Olsen, A., Klein, R. R., Kumari, S., Ware, D., et al. (2011). Functional annotation of the transcriptome of *Sorghum bicolor* in response toosmotic stress and abscisic acid. *BMC Genomics, 12*, 514.
Elhindi, K. M., El-Din, A. S., & Elgorban, A. M. (2016). The impact of arbuscular mycorrhizal fungiin mitigating salt-induced adverse effects in sweetbasil (*Ocimum basilicum* L.). *Saudi Journal of Biological Sciences* (in press). doi:10.1016/j.sjbs.2016.02.010.
Estrada, B., Aroca, R., Maathuis, F. J. M., Barea, J. M., & Ruiz-Lozano, J. M. (2013a). Arbuscular mycorrhizal fungi native from a Mediterranean saline area enhance maize tolerance to salinity through improved ion homeostasis. *Plant Cell & Environment, 36*, 1771–1782.
Estrada, B., Aroca, R., Barea, J. M., & Ruiz-Lozano, J. M. (2013b). Native arbuscular mycorrhizal fungi isolated from a saline habitat improved maize antioxidant systems and plant tolerance to salinity. *Plant Science, 201–202*, 42–51.

Fleury, D., Jefferies, S., Kuchel, H., & Langridge, P. (2010). Genetic and genomic tools to improve drought tolerance in wheat. *Journal of Experimental Botany, 61*, 3211–3222.

Flexas, J., Bota, J., Loreto, F., Cornic, G., & Sharkey, T. D. (2004). Diffusive and metabolic limitations to photosynthesis under drought and salinity in C3 plants. *Plant Biology, 6*, 269–279.

Foley, J. A., Ramankutty, N., Brauman, K. A., Cassidy, E. S., Gerber, J. S., et al. (2011). Solutions for a cultivated planet. *Nature, 478*, 337–342.

Galmés, J., Conesa, M. À., Ochogavía, J. M., Perdomo, J. A., Francis, D., et al. (2011). Physiological and morphological adaptations in relation to water use efficiency in Mediterranean accessions of *Solanum lycopersicum*. *Plant, Cell & Environment, 34*, 245–260.

Galmés, J., Perdomo, J. A., Flexas, J., & Whitney, S. M. (2013). Photosynthetic characterization of Rubisco transplantomic lines reveals alterations on photochemistry and mesophyll conductance. *Photosynthesis Research, 115*, 153–166.

Gilbert, M. E., & Medina, V. (2016). Drought adaptation mechanisms should guide experimental design. *Trends in Plant Science, 21*, 639–647.

González-Guzmán, M., Rodríguez, L., Lorenzo-Orts, L., Pons, C., Sarrión-Perdigones, A., et al. (2014). Tomato PYR/PYL/RCAR abscisic acid receptors show high expression in root, differential sensitivity to the abscisic acid agonist quinabactin, and the capability to enhance plant drought resistance. *Journal of Experimental Botany, 65*, 4451–4464.

Hazzoumi, Z., Moustakime, Y., El hassan, E., & Joutei, K. A. (2015). Effect of arbuscular mycorrhizal fungi (AMF) and water stress on growth, phenolic compounds, glandular hairs, and yield of essential oil in basil (*Ocimum gratissimum* L). *Chemical and Biological Technologies in Agriculture, 2*, 10.

Hirayama, T., & Shinozaki, K. (2010). Research on plant abiotic stress responses in the post-genome era: Past, present and future. *The Plant Journal, 61*, 1041–1052.

Iovieno, P., Punzo, P., Guida, G., Mistretta, C., Van Oosten, M. J., et al. (2016). Transcriptomic changes drive physiological responses to progressive drought stress and rehydration in tomato. *Frontiers in Plant Science, 7*, 371.

Juniper, S., & Abbott, L. K. (2006). Soil salinity delays germination and limits growth of hyphae from propagules of arbuscular mycorrhizal fungi. *Mycorrhiza, 16*, 371–379.

Kamara, A. Y., Menkir, A., Badu-Apraku, B., & Ibikunie, O. (2003). The influence of drought stress on growth, yield and yield components of selected maize genotypes. *Journal of Agricultural Science, 141*, 43–50.

Kakumanu, A., Ambavaram, M. M., Klumas, C., Krishnan, A., Batlang, U., et al. (2012). Effects of drought on gene expression in maize reproductive and leaf meristem tissue revealed by RNA-Seq. *Plant Physiology, 160*, 846–867.

Kaushal, M., & Wani, S. P. (2016). Plant-growth-promoting rhizobacteria: drought stress alleviators to ameliorate crop production in drylands. *Annals of Microbiology, 66*, 35–42.

Khalvati, M. A., Mozafar, A., & Schmidhalter, U. (2005). Quantification of water uptake by arbuscular mycorrhizal hyphae and its significance for leaf growth, water relations, and gas exchange of barley subjected to drought stress. *Plant Biology, 7*, 706–712.

Kohler, A., Kuo, A., Nagy, L. G., Morin, E., Barry, K. W., et al. (2015). Convergent losses of decay mechanisms and rapid turnover of symbiosis genes in mycorrhizal mutualists. *Nature Genetics, 47*, 410–415.

Krasensky, J., & Jonak, C. (2016). Drought, salt, and temperature stress-induced metabolic rearrangements and regulatory networks. *Journal of Experimental Botany, 63*, 1593–1608.

Lareen, A., Burton, F., & Schäfer, P. (2016). Plant root-microbe communication in shaping root microbiomes. *Plant Molecular Biology, 90*, 575–587.

Lafitte, H. R., Yongsheng, G., Yan, S., & Li, Z. H. (2007). Whole plant responses, key processes, and adaptation to drought stress: The case of rice. *Journal of Experimental Botany, 58*, 169–175.

Lenoir, I., Fontaine, J., & Sahraoui, A. L. H. (2016). Arbuscular mycorrhizal fungal responses to abiotic stresses: A review. *Phytochemistry, 123*, 4–15.

Lesk, C., Rowhani, P., & Ramankutty, N. (2016). Influence of extreme weather disasters on global crop production. *Nature, 529*, 84–87.

Li, T., Hu, Y. J., Hao, Z. P., Li, H., & Chen, B. D. (2013a). Aquaporin genes *GintAQPF1* and *GintAQPF2* from *Glomus intraradices* contribute to plant drought tolerance. *Plant Signal & Behavior, 8*, e24030.

Li, T., Hu, Y. J., Hao, Z. P., Li, H., Wang, Y. S., & Chen, B. D. (2013b). First cloning and characterization of two functional aquaporin genes from an arbuscular mycorrhizal fungus *Glomus intraradices*. *New Phytologist, 197*, 617–630.

Martin, F., Aerts, A., Ahren, D., Brun, A., Danchin, E. G. J., Duchaussoy, F., et al. (2008). Symbiosis insights from the genome of the mycorrhizal basidiomycete *Laccaria bicolor*. *Nature, 452*, 88–92.

Martin, F., Kohler, A., Murat, C., Balestrini, R., Coutinho, P. M., Jaillon, O., et al. (2010). Périgord black truffle genome uncovers evolutionary origins and mechanisms of symbiosis. *Nature, 464*, 10331038.

Mastouri, F., Bjorkman, T., & Harman, G. E. (2010). Seed treatment with *Trichoderma harzianum* alleviates biotic, abiotic, and physiological stresses in germinating seeds and seedlings. *Phytopathology, 100*, 1213–1221.

Mastouri, F., Bjorkman, T., & Harman, G. E. (2012). *Trichoderma harzianum* enhances antioxidant defense of tomato seedlings and resistance to water deficit. *Molecular Plant-Microbe Interactions, 25*, 1264–1271.

Maurel, C., Boursiac, Y., Luu, D. T., Santoni, V., Shahzad, Z., & Verdoucq, L. (2015). Aquaporins in plants. *Physiology Review, 95*, 1321–1358.

Mayak, S., Tirosha, T., & Glick, B. R. (2004). Plant growth-promoting bacteria confer resistance in tomato plants to salt stress. *Plant Physiology and Biochemistry, 42*, 565–572.

McAdam, S. A. M., Brodribb, T. J., & Ross, J. J. (2016). Shoot-derived abscisic acid promotes root growth. *Plant Cell and Environment, 39*, 652–659.

Meier, S., Cornejo, P., Cartes, P., Borie, F., Medina, J., & Azcón, R. (2015). Interactive effect between Cu-adapted arbuscular mycorrhizal fungi and biotreated agrowaste residue to improve the nutritional status of *Oenothera picensis* growing in Cu-polluted soils. *Journal of Plant Nutrition and Soil Science, 178*, 126–135.

Millar, N. S., & Bennet, A. E. (2016). Stressed out symbiotes: hypotheses for the influence of abiotic stress on arbuscular mycorrhizal fungi. *Oecologia, 182*, 625–641.

Mo, Y., Wang, Y., Yang, R., Zheng, J., Liu, C., Li, H., Ma, J., Zhang, Y., Wei, C., & Zhang, X. (2016). Regulation of plant growth, photosynthesis, antioxidation and osmosis by an arbuscular mycorrhizal fungus in watermelon seedlings under well-watered and drought conditions. *Frontiers in Plant Science, 7*, 644.

Morán-Diez, E., Rubio, B., Domïnguez, S., Hermosa, R., Monte, E., & Nicolás, C. (2012). Transcriptomic response of *Arabidopsis thaliana* after 24h incubation with the biocontrol fungus *Trichoderma harzianum*. *Journal of Plant Physiology, 169*, 614–620.

Navarro, J. M., Pérez-Tornero, O., & Morte, A. (2014). Alleviation of salt stress in citrus seedlings inoculated with arbuscular mycorrhizal fungi depends on the rootstock salt tolerance. *Journal of Plant Physiology, 171*, 76–85.

Navarro-Ródenas, A., Ruíz-Lozano, J. M., Kaldenhoff, R., & Morte, A. (2012). The aquaporin TcAQP1 of the desert truffle Terfezia claveryi is a membrane pore for water and CO(2) transport. *Molecular Plant-Microbe Interaction, 25*, 259–266.

Ortiz, N., Armada, E., Duque, E., Roldán, A., & Azcón, R. (2015). Contribution of arbuscular mycorrhizal fungi and/or bacteria to enhancing plant drought tolerance under natural soil conditions: Effectiveness of autochthonous or allochthonous strains. *Journal of Plant Physiology, 174*, 87–96.

Osakabe, Y., Osakabe, K., Shinozaki, K., & Tran, L. S. P. (2014). Response of plants to water stress. *Frontiers in Plant Science, 5*, Article 86.

Osman, M. E. H., Kasim, W. A., Omar, M. N., Abd El-Daim, I. A., Bejai, S., & Meijer, J. (2013). Impact of bacterial priming on some stress tolerance mechanisms and growth of cold stressed wheat seedlings. *International Journal of Plant Biology, 4*, e8.

Pandey, P., Ramegowda, V., & Senthil-Kumar, M. (2015). Shared and unique responses of plants to multiple individual stresses and stress combinations: physiological and molecular mechanisms. *Frontiers in Plant Science, 6*, 723.
Pandey, V., Ansari, M. W., Tula, S., Yadav, S., Sahoo, R. K., et al. (2016). Dose-dependent response of *Trichoderma harzianum* in improving drought tolerance in rice genotypes. *Planta, 243*, 1251–1264.
Pedranzani, H., Rodríguez-Rivera, M., Gutiérrez, M., Porcel, R., Hause, B., & Ruiz-Lozano, J. M. (2016). Arbucular mycorrhizal symbiosis regulates physiology and performance of *Digitaria eriantha* plants subjected to abiotic stresses by modulating antioxidant and jasmonate levels. *Mycorrhiza, 26*, 141–152.
Perrone, I., Gambino, G., Chitarra, W., Vitali, M., Pagliarani, C., Riccomagno, N., et al. (2012). The grapevine root-specific aquaporin VvPIP2;4 n controls root hydraulic conductance and leaf gexchange upon irrigation but not under water stress. *Plant Physiology, 160*, 965–977.
Peter, M., Kohler, A., Ohm, R. A., Kuo, A., Krützmann, J., Morin, E., et al. (2016). Ectomycorrhizal ecology is imprinted in the genome of the dominant symbiotic fungus *Cenococcum geophilum. Nature Communication*. doi:10.1038/ncomms12662.
Poosapati, S., Ravulapalli, P. D., Tippirishetty, N., Vishwanathaswamy, D. K., & Chunduri, S. (2014). Selection of high temperature and salinity tolerant *Trichoderma* isolates with antagonistic activity against *Sclerotium rolfsii. SpringerPlus, 3*, 641.
Porcel, R., Azcón, R., & Ruiz-Lozano, J. M. (2004). Evaluation of the role of genes encoding for Δ1-pyrroline-5-carboxylate synthetase (P5CS) during drought stress in arbuscular mycorrhizal *Glycine max* and *Lactuca sativa* plants. *Physiology and Molecular Plant Pathology, 65*, 211–221.
Porcel, R., Aroca, R., Azcón, R., & Ruiz-Lozano, J. M. (2006). PIP aquaporin gene expression in arbuscular mycorrhizal *Glycine max* and *Lactuca sativa* plants in relation to drought stress tolerance. *Plant Molecular Biology, 60*, 389–404.
Porcel, R., Aroca, R., & Ruíz-Lozano, J. M. (2012). Salinity stress alleviation using arbuscular mycorrhizal fungi. *A review. Agronomy for Sustainable Development, 32*, 181–200.
Porras-Soriano, A., Soriano-Martín, M. L., Porras-Piedra, A., & Azcón, R. (2009). Arbuscular mycorrhizal fungi increased growth, nutrient uptake and tolerance to salinity in olive trees under nursery conditions. *Journal of Plant Physiology, 166*, 1350–1359.
Qi, W., & Zhao, L. (2013). Study of the siderophore-producing *Trichoderma asperellum* Q1 on cucumber growth promotion under salt stress. *Journal of Basic Microbiology, 53*, 355–364.
Rampino, P., Pataleo, S., Gerardi, C., & Perotta, C. (2006). Drought stress responses in wheat: physiological and molecular analysis of resistant and sensitive genotypes. *Plant, Cell & Environment, 29*, 2143–2152.
Rolli, E., Marasco, R., Vigani, G., Ettoumi, B., Mapelli, F., et al. (2015). Improved plant resistance to drought is promoted by the root-associated microbiome as a water stress-dependent trait. *Environmental Microbiology, 17*, 316–331.
Rojas-Tapias, D., Moreno-Galván, A., Pardo-Díaz, S., Obando, M., Rivera, D., & Bonilla, R. (2012). Effect of inoculation with plant growth-promoting bacteria (PGPB) on amelioration of saline stress in maize (Zea mays). *Applied Soil Ecology, 61*, 264–272.
Rubio, M. B., Quijada, N. M., Pérez, E., Domínguez, S., Monte, E., & Hermosa, R. (2014). Identifying beneficial qualities of *Trichoderma parareesei* for plants. *Applied and Environmental Microbiology, 80*, 1864–1873.
Ruiz-Lozano, J. M., & Azcón, R. (1996). Mycorrhizal colonization and drought stress as factors affecting nitrate reductase activity in lettuce plants. *Agriculture, Ecosystems & Environment, 60*, 175–181.
Ruiz-Lozano, J. M., Collados, C., Barea, J. M., & Azcón, R. (2001). Clonig of cDNAs encoding SODs from lettuce plants which show differential regulation by arbuscular mycorhizal symbiosis and by drought stress. *Journal of Experimental Botany, 52*, 2241–2242.
Ruiz-Lozano, J. M. (2003). Arbuscular mycorrhizal symbiosis and alleviation of osmotic stress. New perspectives for molecular studies. *Mycorrhiza, 13*, 309–317.

Ruiz-Lozano, J. M., & Aroca, R. (2010). Host response to osmotic stresses: Stomatal behavior and water use efficiency or arbuscular mycorrhizal plants. In *Arbuscular mycorrhiza: physiology and function* (pp. 239–256). Dordrecht: Springer.

Ruiz-Lozano, J. M., Aroca, R., Zamarreño, A. M., Molina, S., Andreo-Jiménez, B., et al. (2016). Arbuscular mycorrhizal symbiosis induces strigolactone biosynthesis under drought and improves drought tolerance in lettuce and tomato. *Plant, Cell & Environment, 39*, 441–452.

Salazar, C., Hernández, C., & Pino, M. T. (2015). Plant water stress: Associations between ethylene and abscisic acid response. *Chilean Journal of Agricultural Research, 75*(Suppl. 1).

Sánchez-Romera, B., Ruiz-Lozano, J. M., Zamarreño, Á. M., García-Mina, J. M., & Aroca, R. (2016). Arbuscular mycorrhizal symbiosis and methyl jasmonate avoid the inhibition of root hydraulic conductivity caused by drought. *Mycorrhiza, 26*, 111–122.

Sandhya, V., Ali, S. K. Z., Grover, M., Reddy, G., & Venkateswarlu, B. (2009). Alleviation of drought stress effects in sunflower seedlings by the exopolysaccharides producing *Pseudomonas putida* strain GAP-P45. *Biology & Fertility of Soils, 46*, 17–26.

Savvides, A., Ali, S., Tester, M., & Fotopoulos, V. (2016). Chemical priming against multiple abiotic stresses: Mission possible? *Trends in Plant Science, 21*, 329–340.

Schlaeppi, K., & Bulgarelli, D. (2014). The plant microbiome at work. *Molecular Plant-Microbe Interaction, 28*, 212–217.

Sheffield, J., et al. (2012). Little change in global drought over the past 60 years. *Nature, 491*, 435–438.

Singh, A., Ganapathysubramanian, B., Singh, A. K., & Sarkar, S. (2016). Machine learning for high-throughput stress phenotyping in plants. *Trends in Plant Science, 21*, 110–124.

Secchi, F., Pagliarani, C., Zwieniecki, M.A. (2016). The funtional role of xylem parenchyma cells and aquaporins during recovery from severe water stress. *Plant, Cell & Environment* (in press). doi:10.1111/pce.12831.

Secchi, F., Perrone, I., Chitarra, W., Zwieniecka, A. K., Lovisolo, C., & Zwieniecki, M. A. (2013). The dynamics of embolism refilling in abscisic acid (ABA)deficient tomato plants. *International Journal of Molecular Sciences, 14*, 359–377.

Sewelam, N., Oshima, Y., Mitsuda, N., & Ohme-Takagi, M. (2014). A step towards understanding plant responses to multiple environmental stresses: a genome-wide study. *Plant, Cell & Environment, 37*, 2024–1035.

Sheffield, J., Wood, E. F., & Roderick, M. L. (2012a). Little change in global drought over the past 60 years. *Nature, 491*, 435–443.

Soda, N., Wallace, S., & Karan, R. (2015). Omics study for abiotic stress responses in plants. *Advances in Plants & Agriculture Research, 2*, 00037.

Srivastava, P. K., Vaish, A., Dwivedi, S., Chakrabarty, D., Singh, N., & Tripathi, R. D. (2011). Biological removal of arsenic pollution by soil fungi. *Science of the Total Environment, 409*, 2430–2442.

Staudinger, C., Mehmeti-Tershani, V., Gil-Quintana, E., Gonzalez, E. M., Hofhansl, F., Bachmann, G., et al. (2016). Evidence for a rhizobia-induced drought stress response strategy in *Medicago truncatula*. *Journal of Proteomics, 136*, 202–213.

Tamayo, E., Gómez-Gallego, T., Azcón-Aguilar, C., & Ferrol, N. (2014). Genome-wide analysis of copper, iron and zinc transporters in the arbuscular mycorrhizal fungus *Rhizophagus irregularis*. *Frontiers in Plant Science, 5*, 547.

Timmusk, S., & Wagner, E. G. (1999). The plant-growth-promoting rhizobacterium Paenibacillus polymyxa induces changes in *Arabidopsis thaliana* gene expression: a possible connection between biotic and abiotic stress responses. *Molecular Plant Microbe Interaction, 12*, 951–959.

Timmusk, S., Timmusk, K., & Behers, L. (2013). Rhizobacterial plant drought stress tolerance enhancement: Towards sustainable water resource management and food security. *Journal of Food Security, 1*, 6–9.

Timmusk, S., Abd El-Daim, I. A., Copolovici, L., Tanilas, T., Kännaste, A., Behers, L., et al. (2014). Drought-tolerance of wheat improved by rhizosphere bacteria from harsh environments: Enhanced biomass production and reduced emissions of stress volatiles. *PLoS ONE, 9*, e96086.

Tisserant, E., Malbreil, M., Kuo, A., Kohler, A., Symeonidi, A., Balestrini, R., et al. (2013). Genome of an arbuscular mycorrhizal fungus provides insight into the oldest plant symbiosis. *Proceedings of the National Academy of Sciences, USA, 110*, 20117–20122.

Tombesi, S., Nardini, A., Frioni, T., Soccolini, M., Zadra, C., et al. (2015). Stomatal closure is induced by hydraulic signals and maintained by ABA in drought-stressed grapevine. *Scientific Reports, 5*, 12449.

Tripathi, R. D., Srivastava, S., Mishra, S., Singh, N., Tuli, R., et al. (2007). Arsenic hazards: strategies for tolerance and remediation by plants. *Trends Biotechnology, 25*, 158–165.

Tripathi, P., Singh, P. C., Mishra, A., Chaudhry, V., Mishra, S., Tripathi, R. D., & Nautiyal, C. S. (2013). Trichoderma inoculation ameliorates arsenic induced phytotoxic changes in gene expression and stem anatomy of chickpea (*Cicer arietinum*). *Ecotoxicology and Environmental Safety, 89*, 8–14.

van der Heijden, M. G. A., Martin, F. M., Selosse, M. A., & Sanders, I. R. (2015). Mycorrhizal ecology and evolution: The past, the present, and the future. *New Phytologist, 205*, 1406–1423.

Wang, W. X., Vinocur, B., & Altman, A. (2003). Plant responses to drought, salinity and extreme temperatures: Towards genetic engineering for stress tolerance. *Planta, 218*, 1–14.

Yang, S., Zhang, Z., Xue, Y., et al. (2014). Arbuscular mycorrhizal fungi increase salt tolerance of apple seedlings. *Botanical Studies, 55*, 70.

Yildirim, E., Taylor, A.G., Spittler, T.D. (2006). Ameliorative effects of biological treatments on growth of squash plants under salt stress. *Science Horticulture* 111, 1-6.

Zeilenger, S., Gruber, S., Bansal, R., & Mukherjee, P. K. (2016). Secondary metabolism in *Trichoderma* – Chemistry meets genomics. *Fungal Biology Reviews, 30*, 74–90.

Zhang, W. P., Jiang, B., Li, W. G., Song, H., Yu, Y. S., & Chen, J. F. (2009). Polyamines enhance chilling tolerance of cucumber (*Cucumis sativus* L.) through modulating antioxidative system. *Science Horticulture, 122*, 200–208.

Zhao, L., Wang, F., Zhang, Y., & Zhang, J. (2014). Involvement of *Trichoderma asperellum* strain T6 in regulating iron acquisition in plants. *Journal of Basic Microbiology, 54*, S115–S124.

Zhu, X. C., Song, F. B., & Xu, H. W. (2010a). Effects of arbuscular mycorrhizal fungi on photosynthetic characteristics of maize under low temperature stress. *Acta Ecologica Sinica, 21*, 470–475.

Zhu, X. C., Song, F. B., & Xu, H. W. (2010b). Influence of arbuscular mycorrhiza on lipid peroxidation and antioxidant enzyme activity of maize plants under temperature stress. *Mycorrhiza, 20*, 325–332.

Chapter 13
Microbial Communities, Functional Genes, and Nitrogen Cycling Processes as Affected by Tree Species

Relena R. Ribbons, Morag A. Mcdonald, and Lars Vesterdal

Abstract Tree species influence soils through direct and indirect inputs above- and belowground through leaf litter and root inputs. Soil microbial communities can in turn influence tree growth and development through processes such decomposition and chemical transformation of nutrients in soils. In this chapter we will provide an overview of the mechanisms by which trees influence soil microbial communities and nitrogen cycling processes. Specifically, we explore the effects of tree species on ammonification and nitrification processes in forest floor soils, and relate those to functional genetic markers for ammonia-oxidation by archaea and bacteria (*amoA* AOA and AOB) bacterial denitrification (*nirS* and *nirK*). We will cover the use of complementary laboratory methods used to investigate these relationships, including the use of molecular techniques such as quantitative polymerase chain reaction (qPCR) to target gene abundances in soils, and ^{15}N tracing experiments to understand the production and consumption of nitrogen. We will also address some of the benefits and drawbacks of these approaches, with special focus on the types of research questions that can be answered using these approaches. The chapter will wrap up with an example study in a common garden tree species trial in Vancouver, B.C., Canada, which demonstrated tree species effects on soil microbial communities and nitrogen cycling dynamics.

R.R. Ribbons (✉)
Department of Geosciences and Natural Resource Management, University of Copenhagen, Copenhagen, Denmark

School of Environment, Natural Resources and Geography, Bangor University, Bangor, UK
e-mail: rerr@ign.ku.dk; rribbons@gmail.com

M.A. Mcdonald
School of Environment, Natural Resources and Geography, Bangor University, Bangor, UK

L. Vesterdal
Department of Geosciences and Natural Resource Management, University of Copenhagen, Copenhagen, Denmark

M. Lukac et al. (eds.), *Soil Biological Communities and Ecosystem Resilience*, Sustainability in Plant and Crop Protection, DOI 10.1007/978-3-319-63336-7_13

Keywords Tree species effects • Soil microbes • Nitrogen • Functional genes • Forests • qPCR

13.1 Introduction

Global changes, such as increased temperature and the spread of invasive species, are leading to large transitions in both the structure and function of ecosystems (Vitousek et al. 1997; Chapin et al. 2000; Ribbons 2014). A major question in ecology is: How will global change influence community structure and ecosystem function? Forests serve as important buffers for climate change through processes such as carbon sequestration, are significant sources of primary productivity, and provide a variety of major ecosystem services. There is uncertainty in how global change will influence forests, which could lead to declines in primary productivity, and reductions in ecosystem functions. How forests will respond to global change is important to study at both a community and ecosystem level, because changes in forest structure and composition have large influences on ecosystem functions. In this chapter we aim to address the following three questions: (1) How do microbial communities differ between tree species grown on the same forest sites? (2) How do gross rates of nitrogen cycling vary under different tree species grown on the same sites, and do these effects differ when trees are grown in nitrogen-rich and nitrogen-poor environments? (3) How do soil microbial communities and functional genes for nitrogen cycling differ between tree species?

Plant soil-feedbacks play an important role in regulating nutrient cycling and ecosystem processes (Bardgett 2011), with important implications for ecosystem responses to climate change and mitigation. Diversity within a community has been widely recognized as a factor that renders ecosystems resilient to perturbation (Loreau et al. 2001), but the mechanisms of how these processes work are less well understood, especially when scaling to soil and ecosystem nutrient cycling (Van der Putten et al. 2010). Understanding how forests are responding to shifts in species composition, and resulting shifts in ecosystem processes, is crucial to planning global climate change mitigation strategies.

A foundational concept within ecology is the role of species diversity and richness on community dynamics and ecosystem functions (Huston 1979; McCann 2000). More recently, the field of functional trait diversity has blossomed, such that large databases on functional traits and characteristics at the species-level are now commonplace (see TRY database, Kattge et al. 2011). Functional traits are often used to categorize plants based on the characteristics that are considered relevant to their response to the environment and/or its effects on ecosystem functioning (Diaz and Cabido 2001). Recently, research discussing plant functional traits has increased in the ecological literature (Diaz and Cabido 2001; Tilman et al. 1998), due to their applicability in management and climate change mitigation. For example, plants which naturally form symbiotic relationships with root-nodules are able to fix atmospheric nitrogen, and are thus categorized as having the functional trait of nitrogen-fixation. Other examples include physical features such as leaf size, canopy height, and ability to re-sprout. The diversity-stability theory in ecology states that increased diversity leads to increased ecosystem stability (Elton 1958; Tilman et al. 1998) in terms of maintaining ecosystem functions. However, using the litter of 32 species of contrasting litter quality, and categorized into four functional groups, Wardle (1997) found that increasing species richness had no effect on an ecosystem function such as the rate of decomposition.

Functional diversity within an ecosystem can also increase ecosystem stability by providing functional redundancy (e.g. several plants with identical functional traits). Functional diversity may play a more important role than species richness in maintaining ecosystem stability. For example, Isbell et al. (2011) found that high plant diversity maintained ecosystem function over time, and that functional redundancy is required within systems to adapt to a changing planet. Tree species have different functional traits and occupy different functional niches, suggesting that certain combinations of species can enhance specific ecosystem functions, and increase ecosystem resilience to changing climatic conditions. The capacity for different tree species to foster functionally distinct soil microbial communities requires further exploration, but a wealth of evidence and data support the role of plant-soil feedbacks in shaping community structure and ecosystem function.

13.1.1 Plant-Soil Feedbacks

Plant-soil feedbacks, such as biotic interactions, play a large role in how combinations of plants will function. Some species have well-documented negative plant-soil feedbacks, such as *Prunus serotina* (black cherry), which creates a soil microbial community that specifically inhibits the growth of its own seedlings in the direct vicinity of the parent tree (Reinhart et al. 2003, 2010), an example of the Janzen-Connell effect (Janzen 1970; Connell 1971). Other species have positive plant-soil feedbacks, which is especially important when the species in question is an invasive species (Levine et al. 2006), such as *Ailanthus altissima* (Tree of Heaven) which alters soil communities resulting in the suppression of competitor species (Felker-Quinn et al. 2011). By combining specific life history traits, functional traits, and known feedbacks we can better predict how combinations of tree species may persist and maintain high levels of forest productivity over time, thus yielding species of interest for influencing soil nutrient dynamics by fixing abiotic N or increasing soil N availability.

It has long been known that sites with different soils will support plant communities differing in species composition with species selected based on differences in environmental conditions (Binkley and Giardina 1998). Research suggests that there are tight couplings between species and the soils they are grown on, whereby a species may achieve different growth forms, biomass, and nutrient cycling when grown under a range of environmental conditions, such as in monoculture or polyculture (Bardgett and Wardle 2010; Harrison and Bardgett 2010; Orwin et al. 2010). There is growing evidence that plant species can alter the soil through their litter chemistry, and exudation or turnover of root and mycorrhizal symbionts (Bardgett and Wardle 2010), with recent experiments exploring additional soil properties (Prescott and Vesterdal 2013). Soil microbial communities (bacteria and fungi) play critical roles in cycling nutrients and decomposing litter. For example, mycorrhizal infection can have a positive effect on tree growth, as hyphae improve nutrient acquisition of tree roots by enhancing root surface area in exchange for carbohydrates. The next step is to consider how sites with different soils are coupled with plant-mediated alterations in these soil communities. How will this coupled system respond to changes in species composition (such as the removal or addition of species) due to climatic change or forest management practices?

Interactions between plants and soils are widely recognized as important for ecosystem functions. However, species-specific interactions and their effects on ecosystem functions such as nutrient cycling and decomposition remain poorly understood, and results are mixed. Plant diversity positively influenced soil carbon storage in grasslands (Steinbeiss et al. 2008; De Deyn et al. 2011), peatlands (Wardle et al. 2009), forests (De Deyn et al. 2008), and drylands around the globe (Maestre et al. 2012); while plant diversity did not influence carbon storage in New Zealand forests altered by mammal browsing (Wardle et al. 2001), had a modest positive effects in deeper soil layers of temperate forests of Poland (Dawud et al. 2016), and increased diversity had a negative effect on soil carbon storage in forests invaded by lianas (Phillips et al. 2002). Further research is needed to address how species diversity across trophic levels in conjunction with abiotic factors influence ecosystem functions (Midgley 2012). Plant functional composition (e.g. conifer/broadleaves) can play an important role in determining how ecosystems cycle nutrients, as has been demonstrated for carbon and nitrogen accumulation in grasslands (Fornara and Tilman 2008) and forests (Dawud et al. 2016).

Research on plant effects on soil properties has increased recently, due to an increased interest in determining the role of plants in regulating carbon to mitigate climatic change, for example by sequestering atmospheric carbon dioxide. Much of this research on specific examples of above- and belowground linkages in forest and grassland ecosystems has been thoroughly examined and synthesized in Bardgett and Wardle (2010). In addition, a plethora of studies have examined the role of plant species diversity on forest soil properties (Laakso and Setala 1999; Small and McCarthy 2005; Blaško et al. 2015). Understanding the linkages between above- and belowground components within natural systems is an important challenge for predicting climate change effects on ecosystems (Bardgett and Wardle 2010). Examining multiple taxa and functional groups in both the above- and belowground communities will provide insights into how species may respond to climate change and yield practical and applied information.

Exploring belowground responses of soil communities or tree root growth to inter- and intra-specific competition between trees planted in forests has implications for climate change mitigation and carbon sequestration. Soil communities have been found to develop tight relationships with the litter they decompose, across many biomes (Keiser et al. 2011; Makkonen et al. 2012). Soil communities often preferentially decompose the litter of plants that have been growing in that soil compared with foreign litter (from plants grown at other locations), suggesting a soil legacy which influences carbon and nitrogen dynamics (Carrillo et al. 2012), commonly referred to as the "home-field advantage" (Veen et al. 2014), although its generality is still debated. Alternatively, plants can be exerting top-down control over the soil community, effectively encouraging a community that is successful at decomposing litter of a certain species. Plant-soil feedbacks can play an important role in the cycling of nitrogen and carbon. In this review we aim to shed some light on the question of how will changes in the forest species composition and soil type influence ecosystem functions such as nitrogen cycling in the future.

13.2 Tree Species Common Garden Experiments

Trees tend to grow on different sites preferentially. For example, pines can generally tolerate sandy nutrient-poor soils and water-limited conditions, whereas ash or maple are more generally suited to richer soil substrates and less water-stressed conditions (Binkley and Fisher 2013). This seriously hampers separation of tree species and site-related factors on soil and ecosystem functioning (Vesterdal et al. 2012). However, experimental forests can provide an excellent research framework for examining tree species identity effects. Common garden experiments are especially pertinent for examining tree species identity or diversity effects while reducing the influence of confounding factors (such as topography, soil characteristics, hydrology) to a minimum.

An additional series of new common garden experiments have recently been established across Europe to address effects of species diversity (Scherer-Lorenzen et al. 2007), but studies of soil functioning will remain for the future when these experiments have matured. Instead, a targeted well-selected exploratory platform has recently been established and used to study effects of tree species diversity in mature forests (see Baeten et al. 2013). In this review we will focus on single-tree species effects on soil microbial communities and ecosystem functioning within a common garden experiment on Vancouver Island.

The University of British Columbia has conducted several studies in the species trial Experimental Project no. 571 (hereafter referred to as EP571) established by the B.C. Ministry of Forests, Research Branch. The trial is located on Vancouver Island in British Columbia, Canada, and includes four tree species: western redcedar (*Thuja plicata*), western hemlock (*Tsuga heterophylla*), Douglas-fir (*Pseudotsuga menziesii*), and Sitka spruce (*Picea sitchensis*) established in 1961 (Klinka et al. 1996; Prescott et al. 2000), at four sites with contrasting N status, two of which were used for the recent work by Ribbons et al. (2016). These two sites include Fairy Lake which has low N and high C:N ratio in the forest floors, and San Juan which has high N and low C:N ratio in the forest floors. Rates of net nitrogen mineralization in laboratory incubations have previously been assessed at the EP571 experimental forest (Prescott et al. 2000), in addition to PLFA analyses on microbial communities (Grayston and Prescott 2005). In contrast, gross quantification of nitrogen mineralization and nitrification rates was only recently explored in addition to the quantification of forest floor microbial communities and functional genes responsible for primary cycling of nitrogen within these forests (Ribbons et al. 2016). The following case study illustrates how multiple mechanism-based experiments can yield new insights into ecosystem function.

13.3 Tree Species Effects on Forest Floor Microbial Communities and Nitrogen Cycling – A Case from British Columbia

We used a common garden forest experiment to disentangle tree species effects from possible confounding factors, such as soil parent material, C:N ratios, and soil bulk density. This experimental framework was employed to explicitly test the role of tree species identity on soil processes like nitrogen cycling. Two deviating sites within EP571 was selected to explore the mechanistic links between tree species, soil microbial communities, and nitrogen dynamics in the forest floors. These two sites were selected because of their contrasting N status in forest floors, and thus deviating potential for N cycling process rates. We sampled forest floors from underneath each of the four tree species and examined them for physico-chemical properties including C:N ratios, microbial C:N ratios, pH, and microbial biomass C and N. A suite of process rates relating to nitrogen transformations were also assessed using an isotope pool-dilution method. This data was coupled with quantitative characterization of the fungal and bacterial community, broadly assessed with gene markers for fungal biomass abundance (*ITS*) and bacterial biomass abundance (*16S*). We also quantitatively examined the functional microbial community assessed with gene markers for two types of denitrifying bacteria: *nirK* and *nirS*, and ammonia-oxidising bacteria (*amoA* AOB) and archaea (*amoA* AOA).

Site-level differences in C:N ratios played a dominant role in controlling N cycling processes. At the high C:N ratio site, Fairy Lake, we found high C:N ratios in both the organic matter and microbial biomass led to the consumption and mineralization of available NH_4^+ by the plant and microbial communities (Fig. 13.1). At San Juan, the low C:N ratio of organic matter led to a higher proportion of N mineralization, only a portion of which was consumed by plant and microbial communities. The accumulation of NH_4^+ stimulated the soil microbial community, specifically the ammonia-oxidizing archaea and bacteria (*amoA* AOA and *amoA* AOB, respectively). This activity further liberates NO_3^- in the system, some of which is consumed by plants and microbes, and some consumed by denitrifiers (*nirK* and *nirS*).

We found larger tree species effects on NH_4^+ cycling processes compared with NO_3^- cycling processes. The greatest differences were between western red cedar

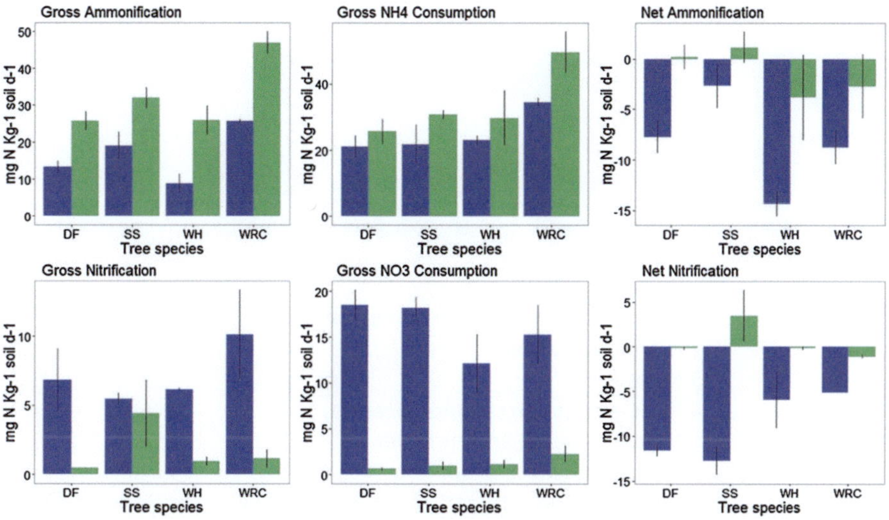

Fig. 13.1 Rates of nitrogen transformations in the forest floors of four tree species at the two sites; (mean ± SE). *Blue bars* Fairy Lake, *green bars* San Juan, *DF* Douglas-fir, *SS* Sitka spruce, *WH* western hemlock, *WRC* western red cedar (Ribbons et al. 2016, reproduced with permission from Soil Biology and Biochemistry)

and the three other tree species in terms of nitrogen transformations and functional genes tied to those microbial processes. We found a positive relationship between gross ammonification rates and the abundance of bacterial gene markers, which indicates the important role of bacteria in the ammonification process. There was a significant positive relationship between AOA *amoA* gene abundances and nitrification rates, which indicates a prominent role of archaeal ammonia-oxidizers over their bacterial counterparts. Taken together, this indicates tree species promote specific soil microbial communities in relation to functional differences in nitrogen cycling processes, as evidenced by both the gross N transformation experiments as well as the quantitative gene abundance data (Fig. 13.2).

Comparing these findings from EP571 to other ecosystems studies can provide valuable information about the functional role of contrasting or complementary tree species in long-term forest soil nutrient availability. Coupling methods for examining fine-temporal resolution of nutrient cycling rates with larger spatial-scale patterns in soil C:N dynamics would be useful for planning forestry and management activities across landscapes.

Specifically, refining our knowledge of the mechanisms by which plants alter soil microbial communities and exert top-down controls on nitrogen cycling has applications for ecosystem restoration and conservation efforts. By increasing our knowledge of the functional roles of individual species we are better able to predict the influence of species gains and losses (Wardle et al. 2011).

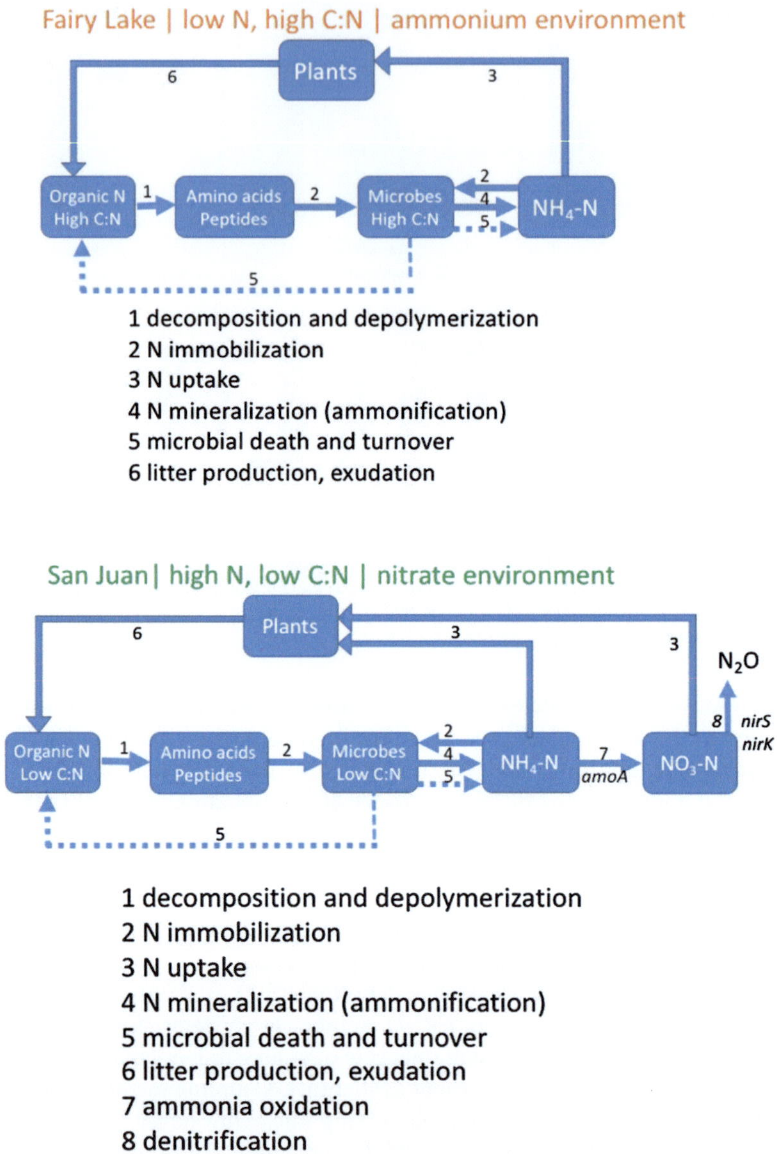

Fig. 13.2 Conceptual model of nitrogen cycling processes at the two study sites with differing N status: Fairy Lake and San Juan (Ribbons et al. 2016, reproduced with permission from Soil Biology and Biochemistry)

13.4 Experimental Considerations

Isotope-based methods, such as tracer experiments, as well as natural abundance, and pool-dilution experiments are examples of tools best applied to research questions that require very precise datasets due to the specialized training for implementing and high cost of analyzing those samples (see Murphy et al. 1999 and Booth et al. 2005 reviews for a further discussion on these experiments). With the EP571 project we designed our study to use ^{15}N pool-dilution experiments to acquire data on the real-time production and consumption of nitrogen in the form of either ammonium of nitrate. This allows us to determine gross process rates and associated information that is typically aggregated in traditional net N mineralization incubations. To collect meaningful data, sample replicates within the experiment, and technical replicates to ensure analytical quality are needed. Furthermore, alternatives such as natural isotopic abundance studies may be considered before enriching samples with isotopes. Would natural ^{15}N abundance be useful for answering the research question, or does it require additional fertilization of an isotope, such as 99% ^{15}N enrichment studies? With an ecosystem-scale N cycling experiment natural abundance would be useful for quantifying ecosystem N inputs and outputs. A ^{15}N enrichment approach would be useful for exploring gross rates of N mineralization under contrasting agricultural fertilization experiments.

The application of ^{15}N enrichment studies is wide-ranging (Murphy et al. 1999) including the fields of agriculture, environmental conservation, ecosystem ecology, and soil biology. The use of isotope-based methods enables us to track real-time production and consumption of isotope-labeled components thus allowing us to calculate gross rates of nitrogen mineralization and nitrification. Non-isotope techniques, such as a standard 28-day incubation of net mineralization and nitrification, can be informative for getting a general picture of how nitrogen is available within soils. In contrast, tracing which forms of nitrogen are being produced and consumed by microbial counterparts is possible using ^{15}N enrichment and gross nitrogen incubations (Booth et al. 2005).

Molecular techniques including DNA and RNA analyses can provide excellent data on the composition, structure, and function of microbial communities. In comparison there is a technique for phospholipid fatty acid analysis (PLFA) which is aimed at discerning general patterns in fatty acids and using those as biomarkers for estimating bacterial and fungal biomass within soil communities (Frostegård and Bååth 1996). While PLFAs have been used with success (Frostegård et al. 1993; Grayston and Prescott 2005), this method does not isolate specific gene-encoding enzymes, which would allow us to track changes in nitrifying bacterial communities, for example. Recent methods have now been refined such that PLFA is growing less common in comparison with DNA or RNA based analyses. Historically, microbial biomass chloroform fumigation extractions (CFE), or variations on chloroform incubations, have also been used to quantify microbial activity levels. CFE relies upon CO_2 production as a proxy for microbial consumption after exposure to chloroform (Brookes et al. 1985), but this is a coarse metric and does not allow us to partition components of the microbial community at a finer resolution.

Recent molecular approaches vary in complexity and associated lab-requirements, but also tend to be a more expensive option, especially if next generation sequencing such as illumina is to be performed, which can be used for whole genome or transcriptome studies. The use of genetic markers and DNA-based technologies to determine abundance of microbial communities has some distinct advantages over alternative methods, such as chloroform-fumigation extraction microbial biomass assays (Philippot et al. 2012). DNA-based methods such as PCR and qPCR enable us to determine not just the abundance of total microbial biomass, but can be used to separate broad functional groups (bacterial vs. fungal) or determine specific functional gene markers (e.g. nitrifying bacteria). Chloroform fumigation extraction, on the other hand, is a proxy for microbial biomass and not a direct measure, as it is calculated from CO_2 production after exposure to chloroform and combined with a study-system-specific constant value (often not adjusted and adds to the lack of reliability with this data). In this example, qPCR can be used to provide a more reliable dataset than chloroform fumigation extraction.

However, DNA-based methods only provide information about the genetic potential for the respective functioning of the microbes. For instance, DNA may persist within the soil long-after a microbe has perished or microbes may be inactive as a result of adverse environmental conditions in the field. More specific information is provided by enzymatic assays which indicate the production of enzymes required in a given sample, RNA-based methods which indicate the active replication of DNA in a given sample, or proteomics-based methods, which can provide real-time information about proteins within cells. Each of these methods provides higher resolution data on the level of activity of specific genes identified via qPCR (Gotelli et al. 2012; Ogunseitan 2006). There are drawbacks to these active methods though, including the challenges in accounting for temporal and spatial variability within these datasets, and the time and labor-intensive nature of the data collection required for sufficient sample sizes for statistics. We suggest careful consideration of pairing research methods required with primary research questions of interest and the required replication and power to be able to make meaningful conclusions from the data.

13.5 Conclusions

In this review we highlight the need for additional experiments to explore tree species identity and diversity effects on soils and plant-soil interactions more broadly taking advantage of emerging analytical methods. We have shown through the EP571 study that multiple complimentary laboratory techniques for assessing ecosystem function can be used to explore the mechanisms behind observed differences in nitrogen transformations between tree species and sites. Microbial techniques for the quantification of functional gene abundances can provide a greater mechanistic understanding of ecosystem process and function, but additional techniques are emerging and show promise to further this field of study.

Acknowledgements Authors acknowledge the University of British Columbia, Vancouver, the University of Copenhagen, and Bangor University for facilities and support during the preparation of this chapter. R. Ribbons acknowledges FONASO for the doctoral research fellowship.

References

Baeten, L., et al. (2013). A novel comparative research platform designed to determine the functional significance of tree species diversity in European forests. *Perspectives in Plant Ecology, Evolution and Systematics, 15*, 281–291.

Bardgett, R. D. (2011). Plant-soil interactions in a changing world. *F1000 Biology Reports, 3*, 16–16.

Bardgett, R. D., & Wardle, D. A. (2010). *Aboveground-Belowground Linkages: Biotic Interactions, Ecosystem Processes, and Global Change*. New York: Oxford University Press. (Oxford Series in Ecology and Evolution).

Binkley, D., & Fisher, R. (2013). *Ecology and management of forest soils*. Hoboken: Wiley.

Binkley, D., & Giardina, C. (1998). Why do tree species affect soils?: The Warp and Woof of tree-soil interactions. *Biogeochemistry, 42*, 89–106.

Blaško, R., Holm Bach, L., Yarwood, S. A., Trumbore, S. E., Högberg, P., & Högberg, M. N. (2015). Shifts in soil microbial community structure, nitrogen cycling and the concomitant declining N availability in ageing primary boreal forest ecosystems. *Soil Biology and Biochemistry, 91*, 200–211.

Booth, M., Stark, J., & Rastetter, E. (2005). Controls on nitrogen cycling in terrestrial ecosystes: A synthetic analysis of literature data. *Ecological Monographs, 72*, 139–157.

Brookes, P. C., et al. (1985). Chloroform fumigation and the release of soil nitrogen: A rapid direct extraction method to measure microbial biomass nitrogen in soil. *Soil Biology & Biochemistry., 17*, 837–842.

Carrillo, Y., Ball, B. A., Strickland, M. S., & Bradford, M. A. (2012). Legacies of plant litter on carbon and nitrogen dynamics and the role of the soil community. *Pedobiologia, 55*, 185–192.

Chapin, F. S., Zavaleta, E. S., Eviner, V. T., Naylor, R. L., Vitousek, P. M., et al. (2000). Consequences of changing biodiversity. *Nature, 405*, 234–242.

Connell, J. H. (1971). On the role of natural enemies in preventing competitive exclusion in some marine animals and in rain forest trees. In P. J. den Boer & G. R. Gradwell (Eds.), *Dynamics of Populations*. Wageningen: Centre for Agricultural Publishing and Documentation.

Dawud, S. M., Raulund-Rasmussen, K., Domisch, T., Finér, L., Jaroszewicz, B., & Vesterdal, L. (2016). Is tree species diversity or tree species identity the more important driver of soil C stocks, C/N ratio, and pH? *Ecosystems, 19*, 645–660.

De Deyn, G. B., Cornelissen, J. H. C., & Bardgett, R. D. (2008). Plant functional traits and soil carbon sequestration in contrasting biomes. *Ecology Letters, 11*, 516–531.

De Deyn, G. B., Quirk, H., Oakley, S., Ostle, N., & Bardgett, R. D. (2011). Rapid transfer of photosynthetic carbon through the plant-soil system in differently managed species-rich grasslands. *Biogeosciences, 8*, 1131–1139.

Diaz, S., & Cabido, M. (2001). Vive la difference: Plant functional diversity matters to ecosystem processes. *Trends in Ecology and Evolution, 16*, 646–655.

Elton, C. S. (1958). *The ecology of invasions by animals and plants*. London: Methuen.

Felker-Quinn, E., Bailey, J. K., & Schweitzer, J. A. (2011). Soil biota drive expression of genetic variation and development of population-specific feedbacks in an invasive plant. *Ecology, 92*, 1208–1214.

Fornara, D. A., & Tilman, D. (2008). Plant functional composition influences rates of soil carbon and nitrogen accumulation. *Journal of Ecology, 96*, 314–322.
Frostegård, Å., & Bååth, E. (1996). The use of phospholipid fatty acid analysis to estimate bacterial and fungal biomass in soil. *Biology and Feritility of Soil, 22*, 59–65.
Frostegård, Å., Bååth, E., & Tunlio, A. (1993). Shifts in the structure of soil microbial communities in limed forests as revealed by phospholipid fatty acid analysis. *Soil Biology and Biochemistry, 25*, 723–730.
Gotelli, N., Ellison, A. M., & Ballif, B. A. (2012). Environmental proteomics, biodiversity statistics, and food-web structure. *Trends in Ecology and Evolution, 27*, 436–442.
Grayston, S. J., & Prescott, C. E. (2005). Microbial communities in forest floors under four tree species in coastal British Columbia. *Soil Biology and Biochemistry, 37*(6), 1157–1167.
Harrison, K. A., & Bardgett, R. D. (2010). Influence of plant species and soil conditions on plant-soil feedback in mixed grassland communities. *Journal of Ecology, 98*, 384–395.
Huston, M. (1979). A general hypothesis of species diversity. *The American Naturalist, 113*, 81–101.
Isbell, F., Calcagno, V., Hector, A., Connolly, J., Harpole, W. S., et al. (2011). High plant diversity if needed to maintain ecosystem services. *Nature, 477*, 199–203.
Janzen, D. H. (1970). Herbivores and the number of tree species in tropical forests. *The American Naturalist, 104*, 501–528.
Kattge, J., Díaz, S., Lavorel, S., et al. (2011). TRY – A global database of plant traits. *Global Change Biology, 17*, 2905–2935.
Keiser, A. D., Strickland, M. S., Fierer, N., & Bradford, M. A. (2011). The effect of resource history on the functioning of soil microbial communities is maintained across time. *Biogeosciences, 8*, 1477–1486.
Klinka, K., Chen, H. Y. H., Wang, Q., & deMontigny, L. (1996). Forest canopies and their influence on understorey vegetation in early seral stands on West Vancouver Island. *Northwest Science, 70*, 193–200.
Laakso, J., & Setala, H. (1999). Sensitivity of primary production to changes in the architecture of belowground food webs. *Oikos, 87*, 57–64.
Levine, J. M., Pachepsky, E., Kendall, B. E., Yelenik, S. G., & Hille Ris Lambers, J. (2006). Plant-soil feedbacks and invasive spread. *Ecology Letters, 9*, 1005–1014.
Loreau, M., Naeem, S., Inchausti, P., Bengtsson, J., Grime, J. P., et al. (2001). Ecology – Biodiversity and ecosystem functioning: Current knowledge and future challenges. *Science, 294*, 804–808.
Maestre, F., et al. (2012). Plant species richness and ecosystem multifunctionality in global drylands. *Science, 335*, 214–217.
Makkonen, M., Berg, M. P., Handa, I. T., Haettenschwiler, S., van Ruijven, J., van Bodegom, P. M., & Aerts, R. (2012). Highly consistent effects of plant litter identity and functional traits on decomposition across a latitudinal gradient. *Ecology Letters, 15*, 1033–1041.
McCann, K. S. (2000). The diversity-stability debate. *Nature, 405*, 228–233.
Midgley, G. (2012). Biodiversity and ecosystem function. *Science, 335*, 174.
Murphy, D. V., Bhogal, A., Shepard, M., Goulding, K. W. T., Jarvis, S. C., Barraclough, D., & Gaunt, J. L. (1999). Comparison of 15 N labelling methods to measure gross nitrogen mineralisation. *Soil Biology and Biochemistry, 31*, 2015–2024.
Ogunseitan, O.A. (2006). Soil proteomics: Extraction and analysis of proteins from soils. In *Nucleic acids and proteins in soil*. London: Springer.
Orwin, K. H., Buckland, S. M., Johnson, D., Turner, B. L., Smart, S., Oakley, S., & Bardgett, R. D. (2010). Linkages of plant traits to soil properties and the functioning of temperate grassland. *Journal of Ecology, 98*, 1074–1083.
Philippot, L., Ritz, K., Pandard, P., Hallin, S., & Martin-Laurent, F. (2012). Standardisation of methods in soil microbiology: Progress and challenges. *FEMS Microbiology Ecology, 82*, 1–10.
Phillips, O. L., Martinez, R. V., Arroyo, L., Baker, T. R., Killeen, T., et al. (2002). Increasing dominance of large lianas in Amazonian forests. *Nature, 418*, 770–774.

Prescott, C. E., & Vesterdal, L. (2013). Tree species effects on soils in temperate and boreal forests: Emerging themes and research needs. *Forest Ecology and Management, 309*, 1–3.

Prescott, C. E., Vesterdal, L., Pratt, J., Venner, K. H., de Montigny, L. M., & Trofymow, J. A. (2000). Nutrient concentrations and nitrogen mineralization in forest floors of single species conifer plantations in coastal British Columbia. *Canadian Journal of Forest Research, 30*, 1341–1352.

Reinhart, K. O., Packer, A., Van der Putten, W. H., & Clay, K. (2003). Plant-soil biota interactions and spatial distribution of black cherry in its native and invasive ranges. *Ecology Letters, 6*, 1046–1050.

Reinhart, K. O., Tytgat, T., Van der Putten, W. H., & Clay, K. (2010). Virulence of soil-borne pathogens and invasion by *Prunus serotina*. *New Phytologist, 186*, 484–495.

Ribbons, R. R. (2014) Red spruce (*Picea rubens*) community dynamics at southern range margins: using dendrochronology to explore climate change. *Peer J.* http://dx.doi.org/10.7717/peerj.293

Ribbons, R. R., Levy-Booth, D. J., Masse, J., Grayston, S., McDonald, M., et al. (2016). Microbial communities, functional genes and nitrogen cycling processes in forest floors under four tree species. *Soil Biology and Biochemistry, 103*, 181–191.

Scherer-Lorenzen, M., Schulze, E. D., Don, A., Schumacher, J., & Weller, E. (2007). Exploring the functional significance of forest diversity: A new long-term experiment with temperate tree species (BIOTREE). *Perspectives in Plant Ecology, Evolution and Systematics, 9*, 53–70.

Small, C. J., & McCarthy, B. C. (2005). Relationship of understory diversity to soil nitrogen , topographic variation, and stand age in an eastern oak forest, USA. *Forest Ecology and Management, 217*, 229–243.

Steinbeiss, S., Bessler, H., Engels, C., Temperton, V. M., Buchmann, N., et al. (2008). Plant diversity positively affects short-term soil carbon storage in experimental grasslands. *Global Change Biology, 14*, 2937–2949.

Tilman, D., Lehman, C. L., & Bristow, C. E. (1998). Diversity-stability realtionships: Statistical inevitability or ecological consequence? *The American Naturalist, 151*, 277–282.

Van der Putten, W. H., Macel, M., & Visser, M. E. (2010). Predicting species distribution and abundance responses to climate change: Why it is essential to include biotic interactions across trophic levels. *Philosophical Transactions of the Royal Society B-Biological Sciences, 365*, 2025–2034.

Veen, C., Freschet, G. T., Ordonez, A., & Wardle, D. (2014). Litter quality and environmental controls of home-field advantage effects on litter decomposition. *Oikos, 124*, 187–195.

Vesterdal, L., Elberling, B., Christiansen, J. R., Callesen, I., & Schmidt, I. K. (2012). Soil respiration and rates of soil carbon turnover differ among six common European tree species. *Forest Ecology and Management, 264*, 185–196.

Vitousek, P. M., Dantonio, C. M., Loope, L. L., Rejmanek, M., & Westbrooks, R. (1997). Introduced species: A significant component of human-caused global change. *New Zealand Journal of Ecology, 21*, 1–16.

Wardle, D. A. (1997). Biodiversity and plant litter: Experimental evidence which does not support the view that enhanced species richness improves ecosystem function. *Oikos, 79*, 247–258.

Wardle, D. A. et al. (2001). Introduced browsing mammals in New Zealand natural forests: aboveground and belowground consequences. *Ecological monographs, 71*(4), 587–614.

Wardle, D. A., Bardgett, R. D., McNamara, N. P., & Ostle, N. J. (2009). Plant functional group identity influences short-term peatland ecosystem carbon flux: Evidence from a plant removal experiment. *Functional Ecology, 23*, 454–462.

Wardle, D. A., Bardgett, R. D., Callaway, R. M., & Van der Putten, W. H. (2011). Terrestrial ecosystem responses to species gains and losses. *Science, 332*, 1273–1277.

Chapter 14
Ectomycorrhizal Fungal Responses to Forest Liming and Wood Ash Addition: Review and Meta-analysis

Rasmus Kjøller, Carla Cruz-Paredes, and Karina E. Clemmensen

Abstract Large-scale liming and wood ash addition are common practices to mitigate soil and water acidification in temperate and boreal forests. In addition, wood ash recycles nutrients removed at harvest to the forest ecosystem. Both liming and wood ash applications typically increase soil pH by 1–2 units. Therefore, they affect a range of soil processes and organisms including ectomycorrhizal (EM) fungi which are vital for the nutrition of many tree species. Here we review field studies reporting the effects of lime and wood ash amendments on EM fungi. We systematically compiled studies where known amounts of ash or lime were distributed to plots paired with comparable control plots, and where mycorrhizal variables were recorded. For a subset of studies meeting explicit criteria, we performed meta-analyses using overall mycorrhizal abundance, species richness or the abundance of specific fungal taxa as response variables. Guided by availability of data, the focus is on Nordic coniferous forests. Although the reviewed field studies varied widely in dosage and experimental setup they clearly demonstrated that liming and wood ash amendments influence EM fungal species composition. Across studies, species belonging to the lineages Cortinarius and Russula-lactarius (Basidiomycota), particularly *Russula ochroleuca*, decreased in relative abundance, while species within the Tuber-helvella (Ascomycota) increased. Particular species within the Amphinema-tylospora lineage responded in opposite directions; *Tylospora fibrillosa* decreased in relation to the control, while *Amphinema byssoides* increased. The significant changes in species or clade abundances were in the range of 5–20% compared to non-treated plots. In contrast, neither the belowground mycorrhizal biomass nor species richness responded to liming or wood ash applications. We conclude that liming and wood ash amendments cause consistent EM fungal species

---

R. Kjøller (✉) • C. Cruz-Paredes
Department of Biology, University of Copenhagen,
Universitetsparken 15, 2100 Copenhagen Ø, Denmark
e-mail: rasmusk@bio.ku.dk

K.E. Clemmensen
Department of Forest Mycology and Plant Pathology, SLU,
Box 7026, SE-750 07 Uppsala, Sweden

M. Lukac et al. (eds.), *Soil Biological Communities and Ecosystem Resilience*,
Sustainability in Plant and Crop Protection, DOI 10.1007/978-3-319-63336-7_14

dominance shifts, but that a high EM fungal biomass and species and phylogenetic richness is maintained on the tree roots. Given the large dispersal potential of many EM fungi, we therefore suggest that these treatments at normal recommended dosages do not pose any immediate threats to EM fungal biodiversity, at least not when applied at relatively small spatial scales. Whether the observed dominance shifts among EM fungal clades have consequences for the functioning of the EM fungal guild, e.g. in relation to nutrient cycling or tree nutrition, is an important question that should be further investigated.

Keywords Ectomycorrhizal fungi • Liming • Wood ash • Biomass • Species richness • Fungal community composition

14.1 Introduction

Maturing forest soils naturally become more acidic because organic matter accumulates and nutrient bases such as Ca, Mg and K are leached. Therefore, liming practices started more than 100 years ago with the aim to improve soil conditions and tree growth in forests (Huettl and Zoettl 1993; Kreutzer 1995; Larsson et al. 2001). With the advent of atmospheric acid deposition arising from anthropogenic pollution in the early 1980s (mainly SO_2, NO_x and NH_3 derived acids), lime and wood ash additions to forest were increasingly used to counteract accelerated acidification of soils and freshwater resources (Huettl and Zoettl 1993; Lundström et al. 2003a; Clair and Hindar 2005). In recent years, there has been increasing interest particularly in applying wood ash because renewable resources such as forest residues, wood chips or wood pills are replacing fossil fuels in power plants (Augusto et al. 2008; Ingerslev et al. 2011). Therefore, returning the resulting wood ash to the forests has been suggested as a means to both counteract acidification and mitigate potential nutrient deficiencies caused by intensive harvesting regimes (Palmberger and Egnell 2006; Augusto et al. 2008).

Forest floor liming with calcite ($CaCO_3$) or dolomite ($CaMg(CO_3)_2$) or wood ash (approx. 16% Ca content) has in most studies generated the anticipated effects of enhanced pH, increased cation exchange capacity and base saturation as well as lower concentrations of exchangeable aluminum in the upper soil layers (Huettl and Zoettl 1993; Kreutzer 1995; Lundström et al. 2003a; Uggla et al. 2003; Löfgren et al. 2008). Surface-applied lime or wood ash also appears to be relatively immobile, as calcium released from the dissolution of granules moves very slowly through the top organic horizons. Effects may therefore be long-lasting depending on site conditions and application method (Lehto and Malkonen 1994; Lundström et al. 2003b; Löfgren et al. 2008).

Concerns about possible unwanted side-effects of forest liming and wood ash applications have also been raised. Some studies reported decreased soil nutrient and C retention, i.e. increased concentrations and leaching of nitrate and dissolved organic C and N, decreased N and C stocks in the humus layer and acidification of the mineral subsoil (Kreutzer 1995; Andersson et al. 1999; Lundström et al. 2003b; Uggla et al. 2003). However, a recent meta-analysis of 67 wood ash and liming studies did not find any changes in forest floor C/N ratios (Reid and Watmough 2014). Also, the relation between pH and tree growth is debated or at least context dependent. An extensive evaluation across 80 sites in Southwest Sweden did not find any correlation between soil acidification and forest production (Nyberg et al. 2001), and Binkley and Högberg (1997) even suggested that the overall effect of forest liming in the Nordic countries is an up to 10% decrease in forest growth especially on N-deficient soils. This view was corroborated by a study by Jacobson et al. (2014) which indicated that wood ash must be supplemented with N in order to promote stem growth at infertile sites. The broader-scale meta-analysis by Reid and Watmough (2014) also found very large variation in the tree growth responses to wood ash and liming additions, although the overall response was positive.

One key element in temperate and boreal forest ecosystems is the ectomycorrhizal (EM) symbiosis where fungi belonging mainly to Basidiomycota and Ascomycota colonize the finest tree roots with fungal mantles (Smith and Read 2008). Fueled by C-supply from the host trees, the EM fungi form a mycelial network that expands the soil volume from which nutrients can be mobilized and also provides access to complexed N and P that plants cannot normally access (Hibbett et al. 2000; Read and Perez-Moreno 2003; Lindahl and Tunlid 2015). EM fungi thus constitute the main pathway of nutrients and water from soil to the trees and create a conduit for recent photosynthates into the soil matrix (Taylor et al. 2000; Smith and Read 2008). The EM fungi not only effectively connect the most active parts of the tree roots with the soil, they also form an important barrier for the roots against changes in the physical-chemical environment. The EM fungal communities are influenced both by natural soil processes such as acidification and accumulation of organic matter (Visser 1995) and by human-induced accelerated acidification and eutrophication (Cairney and Meharg 1999). It is well documented that individual EM species are affected differentially by environmental perturbations such as N deposition (Arnolds 1991; Lilleskov et al. 2002b; Kjøller et al. 2012; Suz et al. 2014). Also experimental acid irrigation (Agerer et al. 1998; Qian et al. 1998) and perturbations intended to increase soil pH (Antibus and Linkins 1992; Taylor and Finlay 2003; Kjøller and Clemmensen 2009; Rineau and Garbaye 2009b; Klavina et al. 2016) have been reported to affect EM community composition, although such effects have not yet been subjected to a comprehensive review.

The several thousands of EM fungal species associated with temperate and boreal forest trees (Taylor and Alexander 2005; Tedersoo et al. 2010) vary widely both structurally and functionally, e. g. with respect to biomass, development and differentiation of their extramatrical mycelium (Agerer 2001), extracellular enzyme production and nutrient uptake capacity (Leake et al. 1997; Bödeker et al. 2014) and weathering capacity (Wallander 2000). The species composition of an EM fungal community is therefore tightly linked with both total EM fungal biomass on the roots and in the soil and EM functioning. A general decline in species richness or complete or partial replacement within the fungal community due to environmental perturbations could potentially have significant implications for forest vitality and production. Therefore, it is crucial to obtain consensus knowledge on effects of liming and wood ash application on overall abundance, species richness and species composition of EM fungi before employing these treatments in large-scale forest management.

Based on a systematic literature search we here review field studies on effects of forest floor liming and wood ash application on abundance of EM roots (biomass or frequency measurements) and community composition, with a focus on Northern European coniferous forests. For studies suitable for meta-analysis we both performed weighted and unweighted meta-analysis depending on the study design.

14.2 Methodologies of the Reviewed Field Studies

14.2.1 Systematic Literature Searches

In April 2016 the Web of Science and Agricola literature databases where searched using the following requests: "ectomycorrhiza" AND liming", "ectomycorrhiza" AND lime", "ectomycorrhiza" AND ash" and "ectomycorrhiza" AND "$CaCO_3$". No attempts were made to systematically search grey literature i.e. institutional reports, PhD theses etc., but these are used in the general discussion when found relevant. All searches were restricted to 1990–2016 as pilot searches indicated that only after 1990 relevant papers accumulated. The four search files where combined and pruned for duplicates which left 140 publications. Titles were then scanned and papers clearly out of focus were removed i.e. papers on ash trees, arbuscular mycorrhizal fungi, in-vitro studies, ash from natural fires etc. This reduced the database to 76 records. After reading the Abstracts, 31 papers were maintained which then all are addressing the effect of lime or wood ash on EM fungi. Of these, the majority of papers addressed the effects on EM root tip abundance or species composition while fewer the effects on mycelia or fruiting bodies. Still, not all downloaded papers were suitable for subsequent meta-analysis. The explicit protocol used to include studies in the meta-analysis were, in addition to the above initial search criteria, that i) studies were using a field design with control and treated plots, ii) studies contained one or more EM fungal species richness or abundance measures, and iii) the pH change, lime or wood ash dosages and host tree were recorded. If a paper included several different study areas or tree species these were treated as separate studies. Thus in total 32 studies from 22 publications were included in the meta-analysis (Table 14.1). In studies where more than one soil horizon was sampled, the top (typically organic) sampled horizon was used. When repeated samplings were done, typically in successive years, data from the latest sampling was used. In cases where sites were sampled several times and published separately by the same or different research groups (Table 14.1), analysis was done both by including/excluding affected publications successively.

Table 14.1 Field studies of effects of lime and wood ash addition on ectomycorrhizal fungal biomass, richness and community composition

		Extracted meta-analysis parameters per species			Site, stand and treatment information						
Study "substudy"	Replication/ type	Abund.	Richness	Changes	Stand (age)	Region (site)	Plot size (m^2)	Dosage and liming agent[a] (Ca dose[b])	Sampled after treatment (years)	pH control/ treated[c]	Direction of effects[d]
Agerer et al. (1998)	No	n.a.	+	+	*Picea abies* (80)	S Germany (Höglwald)	900	4 t dolomite ha^{-1} (0.88 t ha^{-1})	7	4.2/5.5	↑ Species richness (sporocarps) ↓ *Cortinarius* sp. ↓ *Hygrophorus pustulatus* → *Russula ochroleuca*
Andersson and Söderström (1995)	Yes/plot	+	n.a.	n.a.	*P.abies* (seedlings) (50)	SW Sweden (Öringe)	625	3.8 t Ca CO_3 ha^{-1} (1.52 t ha^{-1})	5,6	4.2–4.5 /4.7–5.1	→ no. tips $seedling^{-1}$ ↓ no. ECM root tips mm^{-1} root
Antibus and Linkins (1992)	No	+	+	+	*Pinus resinosa* (60)	USA, MA (Harvard forest)	900	20 t lime ha^{-1} as 0.4 t ha^{-1} $month^{-1}$ for 4 years (8 t ha^{-1})	2,3	3.8/6.2	→ no. ECM root tips $sample^{-1}$ → Species richness ↑ *Piloderma bicolor*, *Piceirhiza* cf. *nigra* (6)

(continued)

Table 14.1 (continued)

Study "substudy"	Replication/ type	Extracted meta-analysis parameters per species			Site, stand and treatment information						Direction of effects[d]
		Abund.	Richness	Changes	Stand (age)	Region (site)	Plot size (m^2)	Dosage and liming agent[a] (Ca dose[b])	Sampled after treatment (years)	pH control/ treated[c]	
Bakker et al. (2000) and	Yes/site	+	n.a.	n.a.	*Quercus* spp. (15–76)	France/ Netherlands	n.a.	0.8^{-1}.6 t CaO ha^{-1} added as Ca CO_3 or CaO (0.57^{-1} • 14 t ha^{-1})	1–27 (depending on stand)	3.7–5.0 /n.a.	↑ no. ECM root tips ha^{-1} → no. ECM root tips m^{-1} root ↑ *Cenococcum geophilum*, morphotypes with extensive mycelia; ↓ smooth morphotypes
Börja and Nilson (2009)	Yes/site	+	n.a.	n.a.	*P. abies* ("old growth")	SE Norway	576	3 t CaO ha^{-1} (2.14)	30–35	3.6/3.8	↓ ECM root tips m^{-2} ↓ *Cenococcum geophilum*

Erland and Söderström (1991), "lime"	Yes/plot	+	n.a.	+	*Pinus sylvestris* (seedl.) (40)	S Sweden (Torrmyra)	400	4 t dolomite ha^{-1} (0.88 t ha^{-1})	1		3.8/5.2	→ no. ECM root tips $sample^{-1}$ ↑ no. ECM root tips m^{-1} root ↑ pink morphotype, *Pinirhiza rosea*; *Piloderma croceum* ↓corraloid morphotype, brown morphotype → *C. geophilum*
Erland and Söderström (1991), "wood ash"	Yes/plot	+	n.a.	+	*P. sylvestris* (seedl.) (40)	S Sweden (Torrmyra)	400	7.5 t wood ash ha^{-1} (1.22 t ha^{-1})	1		3.8/6.4	→ no. ECM root tips $sample^{-1}$ → no. ECM root tips m^{-1} root ↓coralloid morphotype → brown morphotype, pink morphotype

(continued)

Table 14.1 (continued)

Study "substudy"	Replication/ type	Extracted meta-analysis parameters per species			Site, stand and treatment information						Direction of effects[d]
		Abund.	Richness	Changes	Stand (age)	Region (site)	Plot size (m^2)	Dosage and liming agent[a] (Ca dose[b])	Sampled after treatment (years)	pH control/ treated[c]	
Jonsson et al. (1999)	Yes/plot	+	+	+	*P. abies* (50) (50)	S Sweden (Hasslöv)	900	8.75 t dolomite ha^{-1} (1.93 t ha^{-1})	12	4.1 / 5.5	↑ no. ECM root tips m^{-1} root → species richness ↓ *Russula ochroleuca*, *Tyllospora fibrillosa*, *Thelephora terrestris*
Kårén and Nylund (1996) "V"	Yes/plot	+	+	n.a.	*P. abies* (30)	SW Sweden (Skogaby)	2000	48:43:218:46: 75 kg P:K:Ca:Mg:S ha^{-1} (0.218 t ha^{-1})	3–5	3.7/3.8	→ ECM root tips m^{-2} → species richness → *C. geophilum*
Kårén and Nylund, (1996) "PK"	Yes/site	n.a.	n.a.	+	*P. abies* (50–75)	SW Sweden (Backaryd, Hammarby, Olofström)	900	25:62:33:12: 54 kg P:K:Ca:Mg:S ha^{-1} (0.033 t ha^{-1})	3, 4	4.0/ n.a.	→ Species richness → *C. geophilum*, *P. croceum*

Kjøller and Clemmensen (2009)	Yes/site	n.a.	+	+	*P. abies* / *P. sylvestris* (40/60)	S Sweden (Munkedal, Bäckefors, Mullsjö)	900	3 t lime/dolomite ha^{-1} (1.1 t ha^{-1})	16	3.7–4.3 /4.5–5.7	→ Species richness ↑ *Tylospora asterophora*, *Tuber* spp., *Amphinema byssoides*, *Elaphomyces* spp. ↓ *Lactarius rufus*
Klavina et al. (2016), "V"	Yes/plot	n.a.	+	+	*P. sylvestris* (101–125)	Latvia	22	50 t wood ash ha^{-1} (2820 t ha^{-1})	12	4.3/4.9	↓species richness ↑ *Tuber* spp., *A. byssoides*, *Tylospora* spp. ↓ *C. geophilum*, *Lactarius tabidus*
Klavina et al. (2016), "M"	Yes/plot	n.a.	+	+	*P. abies* (76)	Latvia	22	50 t wood ash ha^{-1} (2820 t ha^{-1})	12	4.3/5.1	↓ species richness ↑ *Tuber* spp., *A. byssoides*, *Tylospora* spp. ↓ *C. geophilum*, *L. tabidus*
Klavina et al. (2016), "MC"	Yes/plot	n.a.	+	+	*P. abies*/*P. sylvestris* (76/101–125)	Latvia	22	50 t wood ash ha^{-1} (2820 t ha^{-1})	12	4.8/5.4	↓species richness ↑ *Tuber* spp., *A. byssoides*, *Tylospora* spp. ↓ *C. geophilum*, *L. tabidus*

(continued)

Table 14.1 (continued)

		Extracted meta-analysis parameters per species			Site, stand and treatment information						
Study "substudy"	Replication/type	Abund.	Richness	Changes	Stand (age)	Region (site)	Plot size (m^2)	Dosage and liming agent[a] (Ca dose[b])	Sampled after treatment (years)	pH control/treated[c]	Direction of effects[d]
Lehto (1994a)	n.a.	+	n.a.	+	*P. abies* (59)	Central Finland	900	2 and 4 t dolomite ha^{-1} distributed 1959 & 1980 (0.88 t ha^{-1})	12	3.7/5.2	→ ECM root tip biomass cm^{-3} ↓ *P. croceum* → *C. geophilum*
Lehto (1994b)	Yes/plot	n.a.	n.a.	+	*P. abies* (60)	S Finland	4	2 t Ca CO_3 ha^{-1} (2 t ha^{-1})	n.a.	3.8 / 5.2	↑ % dead EM tips ↑ types with external mycelium; ↓ smooth types, *P. croceum*
Mahmood et al. (2002)	Yes/plot	+	+	+	*P. abies* (40)	SW Sweden (Torup)	900	6 t wood ash ha^{-1} (0.98 t ha^{-1})	7	4.3/4.3	→ no. ECM root tips m^{-1} root ↓ species richness ↑ *Piloderma* spp. and *Thelophora* spp. ↓ *Cortinarius* spp.
Majdi et al. (2008)	Yes/plot	+	n.a.	n.a.	*P. abies* (40)	SW Sweden (Skogeby)	2025	3.2 t wood ash ha^{-1} (0.52 t ha^{-1})	13–15	3.8/n.a.	↑ no. ECM root tips m^{-2}

Majdi and Viebke (2004), "CaMgPK"	Yes/plot	+	n.a.	n.a.	*P. abies* (60)	SW Sweden (Åled)	900	1.5+1.5t limestone and dolomite ha^{-1} (1.14 t ha^{-1})	8	n.a.	↑ no. ECM root tips m^{-2}
Majdi and Viebke (2004), "A"	Yes/plot	+	n.a.	n.a.	*P.abies* (60)	SW Sweden (Åled)	900	4.2 t wood ash ha^{-1} (1.13 t ha^{-1})	8	n.a.	↑ no. ECM root tips m^{-2}
Nowotny et al. (1998)	No	+	n.a.	n.a.	*P. abies* (88)	S Germany (Höglwald)	900	4 t dolomite ha^{-1} (0.88 t ha^{-1})	7	3.4/4.6	↑ ECM root clusters kg ha^{-1}
Qian et al. (1998)	No	n.a.	n.a.	+	*P. abies* (88)	S Germany (Höglwald)	900	4 t dolomite ha^{-1} (0.88 t ha^{-1})	7	n.a.	↑ *Piceirhiza nigra*, *Tuber puberulum*, *A. byssoides* ↓ *Tylospora fibrillosa*, *R. ochroleuca*
Rineau and Garbaye (2009a) "beech"	No	n.a.	n.a.	+	*Fagus sylvatica* (60)	NE France	100	0.76:0.38 t CaO:MgO ha^{-1} (0.54 t ha^{-1})	16	4.0/4.5	↓species richness ↑ *Clavulina cristata*, *Lactarius subdulcis*, *Tomentella* spp. ↓ *C. geophilum*, *Cortinarius* spp., *L. tabidus*, *Xerocomus pruinatus*

(continued)

Table 14.1 (continued)

Study “substudy”	Replication/ type	Extracted meta-analysis parameters per species			Site, stand and treatment information						Direction of effects[d]
		Abund.	Richness	Changes	Stand (age)	Region (site)	Plot size (m^2)	Dosage and liming agent[a] (Ca dose[b])	Sampled after treatment (years)	pH control/ treated[c]	
Rineau and Garbaye (2009a) “spruce”	No	n.a.	n.a.	+	*P. abies* (35)	NE France	100	0.76:0.38 t CaO:MgO ha^{-1} (0.54 t ha^{-1})	16	4.2/4.3	↓species richness ↑ *C. cristata*, *Tomentella sublilacina*; ↓ *C. geophilum*, *L. tabidus*, *X. pruinatus*
Rineau and Garbaye (2009b), “beech”	No	+	n.a.	+	*F. sylvatica* (60)	NE France, Humont, Vosges forest	100	0.76:0.38 t CaO:MgO ha^{-1} (0.54 t ha^{-1})	15	4.0/4.5	↓ no. ECM root tips $sample^{-1}$, changes in ECM root tip distribution in soil horizons ↑ *Clavulina* sp., *L. subdulcis*, *Piceirhiza* sp., ↓ *R. ochroleuca*, *Hygrophorus olivaceoalbus*, *C. geophilum*

Rineau and Garbaye (2009b), "spruce"	No	+	n.a.	+	*P. abies* (35)	NE France, Humont, Vosges forest	100	0.76:0.38 t CaO:MgO ha⁻ (0.54 t ha^{-1})	15	4.2/4.3	→ ECM root tips $sample^{-1}$, changes in ECM root tip distribution in soil horizons ↑ *Clavulina* sp., ↓ *R. ochroleuca*, *C. geophilum*
Rineau et al. (2010), "beech"	No	n.a.	+	+	*F. sylvatica* (60)	NE France, Humont, Vosges forest	400	0.76:0.38 t CaO:MgO ha⁻ (0.54 t ha^{-1})	15	4.0/4.5	↓species richness (ECM root tips and sporocarps) ↑ *C. cristata*, *L. subdulcis*, *Tomentella* spp. ↓ *C. geophilum*, *R. ochroleuca* → *X. pruinatus*
Rineau et al. (2010), "spruce"	No	n.a.	+	+	*P. abies* (35)	NE France, Humont, Vosges forest	400	0.76:0.38 t CaO:MgO ha^{-1} (0.54 t ha^{-1})	15	4.2/4.3	→ species richness (ECM root tips) ↓species richness (sporocarps) ↑ *C. cristata* ↓ *C. geophilum*, *R. ochroleuca* → *X. pruinatus*

(continued)

Table 14.1 (continued)

		Extracted meta-analysis parameters per species			Site, stand and treatment information						
Study "substudy"	Replication/ type	Abund.	Richness	Changes	Stand (age)	Region (site)	Plot size (m^2)	Dosage and liming agent[a] (Ca dose[b])	Sampled after treatment (years)	pH control/ treated[c]	Direction of effects[d]
Rineau and Garbaye (2010)	No	+	+	+	*F. sylvatica* (60)	NE France, Humont, Vosges forest	400	0.76:0.38 t CaO:MgO ha^{-1} (0.54 t ha^{-1})	15	4.0/4.5	↓no. ECM root tips $sample^{-1}$ → species richness ↑ *T. sublilacina*, *L. subdulcis*, *C. cristata* ↓ *C. geophilum*, *R. ochroleuca*, *X. pruinatus*
Taylor and Finlay (2003), "Hörröd-lime"	Yes/plot	+	+	+	*P. abies* (80)	S Sweden (Horröd)	225	3.25 t lime ha^{-1} (1.3 t ha^{-1})	4	2.7/3.0	↑ no. ECM root tips $core^{-1}$ ↓ species richness ↑ *C. geophilum*, *Piloderma* cf. *byssinum* ↓ *P. croceum*, *Russula* sp., *Tomentellopsis submollis* → *Tomentella* sp.

Taylor and Finlay (2003) "Hörröd-ash"	Yes/plot	+	+	+	*P. abies* (80)	S Sweden (Horröd)	225	4.28 t wood ash ha^{-1} (0.70 t ha^{-1})	4		2.7/3.3	↓no. ECM root tips $core^{-1}$ → species richness ↑ *C. geophilum* ↓ *P. croceum*, *Russula* sp., *Tomentella* sp. *T. submollis* → *P.* cf. *byssinum*, *Tylospora* sp.
Taylor and Finlay (2003) "Hasslöv"	Yes/plot	+	+	+	*P. abies* (60)	S Sweden (Hasslöv)	900	8.75 t dolomite ha^{-1} (1.93 t ha^{-1})	15		3.9/5.7	↑ no. ECM root tips $core^{-1}$ ↑ species richness ↑ *A. byssoides*, *P. nigra*, *Thelophora* sp., *Tuber* cf. *puberulum*, *Tylospora* cf. *asterophora* ↓ *Piceirhiza* cf. *gelatinosa*, *Piloderma* sp., *R. ochroleuca*, *Tylospora fibrillosa*

[a]Author's original notations of the used liming agents. Dolomite = $CaMg(CO_3)_2$, Calcite/lime = $CaCO_3$, quick lime = CaO

[b]If not given by authors, Ca% of liming agents was calculated as follows: $CaMg(CO_3)_2$ = 22%, Ca from CaO = 71.47%, Ca from $CaCO_3$ = 40.04%, Ca in wood ash: 16.3% (http://woodash.slu.se/eng/)

[c]When pH was measured in several soil horizons, pH of the top most horizon is given

[d]Only summary for more abundant species are shown. ↑ and ↓ indicate increases or decreases with liming or wood ash in comparison with controls, → indicates no effect

In the studies included, EM fungi were quantified based on the colonized root tips. Specifically, EM fungal abundance was recorded as I) EM root tip biomass per soil volume or area sampled, II) number of EM root tips per soil volume or area sampled, III) number of EM tips per tree seedling (bait seedlings planted into mature forest), or IV) number of EM tips per root length (Table 14.1). If more abundance indexes were used in a study, we used measures per soil volume or area over measures per seedling or root length. Another index of fungal colonization often reported in the (older) literature is the percentage root tips colonized by EM fungi. All but one study (Börja and Nilsen 2009) reported colonization to be 90–100% independent of treatment. Therefore, the percentage colonization level was not used as an abundance index in this study. One single study quantified EM abundance based on the production of external mycelium (Majdi et al. 2008), but since this study also reported the number of ectomycorrhizal root tips this metric was preferred for better comparability with the other studies.

Responses in species richness of EM fungal communities have also been reported in a number of studies. There has been an immense development in methodology especially for describing the below-ground communities during the 26 years from which studies were found (Horton and Bruns 2001; Kjøller and Clemmensen 2009). Therefore, the species richness estimates extracted from the studies are based on a range of different methodologies including I) morphologically identified aboveground sporocarps (1 study), II) morphologically identified belowground EM root tips (morphotyping) (12 studies), III) morphotyping combined with PCR-RFLP-typing (4 studies), IV) morphotyping combined with DNA-sequencing (10 studies), V) DNA-sequencing of randomly sampled EM root tips (1 study). All data types may be assigned to particular fungal species depending on availability of reference material and when possible, data on particularly common fungal species were compared between studies. Lastly, for analysis of fungal phylogenetic lineages, species where grouped into clades based on Tedersoo et al. (2010) except for Boletus, Paxillus-gyrodon, Pisolithus-scleroderma and Suillus-rhizopogon which were grouped into Boletales. It was considered beyond the scope of this study to evaluate the taxonomic validity within the published works and names are used as they are given in the publications.

14.2.2 Meta-analysis

From the studies included in the meta-analysis, spreadsheets were compiled tabulating the means and if possible the sample size and standard deviation for control and treatment. Also other relevant moderators e.g. dosage, stand age, geographical region etc. were recorded. In cases where the variation was expressed graphically e.g. as error bars these were converted to SD manually based on the graphs.

Generally, it is recommended to perform weighted meta-analysis based on standardized mean differences calculated from the mean, standard deviation and sample size for control and treatment of each study.

Based on this, the individual study-based standardized mean difference are then used to calculate a summary across studies. For studies including true replication, either as replicated plots within a site or as replication across several sites, we calculated the Hedges' d as the standardized effect size as this should work down to as few as 5–10 studies (Hedges et al. 1999). Hedges' d range from $-\infty$ to $+\infty$ where 0 is no difference between control and treatment. Responses >0 or < 0 show an increase or decrease of the response variable in comparison with the control. Statistics were calculated using MetaWin 2.0 software (Rosenberg et al. 1999). If the 95% confidence interval of the individual cases as well of the global summary do not overlap with 0 these are judged to be significantly different from the control. Unfortunately, many studies only included one treated and one control plot and some replicated studies fail to report sample size and variation. For EM abundance and species richness we therefore also performed unweighted analyses where effect sizes of each study were calculated as ((treatment/control) − 1) %. In total, 12 or 20 and 10 or 15 studies were applicable to weighted or unweighted meta-analysis of the mycorrhizal abundance and species richness response variables, respectively. Because of scarcity of data available, only unweighted meta-analysis was possible when analyzing abundance of specific fungal taxa (Table 14.1). In the case of specific EM fungal taxa we simply calculated the relative subtractive value (treatment − control). For the unweighted analysis, we also used MetaWin for calculating the summary effect size by setting the variance of all studies to 1 (Rosenberg et al. 1999). Correlations between effect sizes and moderators where calculated using R (R-project 2014).

14.3 Results and Discussion

14.3.1 Effects on Overall Mycorrhizal Abundance

In our weighted meta-analysis, apart from one study (Jonsson et al. 1999), neither EM biomass nor number of EM root tips in treated plots deviated significantly from control plots (Table 14.1, Fig. 14.1). This was further corroborated by including more studies in an unweighted analysis (summary effect size = 7.5 with bootstrap confidence intervals between −7.0 and 22.9). There was a strong and significant ($R = 0.80$; $P < 0.01$) correlation between the weighted Hedges' d and the simple unweighted division metric. Also, analyses were robust when studies from the same site were successively excluded (data not shown). Still, we observed some variation across studies, indicating a stimulation of mycorrhizal biomass in some sites and inhibition in other sites, but effect sizes did not correlate with any of the tested moderators: pH increment, years after treatment, stand age, Ca dosage and plots size (data not shown). EM fungal biomass is influenced by the amount of fine roots available for colonization. Fine root responses to liming or wood ash treatments were not systematically analyzed in this study but additions generally appear to

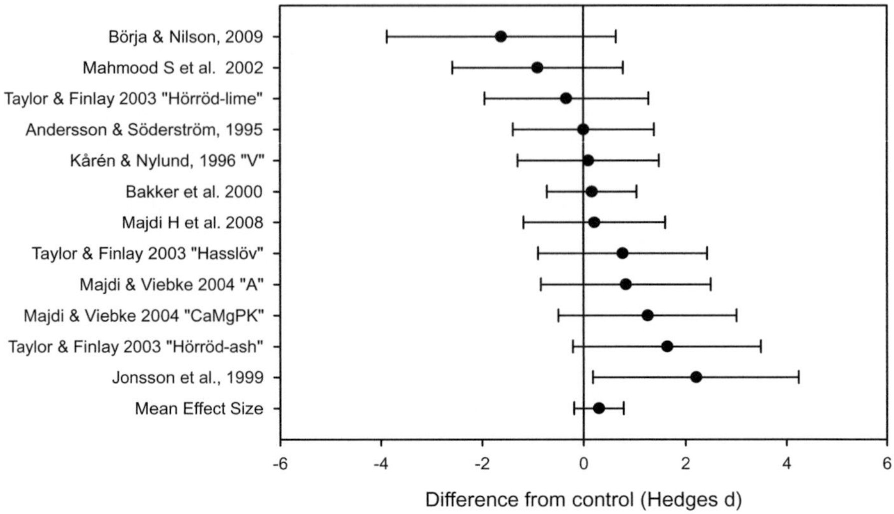

Fig. 14.1 Response of the general EM fungal abundance to liming or wood ash treatments based on a weighted meta-analysis using Hedges' d as effect size with 95% confidence intervals. Positive effect sizes indicate a positive response to the treatment while negative effect sizes indicate decline in the analyzed variable in response to the treatment. Responses are significant only if confidence intervals do not overlap with the vertical zero line. The *bottom* "Mean Effect Size" is the weighted summary statistic calculated from all the individual studies

stimulate fine root development in the uppermost soil layers (Huettl and Zoettl 1993; Persson and Ahlström 1994; Nowotny et al. 1998; Majdi et al. 2008), although there is also reports of neutral or negative responses (Lehto 1994a; Majdi and Viebke 2004). Together these results suggest that forests treated with lime or wood ash can maintain a stable and active fine root and EM fungal biomass, ensuring that nutrients are captured and directed to the growing trees.

In contrast to the below-ground EM root tip abundances, several studies showed large reductions in the range of 40–50% in production of above-ground EM fungal sporocarps (Wästerlund 1982; Agerer et al. 1989; Jacobsson 1993; Wiklund et al. 1995; Brandrud et al. 2001; Brandrud et al. 2003). This was in spite of relatively low liming doses of only 0.2–3 t ha^{-1} but also relatively short time after treatments. For instance, Kårén and Nylund (1996) estimated that *Cortinarius* spp. accounted for 30% of the reduction in sporocarp production but only decreased their relative abundance on roots by 2% with liming. These observations suggest that decreased allocation to fruit-body production may be a first and fast response for some EM fungi, and that mycorrhizal mycelia may be more sensitive than mycorrhizal biomass directly associated with the root tips. However, there was no support for decreased mycelial production in the single study where mycelial production indeed was recorded, as production in mesh bags buried in plots treated with 3.2 t wood ash ha^{-1} was found to be equal to production in control plots (Majdi et al. 2008). Another important factor is that above and belowground EM fungal communities show different patterns, as not all EM fungi produce aboveground structures and because sampling strategies differ. Thus, complex interactions between EM fungal community changes and fruiting patterns may be expected but data are too scarce to pinpoint exact cause-effect relationships yet (but see further discussion below).

14.3.2 Effects on Species Richness and Community Structure

A species rich and phylogenetically diverse EM community is potentially a safeguard against environmental perturbations which allows different nutrient patches to be exploited from the soil e.g. organic versus mineral nutrients. As with general mycorrhizal abundance, the meta-analysis failed to identify any systematic response of EM fungal species richness to liming or wood ash additions (Table 14.1, Fig. 14.2). Also for species richness, there was a strong and significant ($R = 0.98$; $P < 0.01$) correlation between the weighted Hedges' d and the simple unweighted division metric. The summary effect size of the unweighted analysis was −0.4 with bootstrap confidence intervals ranging from −13.6 and 12.4.

Clearly, in some of the earliest studies where diversity was recorded based on only a few EM morphotypes, these were probably pools of many individual species and therefore less likely to show a response in richness or dominance patterns. Reduced species richness after liming, however, was indicated in a few studies with

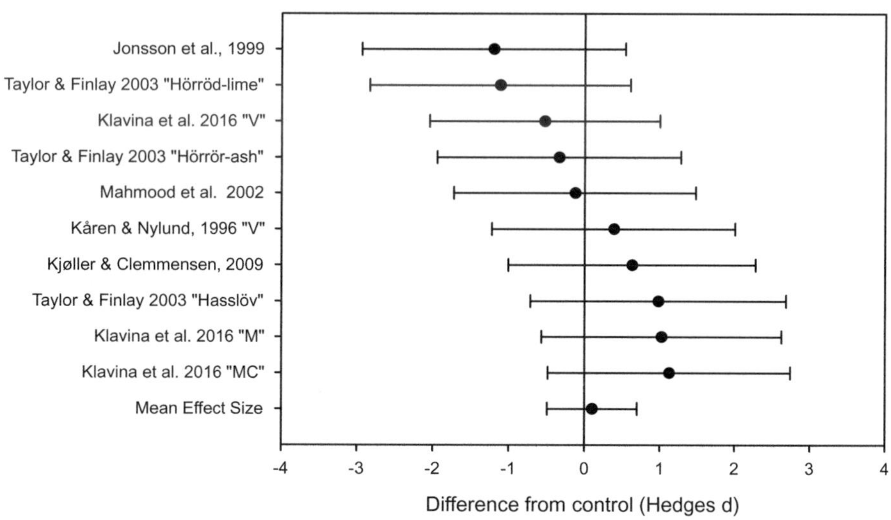

Fig. 14.2 EM fungal species richness responses to liming or wood ash treatments based on a weighted meta-analysis using Hedges' d as effect size with 95% confidence intervals. Positive effect sizes indicate a positive response to the treatment while negative effect sizes indicate decline in the analyzed variable in response to the treatment. Responses are significant only if confidence intervals do not overlap with the vertical zero line. The bottom "Mean Effect Size" is the overall response calculated across studies.

high species resolution and relatively large numbers of samples surveyed. In a Norway spruce stand in S Sweden exposed to 3.25 t lime ha^{-1} , belowground richness was reduced from 27 morphotypes in control plots to 16 in limed plots (Taylor and Finlay 2003 – Horröd). This reduction was mainly due to a loss of rare EM morphotypes as the community structure changed towards an increased dominance of common species. Also, other studies have reported increased dominance of a few species within EM communities with liming (Kåren and Nylund 1996; Kjøller and Clemmensen 2009). In another study, liming with 3 t of dolomite ha^{-1} in a Scots pine stand in Norway, reduced species richness of the aboveground EM sporocarp community from 25–26 species in control areas to 10–14 species in limed areas within 3 years after the lime application (Brandrud et al. 2001). In contrast, other studies showed a tendency to increases in species richness with liming or wood ash application (Taylor and Finlay 2003 – Hasslöv; Klavina et al. 2016). As for mycorrhizal abundance, the range of species richness responses did not correlate with any of our recorded moderators across studies (data not shown).

14.3.3 Effects on EM Community Composition and Individual Taxa

Most of the individual studies from our systematic literature review reported EM fungal community changes with liming or wood ash amendments i. e. increased relative abundance of some EM species or types on the expense of others (Table 14.1). The high species richness of many EM fungal communities combined with the low resolution of many of the older studies reduced the number of cases where species-level meta-analysis was possible. However, even with the scarce data available several species responded significantly to liming or wood ash additions (Fig. 14.3). *Tylospora fibrillosa*, *Piloderma croceum* and *Russula ochroleuca* decreased while *Amphinema byssoides* increased in response to liming or wood ash amendment (Fig. 14.3). There are most likely other species especially within the Tuber-helvella clade (Klavina et al. 2016) and possibly also within Tomentella-thelophora (Taylor and Finlay 2003) that given more data would show a significant positive response to liming or wood ash addition as also indicated in the analysis below (Fig. 14.4).

Furthermore, our analysis suggests that some species, i. e. *Cenococcum geophilum*, *Tylospora asterophora* and *Tomentellopsis submollis*, do not respond to liming or wood ash amendment. One of these species *C. geophilum* seems to be a true generalist interacting with many different soil types and hosts trees including both conifers and broadleaved trees (LoBuglio 1999). This species seems capable of thriving along a relatively broad pH range and it is also known from other studies to be a disturbance tolerant species (LoBuglio, 1999). *T. asterophora* and *T. submollis* are in contrast confined almost exclusively to coniferous hosts and only recorded from the Northern hemisphere (https://unite.ut.ee/index.php), but still these species

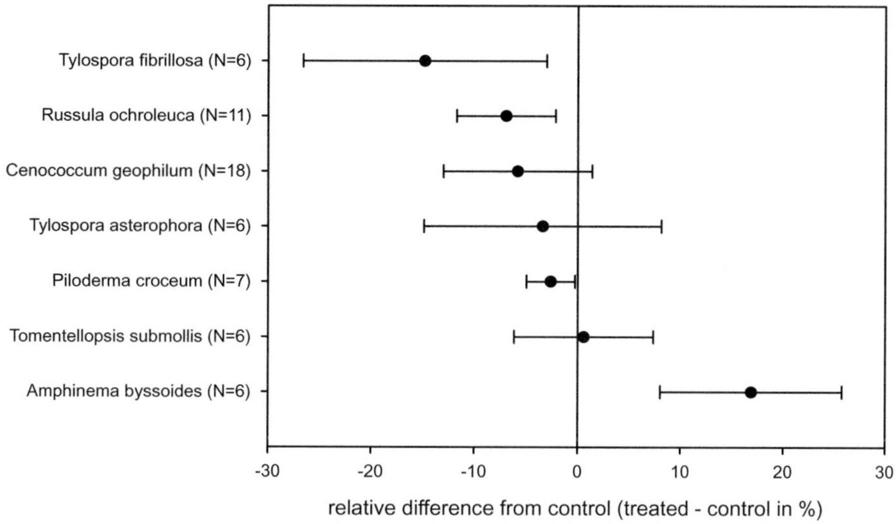

Fig. 14.3 Results of unweighted summary analysis of individual EM fungal species across different studies (number of studies included is indicated by "N="). The unweighted metrics were: treated – control in percent. 95% confidence intervals are calculated with a bootstrapping method (Rosenberg et al. 1999). Positive effect sizes indicate a positive response to the treatment while negative effect sizes indicate decline in the analyzed variable in response to the treatment. Responses are significant only if confidence intervals do not overlap with the zero vertical line.

seem to be resistant to these perturbations and to tolerate the pH ranges experienced in the studies included.

Aggregating species into EM fungal clades according to Tedersoo et al. (2010) provided more observations that we could analyze across studies. This allowed the response of the notoriously species rich clade Cortinarius to be analyzed which showed an overall significantly negative response to liming or wood ash amendment (Fig. 14.4). Also Russula-lactarius showed an overall negative response, while Tuber-helvella responded positively to the treatments (Fig. 14.4). Clearly not all species within a clade may respond in the same direction as illustrated for the Amphinema-tylospora clade, where *Tylospora fibrillosa* and *Piloderma croceum* responded negatively, *Amphinema byssoides* was stimulated, while another abundant conifer species *Tylospora asterophora* was unresponsive. Overall this summed up to a non-significant response of the Amphinema-tylospora clade (Fig. 14.4).

Most of the significant species and clade level effects were robust against inclusion/exclusion of replicated study sites, but in some cases exclusion of particular studies changed the overall response markedly. For the order Boletales, nine out of the ten included studies showed a negative response to liming or wood ash addition but a single study (Jonsson et al. 1999) showed the opposite trend. Here, *Tylopilus felleus* (Boletales) increased from 3 to 18% of colonized root tips with liming which blurred the overall response of the clade (Fig. 14.4). Also for the Russula-lactarius clade the negative response with liming and wood ash became more significant if

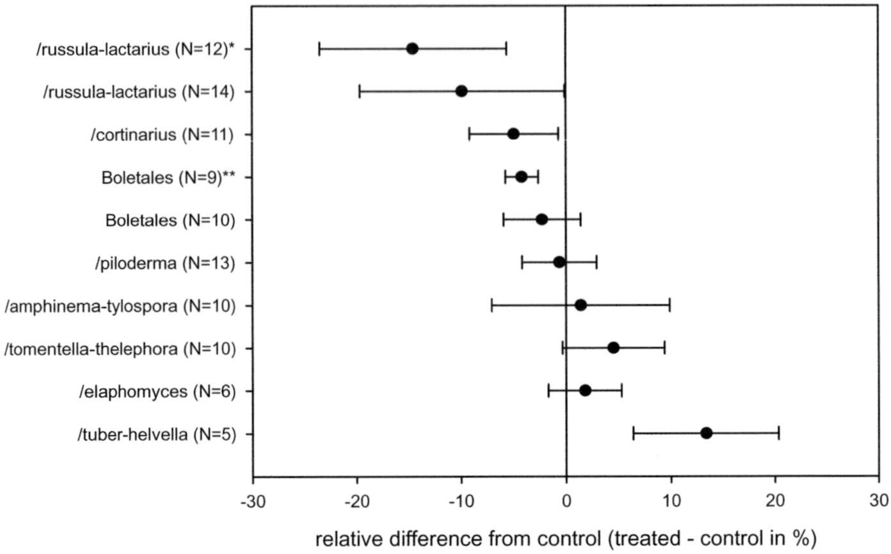

Fig. 14.4 Results of unweighted summary analysis of EM fungal clades across different studies ("N=": number of studies included). Clades were assigned according to Tedersoo et al. (Tedersoo et al. 2010). Boletales is composed of fungi from Boletus, Paxillus-gyrodon, Pisolithus-scleroderma and Suillus-rhizopogon. The unweighted metric was: treated – control in %. 95% confidence intervals are calculated with a bootstrapping method (Rosenberg et al. 1999). Positive effect sizes indicate a positive response to the treatment while negative effect sizes indicate decline in the analyzed variable in response to the treatment. Responses are significant only if confidence intervals do not overlap with the zero vertical line. For Russula-lactarius data are shown with or without (*) two studies with *Fagus sylvatica* as host trees, and for Boletales with and without (**) the study by Jonnsson et al. (1999).

studies were restricted to coniferous forests (Fig. 14.4). This was caused by a very strong stimulation of a single species, *Lactarius subdulcis*, with liming in the *Fagus sylvatica* plots of the French Humont experiment (Rineau and Garbaye 2009a; Rineau et al. 2010).

As an example of these general trends the Norway spruce experiments in S Sweden, Hasslöv (8.75 dolomite ha^{-1}), and S Germany, Högwald (4 t dolomite ha^{-1}), are illustrative. In Hasslöv *Russula ochroleuca* and *Tyllospora fibrillosa* almost disappeared from the root tips in limed plots (Jonsson et al. 1999; Taylor and Finlay 2003), and the same two species decreased from 32 to 4% and 30 to 3% with liming in the Högwald experiment (Taylor et al. 1992; Qian et al. 1998). Also in Högwald, *R. ocroleuca* decreased as assessed by fruit body observations (Agerer et al. 1998). *Russula ocroleuca* and *T. fibrillosa* are common in nutrient poor Norway spruce forests, and they made up 15 and 45%, respectively, of all mycorrhizal root tips in control plots at the Swedish site and 32 and 29%, respectively, at the German site. In the study by Kjøller and Clemmensen (Kjøller and Clemmensen 2009) the Russula-lactarius clade made up about 30% of the root tip community in control plots but was reduced to 5% after liming, in this case mainly driven by a reduction of the *Lactarius rufus*.

In spite of the relatively high liming dose (8.75 t dolomite ha^{-1}), EM fungal species richness in the Norway spruce forest studied by Taylor and Finlay (2003) did not change although community composition was almost completely replaced. Control and limed areas had only 3 species in common out of the 12 and 15 species found in control and treated areas, respectively. Taylor and Finlay (2003) also noted that some of the mycorrhizal types, *A. byssoides* and *Tuber puberulum*, that thrived after the high liming dose were species that are normally associated with more nutrient-rich or calcareous soils. The two species were not recorded in control plots but made up 38 and 9%, respectively, of the mycorrhizal root tips in the limed plots. Similarly, the morphotype *Piceirhiza nigra* was not present in control plots but made up 20% of the community in the limed plots. *A. byssoides*, *T. puberulum* and *P. nigra* also showed large increases in abundance at the German spruce forest, Högwald, limed with 4 t dolomite ha^{-1} (Taylor et al. 1992; Qian et al. 1998).

In the three studies reporting decreased production of aboveground sporocarps with liming or ash additions, the changes were partly attributed to large decreases in the fruit-body production by *Cortinarius* species. Brandrud et al. (2001, 2003) reported 75–90 % reduction in sporocarp production by *Cortinarius* species 2½-6 years after liming with 3 t dolomite ha^{-1} in two coniferous forests in Norway. A lower liming dose of 218 kg Ca ha^{-1} (equivalent to the Ca content of 1 t dolomite) combined with additions of P, K, Mg and S to a Norway spruce stand in S Sweden reduced sporocarp production in almost all of the common species including *Cortinarius brunneus* (Wiklund et al. 1995). Belowground in the same experiment, Kårén and Nylund (1996) observed significantly decreased frequency (from 3.4 to 1.2 % of EM tips) of a white morphotype probably including *C. brunneus*. They calculated that fungal species included in this morphotype could account for 30% of the reduction in sporocarp production with liming at the site. Also in Högwald a *Cortinarius* species (*Dermocybe cinnamomea*) was in some years a common fruiting species in control plots but disappeared in limed plots (Agerer 1990; Agerer et al. 1998). Liming resulted in an increased abundance of species that produce inconspicuous resupinate sporocarps including *Amphinema byssoides* and most of the tomentelloid fungi recorded (Taylor and Finlay 2003). Therefore, positive liming effects on these mycorrhizal fungi may generally have been underestimated in previous sporocarp-based field studies.

Only a few studies reported no detected effects on community composition with liming or wood ash addition. One of these studies investigated effects of a low application rate (1.55 t dolomite ha^{-1}) in the Norway spruce experiment at Hasslöv by Jonsson et al. (1999). This application rate may have been below the critical load for the belowground EM community at this site. Kårén and Nylund (1996) also concluded that in the short-term (<5 years) the use of moderate amounts of N-free fertilizer (with a Ca content corresponding to that in 0.15–1 t dolomite) is not likely to drastically affect dominant EM fungi on roots. Also, the critical load for sporocarp communities in nutrient-poor coniferous forests in western Norway seems to be below 2 t dolomite ha^{-1} according to a field study of short-term effects (Brandrud et al. 2001; TE Brandrud, pers. com.). The addition of moderate levels of wood ash additions (3–7.5 t ha^{-1}) also seems to provoke lesser changes to the EM fungal community than similar dosages of lime (Erland and Söderström 1991; Mahmood et al. 2002; Taylor and Finlay 2003) (Table 14.1). This is in spite of that wood ash addition generate some of the same effects on forest soils as liming, e.g. increased pH. An exception was the study by Klavina et al. (2016) where a very significant community shift between wood ash treated and control plots was found, but it is noticeable that in this experiment a rather extreme dose (50 t ha^{-1}) was used.

14.3.4 Possible Drivers of EM Fungal Community Shifts

Several of the reviewed studies pointed to increased availability of inorganic N after liming as the primary mechanism behind reduced production of EM sporocarps (Wiklund et al. 1995; Brandrud et al. 2003) and shift in the belowground EM community (Taylor and Finlay 2003). Most boreal forests are considered N limited (Tamm 1991), and it is well documented that increased N inputs to forest ecosystems affect EM communities both aboveground (Arnolds 1991; Wiklund et al. 1995; Lilleskov et al. 2001) and belowground (Lilleskov et al. 2002a; Cox et al. 2010; Kjøller et al. 2012). Since neither lime nor wood ash contains any N, increased N availability with these amendments must then be a consequence of pH driven increased mineralization and nitrification rates which has indeed been reported from some studies (Kreutzer 1995; Zimmermann and Frey 2002; Jandl et al. 2007). The observation that liming decreased the abundance of species adapted to N limited and acidic conditions, like many *Cortinarius* species (Brandrud et al. 2003), while species associated with more rich sites e.g. *Lactarius subdulcis* or *Tuber* spp. increased (Taylor and Finlay 2003; Rineau et al. 2010), may support this hypothesis.

On the other hand, the maintenance of species richness after liming and wood ash amendment as seen in Fig. 14.2 is not indicative of increased N availability as EM richness normally is reduced under such circumstances (Lilleskov et al. 2002a; Kjøller et al. 2012). Indeed, none of the studies captured in the systematic literature search reported any stimulation of the N status with liming or wood ash amendment but rather indications of lower N availability or no response (Wiklund et al. 1995; Börja and Nilsen 2009; Kjøller and Clemmensen 2009; Rineau and Garbaye 2009b; Klavina et al. 2016). Therefore, increased mineralization and nitrification rates with liming or wood ash additions may be context dependent and are most likely to occur at more fertile sites (Jacobson et al. 2014).

Soil pH is often the best predictor of microbial community changes (Högberg et al. 2007; Dumbrell et al. 2010; Rousk et al. 2010), and species turnover between EM fungal communities often correlates with pH changes (Kjøller and Clemmensen 2009; Rineau and Garbaye 2009b; Tedersoo et al. 2014). Indeed, laboratory tests have shown that pH of the growth medium appears to have differential effects on the growth of different EM fungal isolates in pure culture (Hung and Trappe 1983). EM fungal species produce varying types and amounts of extracellular enzymes, and the activities of these have been shown to have different pH optima (Leake et al. 1997). For instance, Bending and Read (1995) found that activities of proteases involved in the mobilization of N from humus declined above pH 4.5. Thus the effects of increased pH on enzyme activity probably vary among fungal species and ecotypes, which in turn could lead to changed growth of individual EM fungi and cascading changes in composition of EM communities.

The availabilities of a range of nutrients including heavy metals are tightly linked to changes in soil pH. The solubility of many essential elements such as P, N, K and Mg increases with increased pH from 4 to 5, while the solubility e.g. Fe and Al decreases. Deficiencies of particularly manganese (Mn) and boron (B) can be induced by liming due to enhanced adsorption of these ions to organic particles at higher pH (Lehto 1994a). Boron supply is commonly low in the interior of northern Sweden and Finland because these areas receive little sea spray, and liming has caused declined growth of pine and spruce through decreased solubility and plant access to borate in both peatlands and podzolic soils (Gupta et al. 1985; Lehto and Mälkonen 1994).

In the limed spruce forests in S Finland studied by Lehto (1994a), uptake of B was probably also impaired by lime-induced injuries on the mycorrhizal fungi, as the proportion of dead mycorrhizas increased from 10 to 30% with liming. Boron application at the contrary doubled the number of root tips in the humus layer and in practice, B fertilization turned out to be a way to counteract the negative effects of liming on EM fungi (Lehto 1994a). Later growth chamber study established a positive relationship between the internal B concentration in spruce seedlings and the numbers of root tips and mycorrhizas as well as mycorrhizal percentage and root dry weight (Mottonen et al. 2001). This suggested a direct effect of B status in fine root and mycorrhizal development in the investigated soil, but the generality of this phenomenon should be further studied.

For wood ash, the matter is even more complicated as wood ash both increases pH while at the same time introduce a burst of elements most notably K, Ca, Mg, Fe and P. Disentangling the effects of pH and changed availabilities of the introduced as well as the native elements can only be achieved in experiments with factorial additions of lime and individual elements (Lehto 1994a, b).

14.3.5 Possible Consequences on EM Communities and Unique Species with Liming and Wood Ash

From this analysis we can conclude that EM fungal community composition is generally affected by lime and wood ash amendments, while total EM fungal biomass and species richness appear resistant. In general, EM fungal communities have been found to adjust to various changes in their environment (Stendell et al. 1999; Toljander et al. 2006; Kjøller and Clemmensen 2009; Suz et al. 2014; Klavina et al. 2016). Depending on the magnitude of the environmental change, the response by an EM community probably starts by physiological responses in the species present, then by altered relative abundances between species within the community and ultimately by species losses or replacements.

Compared to clear-cutting or stand replacing fires, liming and wood ash amendments may seem to constitute relatively moderate disturbances in the forest ecosystems. However, lime acts over many years and especially the study by Taylor and Finlay (2003) showed that liming can result in an almost complete replacement of an EM community, in parallel to observations following clear-cutting or stand-replacing fires (Baar et al. 1999; Clemmensen et al. 2015; Kyaschenko et al. 2017). The study by Taylor and Finlay (2003) however may be extreme, as the liming dosage was relatively high (8.75 t ha^{-1}) causing a large pH increment (1.8 units). The other included studies showed significant, but still more moderate, community responses with many species overlapping between treated and non-treated areas, although at different relative abundances. As species richness and fungal biomass seems to be unaffected, the moderate dosages normally used (often 3–4 t ha^{-1}) appears to be a safe forest management application rate in production forests. Also, soil borne resistant propagules and re-colonization from surrounding areas will in the longer run act to limit the risk of species losses, although the larger the managed areas, the higher the risk for geographically isolated sensitive species to go extinct. Whether the observed EM fungal community changes with liming and wood ash amendments have consequences in relation to nutrient cycling and tree nutrition, is an important question that needs to be clarified in future studies. For example, the reduction of *Cortinarius* abundance with lime and wood ash treatments observed in several studies, may indicate a shift in mode of nutrient cycling away from organic N mobilization (Bödeker et al. 2014). As *Cortinarius* species recolonize forest areas only slowly after eradication (Kyaschenko et al. 2017), this could have long term consequences for the forest ecosystem.

Acknowledgements This work was supported by a grant from the Swedish Forest Agency, (grant no 91/06 4.43/HK) and by the "Center for Bioenergy Recycling- ASHBACK" project, funded by the Danish Council for Strategic Research (grant no 0603-00587B).

References

Agerer, R. (1990). Impacts of artificial acid-rain and liming on fruitbody production of ectomycorrhizal fungi. *Agriculture, Ecosystems and Environment, 28*, 3–8.

Agerer, R. (2001). Exploration types of ectomycorrhizae. A proposal to classify ectomycorrhizal mycelial systems according to their patterns of differentiation and putative ecological importance. *Mycorrhiza, 11*, 107–114.

Agerer, R., Brand, F., & Treu, R. (1989). Systematic aspects of ectomycorrhizas. In Proceedings of the 10th Congress of European Mycologists, Tallinn, Estonia.

Agerer, R., Taylor, A. F. S., & Treu, R. (1998). Effects of acid irrigation and liming on the production of fruit bodies by ectomycorrhizal fungi. *Plant and Soil, 199*, 83–89.

Andersson, S., Nilsson, I., & Valeur, I. (1999). Influence of dolomitic lime on DOC and DON leaching in a forest soil. *Biogeochemistry, 47*, 297–317.

Antibus, R. K., & Linkins, A. E. (1992). Effects of liming a red pine forest floor on mycorrhizal numbers and mycorrhizal and soil acid–phosphatase activities. *Soil Biology and Biochemistry, 24*, 479–487.

Arnolds, E. (1991). Decline of ectomycorrhizal fungi in Europe. *Agriculture, Ecosystems and Environment, 35*, 209–244.

Augusto, L., Bakker, M. R., & Meredieu, C. (2008). Wood ash applications to temperate forest ecosystems – Potential benefits and drawbacks. *Plant and Soil, 306*, 181–198.

Baar, J., Horton, T. R., Kretzer, A. M., & Bruns, T. D. (1999). Mycorrhizal colonization of Pinus muricata from resistant propagules after a stand–replacing wildfire. *New Phytologist, 143*, 409–418.

Bakker, M. R., Garbaye, J., & Nys, C. (2000). Effect of liming on the ectomycorrhizal status of oak. *Forest Ecology & Managment, 126*, 121–131.

Bending, G. D., & Read, D. J. (1995). The structure and function of the vegetative mycelium of ectomycorrhizal plants VI. Activities of nutrient mobilizing enzymes in birch litter colonized by *Paxillus involutus* (Fr.) Fr. *New Phytologist, 130*, 411–417.

Binkley, D., & Hogberg, P. (1997). Does atmospheric deposition of nitrogen threaten Swedish forests? *Forest Ecology and Management, 92*, 119–152.

Bödeker, I. T. M., Clemmensen, K. E., de Boer, W., Martin, F., Olson, A., & Lindahl, B. D. (2014). Ectomycorrhizal *Cortinarius* species participate in enzymatic oxidation of humus in northern forest ecosystems. *New Phytologist, 203*, 245–256.

Börja, I., & Nilsen, P. (2009). Long term effect of liming and fertilization on ectomycorrhizal colonization and tree growth in old Scots pine *(Pinus sylvestris* L.) stands. *Plant and Soil, 314*, 109–119.

Brandrud, T. E., Bakkestuen, V., Aarrestad, P. A., & Direktoratet for naturforvaltning. (2001). Terrengkalking i Suldal, Rogaland – Effekter på vegetasjon og sopp. In: Terrengkalkningsprosjektet. Årsrapport 2000. Terrengkalking for å avgifte surt overfladevann. *DN Notat, 2001–2004*, 57–76.

Brandrud, T.E., Bakkestuen, V., Bendiksen, E., Eilertsen, O., Aarrestad, P.A. (2003). Terrengkalking i Gjerstad, Aust–Agder. Effekter på skogvegetasjon og sopp, *NINA Fagrapport*, p. 75.

Cairney, J. W. G., & Meharg, A. A. (1999). Influences of anthropogenic pollution on mycorrhizal fungal communities. *Environmental Pollution, 106*, 169–182.

Clair, T. A., & Hindar, A. (2005). Liming for the mitigation of acid rain effects in freshwaters: A review of recent results. *Environmental Reviews, 13*, 91–128.

Clemmensen, K. E., Finlay, R. D., Dahlberg, A., Stenlid, J., Wardle, D. A., & Lindahl, B. D. (2015). Carbon sequestration is related to mycorrhizal fungal community shifts during long–term succession in boreal forests. *New Phytologist, 205*, 1525–1536.

Cox, F., Barsoum, N., Lilleskov, E. A., & Bidartondo, M. I. (2010). Nitrogen availability is a primary determinant of conifer mycorrhizas across complex environmental gradients. *Ecology Letters, 13*, 1103–1113.

Dumbrell, A. J., Nelson, M., Helgason, T., Dytham, C., & Fitter, A. H. (2010). Relative roles of niche and neutral processes in structuring a soil microbial community. *ISME Journal, 4*, 337–345.

Erland, S., & Söderström, B. (1991). Effects of lime and ash treatments on ectomycorrhizal infection of *Pinus sylvestris* L. seedlings planted in a pine forest. *Scandinavian Journal of Forest Research, 6*, 519–526.

Gupta, U. C., Jame, Y. W., Campbell, C. A., Leyshon, A. J., & Nicholaichuk, W. (1985). Boron toxicity and deficiency – A review. *Canadian Journal of Soil Science, 65*, 381–409.

Hedges, L. V., Gurevitch, J., & Curtis, P. S. (1999). The meta–analysis of response ratios in experimental ecology. *Ecology, 80*, 1150–1156.

Hibbett, D. S., Gilbert, L. B., & Donoghue, M. J. (2000). Evolutionary instability of ectomycorrhizal symbioses in basidiomycetes. *Nature, 407*, 506–508.

Högberg, M. N., Högberg, P., & Myrold, D. D. (2007). Is microbial community composition in boreal forest soils determined by pH, C–to–N ratio, the trees, or all three? *Oecologia, 150*, 590–601.
Horton, T. R., & Bruns, T. D. (2001). The molecular revolution in ectomycorrhizal ecology: peeking into the black–box. *Molecular Ecology, 10*, 1855–1871.
Huettl, R. F., & Zoettl, H. W. (1993). Liming as a mitigation tool in Germany declining forests – Reviewing results from former and recent trials. *Forest Ecology and Management, 61*, 325–338.
Hung, L., & Trappe, J. M. (1983). Growth variation between and within species of ectomycorrhizal fungi in response to pH in vitro. *Mycologia, 75*, 234–241.
Ingerslev, M., Skov, S., Sevel, L., & Pedersen, L. B. (2011). Element budgets of forest biomass combustion and ash fertilisation – A Danish case–study. *Biomass Bioenergy, 35*, 2697–2704.
Jacobson, S., Lundstrom, H., Nordlund, S., Sikstrom, U., & Pettersson, F. (2014). Is tree growth in boreal coniferous stands on mineral soils affected by the addition of wood ash? *Scandinavian Journal of Forest Research, 29*, 675–685.
Jacobsson, S. (1993). Inventering av svampfloran i kalkade provrutor på Hallandsåsen. *Jordstjärnan, 14*, 13–22.
Jandl, R., Lindner, M., Vesterdal, L., Bauwens, B., Baritz, R., Hagedorn, F., et al. (2007). How strongly can forest management influence soil carbon sequestration? *Geoderma, 137*, 253–268.
Jonsson, T., Kokalj, S., Finlay, R., & Erland, S. (1999). Ectomycorrhizal community structure in a limed spruce forest. *Mycological Research, 103*, 501–508.
Kårén, O., & Nylund, J. E. (1996). Effects of N–free fertilization on ectomycorrhiza community structure in Norway spruce stands in Southern Sweden. *Plant and Soil, 181*, 295–305.
Kjøller, R., & Clemmensen, K. E. (2009). Belowground ectomycorrhizal fungal communities respond to liming in three southern Swedish coniferous forest stands. *Forest Ecology and Management, 257*, 2217–2225.
Kjøller, R., Nilsson, L. O., Hansen, K., Schmidt, I. K., Vesterdal, L., & Gundersen, P. (2012). Dramatic changes in ectomycorrhizal community composition, root tip abundance and mycelial production along a stand–scale nitrogen deposition gradient. *New Phytologist, 194*, 278–286.
Klavina, D., Pennanen, T., Gaitnieks, T., Velmala, S., Lazdins, A., Lazdina, D., et al. (2016). The ectomycorrhizal community of conifer stands on peat soils 12 years after fertilization with wood ash. *Mycorrhiza, 26*, 153–160.
Kreutzer, K. (1995). Effects of forest liming on soil processes. *Plant and Soil, 169*, 447–470.
Kyaschenko, J., Clemmensen, K. E., Hagenbo, A., Karltun, E., & Lindahl, B. D. (2017). Shift in fungal communities and associated enzyme activities along an age gradient of managed *Pinus sylvestris* stands. *ISME Journal, 11*, 863–874.
Larsson, P. E., Uggla, E., & Westling, O. (2001). Långsiktige effekter av skogsmarkskalkning på mark– och markvattenkemi. In: *IVL Rapport*. IVL Svenska Miljöinstitutet AB, p. 52.
Leake, J. R., Read, D. J., Wicklow, D. T., & Söderström, B. (1997). Mycorrhizal fungi in terrestrial habitats. In *Environmental and Microbial Relationships* (pp. 281–301). Berlin: Springer.
Lehto, T. (1994a). Effects of liming and boron fertilization on mycorrhizas *of Picea abies*. *Plant and Soil, 163*, 65–68.
Lehto, T. (1994b). Effects of soil pH and calcium on mycorrhizas of *Picea abies*. *Plant and Soil, 163*, 69–75.
Lehto, T., & Malkonen, E. (1994). Effects of liming and boron fertilization on boron uptake of *Picea abies*. *Plant and Soil, 163*, 55–64.
Lilleskov, E. A., Fahey, T. J., & Lovett, G. M. (2001). Ectomycorrhizal fungal aboveground community change over an atmospheric nitrogen deposition gradient. *Ecological Applications, 11*, 397–410.
Lilleskov, E. A., Fahey, T. J., Horton, T. R., & Lovett, G. M. (2002a). Belowground ectomycorrhizal fungal community change over a nitrogen deposition gradient in Alaska. *Ecology, 83*, 104–115.

Lilleskov, E. A., Hobbie, E. A., & Fahey, T. J. (2002b). Ectomycorrhizal fungal taxa differing in response to nitrogen deposition also differ in pure culture organic nitrogen use and natural abundance of nitrogen isotopes. *New Phytologist, 154*, 219–231.

Lindahl, B. D., & Tunlid, A. (2015). Ectomycorrhizal fungi – Potential organic matter decomposers, yet not saprotrophs. *New Phytologist, 205*, 1443–1447.

LoBuglio, K. F. (1999). *Cenococcum*. In *Ectomycorrhizal fungi. Key general in profile* (pp. 287–309). Berlin: Springer.

Löfgren, S., Zetterberg, T., Larsson, P.-E., Cory, N., Klarqvist, M., Kronnäs, V., et al. (2008). *Skogsmarkskalkningens effekter på kemin i mark, grundvatten och ytvatten i SKOKAL–områdena 16 år efter behandling, Rapport 16*. Jönköping: Skogsstyrelsens förlag.

Lundström, U. S., Bain, D. C., Taylor, A. F. S., & van Hees, P. A. W. (2003a). Effects of acidification and its mitigation with lime and wood ash on forest soil processes: A review. *Water, Air, Soil Pollution Focus, 3*, 5–28.

Lundström, U. S., Bain, D. C., Taylor, A. F. S., van Hees, P. A. W., Geibe, C. E., Holmström, S. J. M., et al. (2003b). Effects of acidification and its mitigation with lime and wood ash on forest soil processes in southern Sweden. A joint multidisciplinary study. *Water, Air Soil Pollution Focus, 3*, 167–188.

Mahmood, S., Finlay, R. D., Wallander, H., & Erland, S. (2002). Ectomycorrhizal colonisation of roots and ash granules in a spruce forest treated with granulated wood ash. *Forest Ecology and Management, 160*, 65–74.

Majdi, H., & Viebke, C. R. (2004). Effects of fertilization with dolomite lime plus PK or wood ash on root distribution and morphology in a Norway spruce stand in Southwest Sweden. *Forest Science, 50*, 802–809.

Majdi, H., Truus, L., Johansson, U., Nylund, J. E., & Wallander, H. (2008). Effects of slash retention and wood ash addition on fine root biomass and production and fungal mycelium in a Norway spruce stand in SW Sweden. *Forest Ecology and Management, 255*, 2109–2117.

Mottonen, M., Lehto, T., & Aphalo, P. J. (2001). Growth dynamics and mycorrhizas of Norway spruce (*Picea abies*) seedlings in relation to boron supply. *Trees–Structure and Function, 15*, 319–326.

Nowotny, I., Dahne, J., Klingelhofer, D., & Rothe, G. M. (1998). Effect of artificial soil acidification and liming on growth and nutrient status of mycorrhizal roots of Norway spruce (*Picea abies* [L.] Karst.) *Plant and Soil, 199*, 29–40.

Nyberg, L., Lundstrom, U., Söderberg, U., Danielsson, R., & van Hees, P. (2001). Does soil acidification affect spruce needle chemical composition and tree growth? *Water, Air, Soil Pollution Focus, 1*, 241–263.

Palmberger, B., Egnell, G. (2006). Miljöeffekter av skogsbränsleuttag och askåterföring i Sverige. En syntes av Energimyndighetens forskningsprogram 1997 till 2004. Statens Energimyndighet.

Persson, H., & Ahlström, K. (1994). The effects of alkalizing compounds on fine–root growth in a Norway spruce stand in Southwest Sweden. *Journal of Environmental Science and Health, 29*, 803–820.

Qian, X. M., Kottke, I., & Oberwinkler, F. (1998). Influence of liming and acidification on the activity of the mycorrhizal communities in a *Picea abies* (L.) Karst. stand. *Plant and Soil, 199*, 99–109.

Read, D. J., & Perez-Moreno, J. (2003). Mycorrhizas and nutrient cycling in ecosystems: A journey towards relevance? *New Phytologist, 157*, 475–492.

Reid, C., & Watmough, S. A. (2014). Evaluating the effects of liming and wood–ash treatment on forest ecosystems through systematic meta–analysis. *Canadian Journal of Forest Research, 44*, 867–885.

Rineau, F., & Garbaye, J. (2009a). Does forest liming impact the enzymatic profiles of ectomycorrhizal communities through specialized fungal symbionts. *Mycorrhiza, 19*, 493–500.

Rineau, F., & Garbaye, J. (2009b). Effects of liming on ectomycorrhizal community structure in relation to soil horizons and tree hosts. *Fungal Ecology, 2*, 103–109.

Rineau, F., & Garbaye, J. (2010). Effects of liming on potential oxalate secretion and iron chelation of beech ectomycorrhizal root tips. *Microbial Ecology, 60*, 331–339.

Rineau, F., Maurice, J. P., Nys, C., Voiry, H., & Garbaye, J. (2010). Forest liming durably impact the communities of ectomycorrhizas and fungal epigeous fruiting bodies. *Annals of Forest Science, 67*, 1–12.

Rosenberg, M. S., Adams, D. C., & Gurevitch, J. (1999). *MetaWin: Statistical Software for Meta–Analysis, Version 2.0*. Sunderland: Sinauer Associates.

Rousk, J., Baath, E., Brookes, P. C., Lauber, C. L., Lozupone, C., Caporaso, J. G., et al. (2010). Soil bacterial and fungal communities across a pH gradient in an arable soil. *ISME Journal, 4*, 1340–1351.

R–project. (2014). *R: A Language and Environment for Statistical Computing*. Vienna: R Foundation for Statistical Computing.

Smith, S. E., & Read, D. J. (2008). *Mycorrhizal symbiosis*. London: Academic Press.

Stendell, E. R., Horton, T. R., & Bruns, T. D. (1999). Early effects of prescribed fire on the structure of the ectomycorrhizal fungus community in a Sierra Nevada ponderosa pine forest. *Mycological Research, 103*, 1353–1359.

Suz, L. M., Barsoum, N., Benham, S., Dietrich, H. P., Fetzer, K. D., Fischer, R., et al. (2014). Environmental drivers of ectomycorrhizal communities in Europe's temperate oak forests. *Molecular Ecology, 23*, 5628–5644.

Tamm, C. O. (1991). *Nitrogen in terrestrial habitats*. Berlin: Springer.

Taylor, A. F. S., & Alexander, I. (2005). The ectomycorrhizal symbiosis: Life in the real world. *Mycologist, 19*, 102–112.

Taylor, A. F. S., & Finlay, R. D. (2003). Effects of liming and ash application on below ground ectomycorrhizal community structure in two Norway spruce forests. *Water, Air, Soil Pollution Focus, 3*, 63–76.

Taylor, A. F. S., Brand, F., Read, D. J., Lewis, D. H., Fitter, A. H., & Alexander, I. J. (1992). Reaction of the natural Norway spruce mycorrhizal flora to liming and acid irrigation. In *Mycorrhizas in ecosystems* (p. 404). Wallingford: C.A.B. International.

Taylor, A. F. S., Martin, F., Read, D. J., & Schulze, E. D. (2000). Fungal diversity in ectomycorrhizal communities of Norway spruce (*Picea abies* (L.) Karst.) and Beech (*Fagus sylvatica* L.) along north–south transects in Europe. In *Carbon and nitrogen cycling in European forest ecosystems* (pp. 343–365). Berlin: Springer.

Tedersoo, L., May, T. W., & Smith, M. E. (2010). Ectomycorrhizal lifestyle in fungi: global diversity, distribution, and evolution of phylogenetic lineages. *Mycorrhiza, 20*, 217–263.

Tedersoo, L., Bahram, M., Polme, S., Koljalg, U., Yorou, N. S., Wijesundera, R., et al. (2014). Global diversity and geography of soil fungi. *Science, 346*, 6213.

Toljander, J. F., Eberhardt, U., Toljander, Y. K., Paul, L. R., & Taylor, A. F. S. (2006). Species composition of an ectomycorrhizal fungal community along a local nutrient gradient in a boreal forest. *New Phytologist, 170*, 873–884.

Uggla, E., Larsson, P.E., & Westling, O. (2003). Markkemi i kalkad skog. Lägesrapport 10 år efter kalkning. Årsrapport 2002. Effektoppföljning av Skogsstyrelsens program för åtgärder mot markförsurning. IVL rapport B 1536.

Visser, S. (1995). Ectomycorrhizal fungal succession in jack pine stands following wildfire. *New Phytologist, 129*, 389–401.

Wallander, H. (2000). Uptake of P from apatite by *Pinus sylvestris* seedlings colonised by different ectomycorrhizal fungi. *Plant and Soil, 218*, 249–256.

Wästerlund, I. (1982). Försvinner tallens mykorrhizasvamper vid gödsling? *Botanisk Tidskrift, 76*, 411–417.

Wiklund, K., Nilsson, L. O., & Jacobsen, S. (1995). Effect of irrigation, fertilization, and artificial drought on basidioma production in a Norway spruce stand. *Canadian Journal of Botany, 73*, 200–208.

Zimmermann, S., & Frey, B. (2002). Soil respiration and microbial properties in an acid forest soil: Effects of wood ash. *Soil Biology and Biochemistry, 34*, 1727–1737.

Chapter 15
β-Glucosidase Activity of Forest Soil as an Indicator of Soil Carbon Accumulation

Ewa Błońska and Jarosław Lasota

Abstract In this research study, β-glucosidase activity was used to assess differences occurring in soils as a result of the influence of different tree species. The aim of the study was to assess the effects of Scots pine (*Pinus sylvestris*) and pedunculate oak (*Quercus robur*) on the enzymatic activity of β-glucosidase, selected biochemical properties, as well as physical and chemical characteristics of soil. Sample plots were located in central Poland, in the Przedbórz forest district (51.09.59.50 °N, 20.00.24.25 °E). The test area was dominated by Cambisols (WRB 2002). Twenty research plots were established: 10 plots under pine and 10 plots under oak. The pH, soil texture, and organic carbon, nitrogen and base cation contents, β-glucosidase activity, and microbial biomass carbon, respiration were determined in the soil samples. The highest activity of β-glucosidase in this study was reported in the soils of pine stands. In the case of pine trees, the enzymatic activity is probably due to a simultaneous interaction of mycorrhizal fungi in the roots of pine trees with fungi present in genetic horizons which are rich in acidic decomposition products of organic matter derived from surface humus. The rate of organic substances transformation depends largely on the quality of soil organic matter, which is related to species composition of forest stands. We noted a significant correlation between β-glucosidase activity and carbon content, C/N ratio and acidity of soil. The profile distribution of β-glucosidase activity in the estimation of soil profiles decreased rapidly with soil depth.

Keywords Forest soil • Soil organic matter • Scots pine • Pedunculate oak

---

E. Błońska (✉) • J. Lasota
Department of Forest Soil Science, Faculty of Forestry, University of Agriculture in Krakow, Al. 29-go Listopada 46, 31-425 Kraków, Poland
e-mail: eblonska@ar.krakow.pl

M. Lukac et al. (eds.), *Soil Biological Communities and Ecosystem Resilience*, Sustainability in Plant and Crop Protection, DOI 10.1007/978-3-319-63336-7_15

15.1 β -Glucosidase Activity in Soils – Introduction

Through diverse composition of tree species the forest stand may modify various soil properties (physical, chemical and biochemical). Tree species affect soil organic matter (SOM) accumulation (Hobbie et al. 2006; Mueller et al. 2012), pH (Paluch and Gruba 2012; Łabaz et al. 2014) and the cation exchange capacity (Gruba and Mulder 2015). They may also affect soil microbial biomass and soil enzyme activities in a variety of ways, e.g. by changing the properties of SOM, pH modification or nutrient availability (Adamczyk et al. 2014; Kotroczó et al. 2014).

According to Janssens et al. (2002), roots are the key components of the forest ecosystems and the primary source of soil organic matter. Not only forest litter but also roots deliver organic matter that contains various components such as soluble sugars, organic acids, cellulose and lignin (Baldrian and Šnajdr 2011). The decomposition of soil organic matter depends on soil microorganisms and their sets of extracellular enzymes (Baldrian 2014). In comparison with arable soils, forest soils contain larger fungal biomass. The fungi in the forest environment are thought to play a major role in the early stages of decomposing substances such as lignin and cellulose. Kotroczó et al. (2014) suggested that plants change enzymatic activity to a longer extent through roots and their exudates than through the fallout of detritus. Decomposition of organic matter in soils is mediated by a complex set of extracellular enzymes that are produced by soil fungi and bacteria. β-glucosidase is an enzyme taking part in the decomposition of cellulose to glucose by the hydrolysis of glucosides. Cellulose is quantitatively the most important organic compound in the biosphere, therefore the products of its enzymatic hydrolysis are important as an energy sources for soil microorganisms. β-glucosidase has been detected in microorganisms, plants and animals. It is of significant importance in the carbon cycle (Sinsabaugh et al. 1991). β-glucosidase is very sensitive to different factors, therefore the determination of its activity might be helpful in soil quality monitoring (Piotrowska and Koper 2010). β-glucosidase activity has been found to be sensitive to soil management and proposed as a soil quality indicator (Dick et al. 1996; Naidja et al. 2000).

Only scattered information exists concerning the complexity of the relationships between tree species, soil properties and β-glucosidase activity. In this research study, β-glucosidase activity was used to assess differences occurring in soil as a result of the influence of different tree species. The aim of the study was to assess the effects of Scots pine (*Pinus sylvestris*) and pedunculate oak (*Quercus robur*) on the β-glucosidase activity and chemical properties of soil. We expect that pine forests cause a high organic matter accumulation and decrease decomposition rates. We hypothesized that in pine forests the simultaneous interactions of mycorrhizal fungi in the roots and fungi in the genetic horizon of soil stimulate β-glucosidase activity.

15.2 Materials and Methods

Sample plots were located in central Poland, in the Przedbórz Forest District (51.09.59.50°N; 20.00.24.25°E). The study area is characterized by the following conditions of climate: the average precipitation in the year amounts 649 mm, the average temperature is 7.4 °C. The test area is dominated by Brunic Arenosols (WRB 2014). Sample plots were located in the area with a predominance of glacio-fluvial sands and boulder clay. Twenty plots (1000 m^2) were established in pine (*P. sylvestris*) and oak (*Q. robur*) forest stands.

The samples for laboratory testing were collected from the organic horizon (Ofh), the first mineral horizon down to 20 cm depth (A) and the second mineral horizon (B) (from 20 to 35 cm depth horizon). Soil samples were taken during vegetation season (June – July). From each sampling point, 5 subsamples were taken to form a composite sample. For the determination of enzymatic activity one part of fresh samples of natural moisture were sieved through a sieve (diam: 2 mm) and stored at 4 °C before the analysis. Samples for physico-chemical properties were first dried at room temperature to an air-dry condition and then sieved. The pH of the samples was analysed in H_2O and KCl using the potentiometric method (Ostrowska et al. 1991). The soil texture was determined by laser diffraction (Analysette 22, Fritsch, Idar-Oberstein, Germany). The content of carbon (C) and nitrogen (N) were measured by an elemental analyzer (LECO CNS TrueMac Analyzer, St. Joseph, MI, USA). Base cations (BC: Ca^{2+}, Mg^{2+}, K^+, Na^+) concentrations were determined by inductivey coupled plasma-optical emission spectrometry (ICP-OES) (iCAP 6500DUO, Thermo Fisher Scientific, Cambridge, UK).

The activity of β-glucosidase (AB) was determined with by the method of Eivazi and Tabatabai (1988) using of p-nitrophenyl-β-D-glucopyranoside (PNG Sigma Chem. Co.) as a substrate. The enzyme activity was given in μg pNP (paranitrophenol) per 1 g of soil within 1 hour. In order to determine respiration, 50 g amount moist field samples were placed in 1liter plastic jars. Production of CO_2 was estimated using NaOH traps. Fifty ml polyethylene containers filled with 30 ml of 1.0 M NaOH were placed in the 1 liter jars on the soil surface. The jars were tightly sealed for one week. To determine microbial biomass carbon, 5 g of soil were weighed and fumigated with $CHCl_3$ in a glass desiccator (vacuum type) for 24 h at 25 C. Fumigated and non-fumigated samples were extracted with 0.5 M K_2SO_4 and then filtered with the Whatman filters (84 g/m^2) (Vance et al. 1987). The amount of organic C was determined quantitatively by oxidation of potassium dichromate (Jenkinson and Powlson 1976). Microbial biomass carbon was calculated from the difference between the quantity of total C extracted from fresh samples fumigated with $CHCl_3$ and the amount extracted in the non-fumigated control samples.

Statistical analyses were performed using STATISTICA 10 software. The U Mann-Whitney test was used to assess statistically significant differences between the mean values for soil properties under the study. The strength of the relationship between variables was established using the Pearson linear correlation coefficient (product-moment correlation coefficient). Principal component analysis (PCA) was used to reduce the number of variables in statistical datasets and visualize multivariate datasets. The PCA method was also applied in order to interpret factors depending on the kind of data set.

15.3 Depth Trends in β-Glucosidase Activities

The activities of β-glucosidase depending from soil profiles are shown in Fig. 15.1. Among the studied horizons, the highest β-glucosidase activities were detected in the organic horizon (Ofh) of soil under pine and oak stands.

The average activity was 375.56 mmol pNP·g^{-1}·h^{-1} in Ofh horizon under pine stands, and 223.81 mmol pNP·g^{-1}·h^{-1} in Ofh horizon under oak stands. In the humus-mineral horizon (A), the β-glucosidase activity decreased (approx. 70% regardless of the stand type), while in the B horizon it decreased about 90% (Table 15.1). The organic horizon (Ofh) could be characterized with the highest carbon and the base cation content (Table 15.2).

The tree stands had four basic types of impact the tree stands have on the soil, through: plant litter, the impact on microclimate, root systems, and the impact of rainfall which infiltrates treetops and flows down the trunks. The surface horizon could be characterized by the highest roots biomass at all sites (Stone et al. 2014). Roots provide root exudates and dead organic material. The remains and

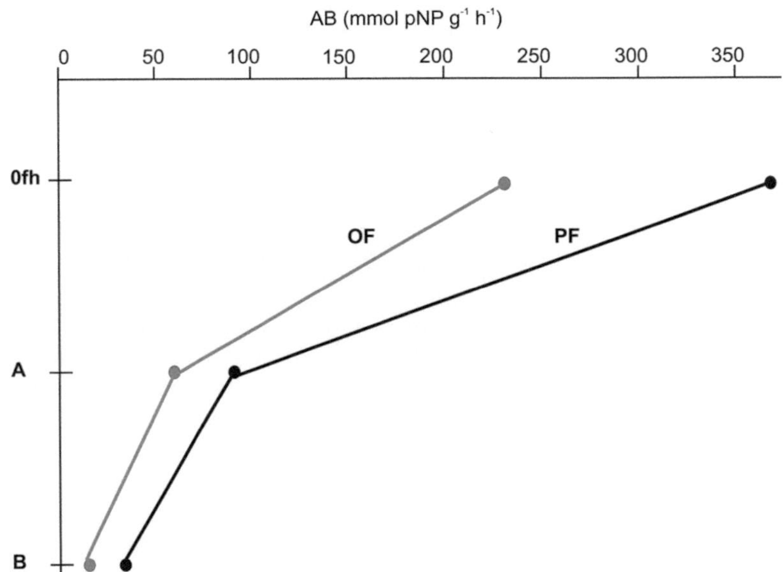

Fig. 15.1 Declines of β-glucosidase activity as a function of soil depth (*Ofh* humus-organic horizon, *A* humus mineral horizon, *B* mineral horizon, *OF* oak stands, *PF* pine stands)

Table 15.1 Comparison of biochemical properties of the examined soils under different tree stands (mean ± SD)

Stands*	AB	RESP	MBC
	Ofh		
PS	[a]362.21 ± 63.31	8.38 ± 1.57	306.16 ± 20.44
OS	[b]247.36 ± 91.41	22.98 ± 2.27	432.55 ± 23.98
	A		
PS	83.70 ± 53.41	n.d.	n.d.
OS	56.03 ± 41.37	n.d.	n.d.
	B		
PS	32.32 ± 22.04	n.d.	n.d.
OS	27.06 ± 11.21	n.d.	n.d.

**PS* pine stands, *OS* oak stands, *AB* β-glucosidase activity, *RESP* respiration, *MBC* microbial biomass carbon. *Ofh* organic horizon, *A* humus mineral horizon, *B* mineral horizon. Different lower case letters in the upper index of mean values indicate significant differences ($P \leq 0.05$)

exudates of roots are important sources of nutrients for microorganisms, and increase their numbers and activity, which leads to a rise in the enzymatic activity (Kumar et al. 2006). Ectomycorrhizal fungi show a considerable enzymatic activity (Colpaert and Laere 2006), are responsible for mobilizing nutrients required by plants by decomposing soil organic matter, which stimulates the enzymatic activity (Courty et al. 2006).

According to Esen (1993), β-glucosidase is produced by a variety of organisms (plants, animals, fungi, and bacteria), but comes mostly from fungi, such as Actinomucor and Mortierella. High concentrations of β-glucosidase are present on the surface of organic debris and fine fraction (< 2mm) where fungi predominate (Hayano and Tubaki 1985). The distribution of β-glucosidase activity in the soil is associated with the profile distribution of soil organic matter. The high activity of enzymes in upper horizons of the soil is associated with the presence of microorganisms and their activity and the content of the organic matter constituting the feeding base. This patterns are linked mainly to the location of humus in the soil and the amount of available carbon substrates for the microorganisms present in the soil as well as the decreasing quantity of enzymes with the increasing depth of the soil profile. The observed decline in enzyme activity in the soil profile is fully consistent with the observations of Eivazi and Tabatabai (1990), Trasar-Cepeda et al. (1998), Kandeler et al. (1999) and Wang and Lu (2006). Additionally, microbial biomass and substrate pools generally decline with soil depth (Kumar et al. 2006) and the microbial community structure often changes with depth (Eilers et al. 2012). Not only soil organic matter affects the enzyme activity but numerous soil properties can influence microbial communities, including pH (Rousk et al. 2010), temperature and moisture (Brzostek and Finzi 2012; Wallenstein et al. 2012), redox status (DeAngelis et al. 2010) and texture (Sessitsch et al. 2001). All of these properties can change with depth.

Table 15.2 Comparison of soil properties under different tree stands* (mean ± SD)

Stand	pH_{H2O}	pH_{KCl}	C	N	C/N	Al.	Ca	K	Mg	Na	Sand	Silt	Clay
	Ofh												
PS	^{a}3.69 ± 0.22	^{a}2.79 ± 0.23	^{a}19.76 ± 11.2	^{a}0.68 ±0.4	^{a}27.24 ± 3.5	^{a}4.56 ± 2.43	^{a}85.35 ± 33.4	^{a}19.15 ± 7.7	^{a}6.53 ± 3.91	^{a}2.16 ±0.56	n.d.	n.d.	n.d.
OS	^{b}4.23 ± 0.30	^{b}3.50 ± 0.35	^{b}2.79 ± 2.22	^{b}0.16 ± 0.11	^{b}17.43 ± 2.23	^{b}1.68 ± 0.89	^{b}11.45 ± 2.96	^{b}3.53 ± 1.53	^{b}1.32 ± 0.53	^{b}0.80 ±0.33	n.d.	n.d.	n.d.
	A												
PS	^{a}3.85 ± 0.21	^{a}3.08 ± 0.21	2.33 ± 0.87	0.09 ± 0.03	^{a}23.6± 4.17	1.26± 0.93	6.27 ± 2.87	1.24 ± 0.75	0.52 ± 0.21	0.7± 0.23	^{a}92± 2.01	5 ± 1.66	2 ± 0.91
OS	^{b}4.36± 0.40	^{b}3.84 ± 0.23	1.4 ± 0.88	0.09 ± 0.04	^{b}17.27± 2.79	1.53 ± 0.51	6.12 ± 1.56	1.44 ± 0.74	0.66 ± 0.17	0.78 ±0.23	^{b}84 ±11	15 ± 10	2 ± 1.67
	B												
PS	^{a}4.65 ± 0.14	^{a}4.35 ± 0.16	1.01 ± 0.67	0.06 ± 0.04	19.67 ± 11.78	0.15 ±0.03	^{a}2.23 ± 0.23	^{a}0.24 ± 0.45	^{a}0.24 ± 0.02	^{a}0.40 ±0.16	^{a}95± 2.16	5 ± 2.30	1 ± 1.03
OS	^{b}5.15 ± 0.68	^{b}3.75 ± 0.29	0.76 ± 0.24	0.04 ± 0.02	17.21 ± 2.28	0.92 ± 0.53	^{b}33.59 ± 10.69	^{b}4.18 ± 2.57	^{b}2.87 ± 5.82	^{b}0.97 ±0.35	^{b}88± 12.0	9 ± 11.6	2 ± 1.5

*PS pine stands, OS oak stands, C carbon content, N nitrogen content. Lower case letters in the upper index of mean values indicate significant differences ($\underline{P} < 0.05$; *Ofh* humus-organic horizon)

15.4 The β-Glucosidase Activity and Its Relationship with Other Soil Properties

In this research study, β-glucosidase activity was used to assess differences occurring in soils as a result of the influence of tree species. The highest β-glucosidase activity was found in soils under pine stand. Significant differences in β-glucosidase activity were noted between the soils of pine and oak stands (Table 15.1). The smallest difference between the soils of pine and oak stand in β-glucosidase activity (30%) was observed in the mineral horizon (A and B), while a greater difference (60%) was recorded in the organic horizon (Ofh) (Table 15.1). A projection of the variables on the factor plane clearly demonstrated correlation between β-glucosidase activity and other soil properties of organic horizon and the type of the forest stand. Two main factors had a significant total impact (83.4%) on the variance of the variables. Factor 1 explained 74.05% of the variance of the examined properties, and factor 2 explained 9.39% of the variance (Fig. 15.2). In contrast, the soil of oak stands was characterized by a high pH, respiration and microbial biomass carbon.

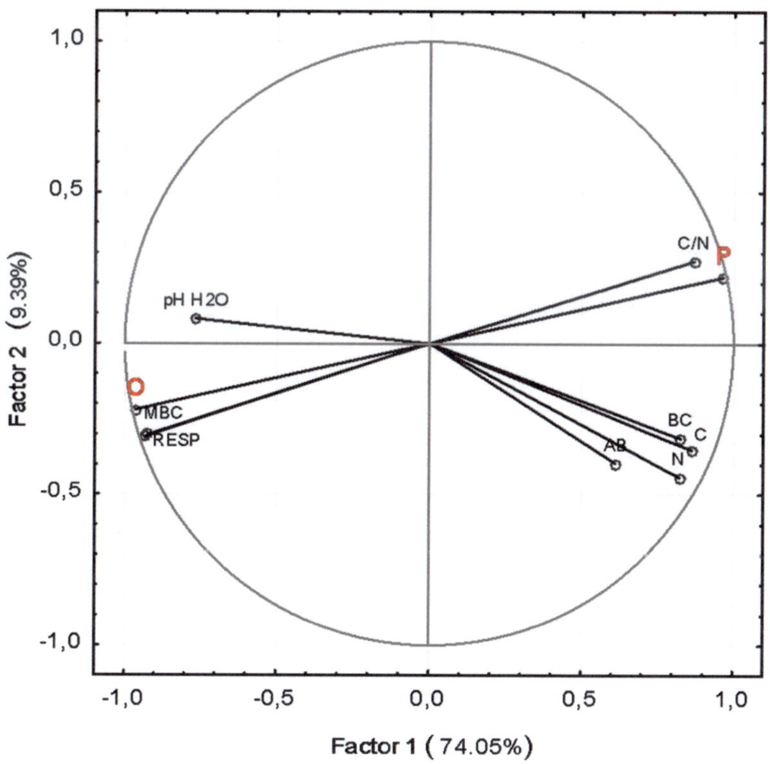

Fig. 15.2 Projection of the variables on the factor-plane in soils under different tree stands (*O* oak stand, *P* pine stand, *AB* β-glucosidase activity, *MBC* microbial biomass carbon, *RESP* respiration, *BC* base cation content, *N* nitrogen content, *C* carbon content)

Table 15.3 Correlation between β-glucosidase activity (AB) and soil characteristics

	pH_{H2O}	pH_{KCl}	C	N	Ca	K	Mg	Na
AB	Ofh							
	−0.503*	−0.501*	0.542*	0.507*	0.218	0.291	0.224	0.575
	A							
	−0.099	−0.273	0.646*	0.474	0.630*	−0.013	0.355	0.327
	B							
	−0.001	−0.321	0.257	−0.268	−0.278	−0.308	−0.240	−0.427
	All horizons							
	−0.513*	−0.603*	0.758*	0.703*	0.550*	0.676*	0.390*	0.683*

*= Significant for $P \leq 0.05$. *Ofh* organic horizon, *A* humus mineral horizon, *B* mineral horizon

The soil of pine stands contained a lot of weakly decomposed organic matter (high C/N ratio). β-glucosidase activity was associated with the carbon and nitrogen content (Table 15.3). The β-glucosidase activity increased with the increase of C content. Generally, coniferous forests have a more acidifying effect on the soil than deciduous forest (Augusto et al. 1998; Gruba and Mulder 2008; Mueller et al. 2012; Paluch and Gruba 2012).

According to Błońska et al. (2016) decrease in pH was probably caused by tree species related with modification of the relationship between H^+ and Al^{3+}, which affected the composition of cation exchange capacity (CEC), thereby leading to shifts in microbial community composition and size, as well as enzyme dynamics. In the case of pine trees, the β-glucosidase activity is probably due to a simultaneous interaction of mycorrhizal fungi in the root of pine trees with fungi present in the horizons which are rich in acidic decomposition products of organic matter derived from surface humus.

Myśków (1987) proved that a decrease in pH leads to an increase in the percentage of fungi and enzymatic activity associated with them in relation to the activity of soil bacteria. Januszek (1993) presented the number of bacteria and fungi in different types of forest humus. In mull, a type of humus typical of deciduous stands, the total number of fungi amounted to 150 thousand per cm^3 of soil, and the total number of bacteria to over 7.5 million per cm^3 of soil. In moder, a type of humus found at sites with pine stands, the number of fungi amounted to over 1.2 million per cm^3 of soil, and the number of bacteria to over 1 million per cm^3 of soil. In an early study Błońska (2015) revealed relation between the β-glucosidase activity and root system. The highest biomass of fine roots was associated with ash trees, in the soils of which the highest β-glucosidase activity was reported. The lowest biomass of fine roots was observed in the beech and pine tree stands. The roots of beech and pine trees are colonized by ectomycorrhizal fungi, whereas the roots of ash trees by arbuscular mycorrhizal fungi, which accounts for their differing strategies of nutrient acquisition. Additionally, the author noted that soil samples under the ash stand had the best aggregate structure, as evidenced by the number of aggregates with a diameter of 1–2 mm and 2–5 mm, which was much higher than the number of these aggregates in soils under pine, beech and oak stands.

Other parameters expressing the biochemical activity of soil arranged inversely than the activity of β-glucosidase. The statistically significantly higher respiration and microbial biomass carbon in soils of oak stands were noted (Fig. 15.2). Respiration in organic horizon of soils under pine stands was 60% lower compared to the organic horizon in soils of oak stands. Raich and Tufekcioglu (2000) reported that the soils of deciduous stands showed higher respiration rate in comparison with coniferous stand. The level of soil respiration is strongly influenced by the quality of organic litterfall as well as the quantity and quality of organic remains and exudates provided by the root system. Microbial biomass C in soils of pine stands was 30% lower than in the soils of oak stands. The microbial biomass C had a strong positive correlation with C and N contents.

According to Peacock et al. (2001), the quantity and quality of organic litter input is one of the most important factors affecting microbial biomass. The activity of β-glucosidase may serve as a soil quality monitoring indicator, since it plays an import ant role in the cycle of organic matter in the soil. This study reported a correlation between β-glucosidase activity and carbon content and microbial biomass carbon. Microbial biomass C is connected with the biomass of bacteria and fungi, regardless of their enzymatic abilities. We confirmed that the forest type would be a primary driver of enzyme activities in soils due to differences in leaf litter chemistry. Different tree species have different impact on the quantity and quality of the supplied organic matter, as wel as on the properties and processes that determine the fertility and on soil quality parameters, such as pH, the amount of N and C, or the circulation of nutrients. Organic matter that comes from deciduous and coniferous species is characterized by varying rates of decomposition. Faster decomposition of organic matter occurs in deciduous or in coniferous tree stands where there is a high proportion of deciduous species (Légaré et al. 2005; Bonifacio et al. 2008).

References

Adamczyk, B., Kilpeläinen, P., Kitunen, V., & Smolander, A. (2014). Potential activities of enzymes involved in N, C, P and S cycling in boreal forest soil under different tree species. *Pedobiologia, 57*, 97–102.

Augusto, L., Bonnaud, P., & Ranger, J. (1998). Impact of tree species on forest soil acidification. *Forest Ecology and Management, 105*, 67–78.

Baldrian, P. (2014). Distribution of extracellular enzymes in soils: spatial heterogeneity and determining factors at various scales. *Soil Science Society of America Journal, 78*, 11–18.

Baldrian, P., & Šnajdr, J. (2011). Lignocellulose-degrading enzymes in soil. In G. Shukla & A. Varma (Eds.), *Soil enzymology* (pp. 167–186). Berlin: Springer-Verlag.

Błońska, E. (2015). Effect of stand species composition on the enzyme activity and organic matter stabilization in forest soil. Scientific papers of University of Agriculture in Krakow No. 527. *Dissertation, Paper No., 404*, 1–05.

Błońska, E., Lasota, J., & Gruba, P. (2016). Effect of temperate forest tree species on soil dehydrogenase and urease activities in relation to other properties of soil derived from loess and glaciofluvial sand. *Ecological Research, 31*, 655–664.

Bonifacio, E., Caimi, A., Falsone, G., Trofimov, S., Zanini, E., & Godbold, D. L. (2008). Soil properties under Norway spruce differ in spruce dominated and mixed broadleaf forests of the Southern Taiga. *Plant and Soil, 308*, 149–159.

Brzostek, E. R., & Finzi, A. C. (2012). Seasonal variation in the temperature sensitivity of proteolytic enzyme activity in temperate forest soils. *Journal of Geophysical Research, 117*, G01018.

Colpaert, J. V., & Laere, A. (2006). A comparison of the extracellular enzyme activities of two ectomycorrhizal and a leaf-saprotrophic basidiomycete colonizing beech leaf litter. *New Phytologist, 134*, 133–141.

Courty, P. E., Pouysegur, R., Buée, M., & Garbaye, J. (2006). Laccase and phosphatase activities of the dominant ectomycorrhizal types in a lowland oak forest. *Soil Biology and Biochemistry, 38*, 1219–1222.

DeAngelis, K. M., Silver, W. L., Thompson, A. W., & Firestone, M. K. (2010). Microbial communities acclimate to recurring changes in soil redox potential status. *Environmental Microbiology, 12*, 3137–3149.

Dick, R.P., Breakwell, D.P., Turco, R.F. (1996). *Soil enzyme activities and biodiversity measurements as integrative microbiological indicators* (Methods for assessing soil quality, Vol. 9, pp. 9–17). Madison: Soil Science Society of America.

Eilers, K. G., Debenport, S., Anderson, S., & Fierer, N. (2012). Digging deeper to find unique microbial communities in soil: The strong effect of depth on the structure of bacterial and archaeal communities in soil. *Soil Biology and Biochemistry, 50*, 58–65.

Eivazi, F., & Tabatabai, M. A. (1988). Glucosidases and galactosidases in soils. *Soil Biology and Biochemistry, 20*, 601–606.

Eivazi, F., & Tabatabai, M. A. (1990). Factors affecting glucosidase and galactosidase activities in soils. *Soil Biology and Biochemistry, 22*, 891–897.

Esen, A. (1993). β-Glucosidases: overview. In: A. Esen (Ed.), *Beta-Glucosidases: Biochemistry and molecular biology* (pp. 1–14). Washington, DC: American Chemical Society.

Gruba, P., & Mulder, J. (2008). Relationship between aluminum in soils and soil water in mineral horizons of a range of acid forest soils. *Soil Science Society of America Journal, 72*, 1150–1157.

Gruba, P., & Mulder, J. (2015). Tree species affect cation exchange capacity (CEC) and cation binding properties of organic matter in acid forest soils. *Science of the Total Environment, 511*, 655–662.

Hayano, K., & Tubaki, K. (1985). Origin and properties of β-glucosidase activity of tomato-field soil. *Soil Biology and Biochemistry, 17*, 553–557.

Hobbie, S. E., Reich, P. B., Oleksyn, J., Ogdahl, M., Zytkowiak, R., Hale, C., & Karolewski, P. (2006). Tree species effects on decomposition and forest floor dynamics in a common garden. *Ecology, 87*, 2288–2297.

Janssens, J. A., Sampson, D. A., Curiel-Yuste, J., Carrara, A., & Ceulemans, R. (2002). The carbon cost of fine root turnover in a Scots pine forest. *Forest Ecology and Management, 168*, 231–240.

Januszek, K. (1993). Seasonal changes of enzyme activity in mor, moder and mull humus of selected forest soils in the Western Beskid Mountains. *Folia Forestalia Polonica, 35*, 59–75.

Jenkinson, D. S., & Powlson, D. S. (1976). The effects of biocidal treatments on metabolism in soil - I. Fumigation with chloroform. *Soil Biology and Biochemistry, 8*, 167–177.

Kandeler, E., Tscherko, D., & Spiegel, H. (1999). Long-term monitoring of microbial biomass, N mineralisation and enzyme activities of a Chernozem under different tillage management. *Biology and Fertility of Soils, 28*, 343–351.

Kotroczó, Z., Veres, Z., Fekete, I., Krakomperger, Z., Tóth, J. A., Lajtha, K., & Tóthmérész, B. (2014). Soil enzyme activity in response to long-term organic matter manipulation. *Soil Biology and Biochemistry, 70*, 237–243.

Kumar, R., Pandey, S., & Pandey, A. (2006). Plant roots and carbon sequestration. *Current Science, 91*, 885–890.

Łabaz, B., Gałka, B., Bogacz, A., Waroszewski, J., & Kabała, C. (2014). Factors influencing humus forms and forest litter properties in the mid-mountains under temperate climate of southwestern Poland. *Geoderma, 230-231*, 265–273.
Légaré, S., Paré, D., & Bergeron, Y. (2005). Influence of aspen on forest floor properties in black spruce-dominated stands. *Plant and Soil, 275*, 207–220.
Mueller, K. E., Eissenstat, D. M., Hobbie, S. E., Oleksyn, J., Jagodzinski, A. M., Reich, P. B., Chadwick, O. A., & Chorover, J. (2012). Tree species effects on coupled cycles of carbon, nitrogen, and acidity in mineral soils at a common garden experiment. *Biogeochemistry, 111*, 601–614.
Myśków, W. (1987). Wpływ głębokiej uprawy i zmianowania roślin na właściwości biologiczne gleby. *Polish Journal of Agronomy, 90*, 7–23.
Naidja, A., Huang, P. M., & Bollag, J. M. (2000). Enzyme-clay interactions and their impact on transformations of natural and anthropogenic organic compounds in soil. *Journal of Environmental Quality, 29*, 677–691.
Ostrowska A, Gawliński S, Szczubiałka Z (1991) *Methods of analysis and assessment of soil and plant properties*. Environmental Protection Institute, Warszawa, PL, p. 334.
Paluch, J. G., & Gruba, P. (2012). Effect of local species composition on topsoil properties in mixed stands with silver fir (*Abies alba* Mill.) *Forestry, 85*, 413–426.
Peacocka, A. D., Mullen, M. D., Ringelberg, D. B., Tyler, D. D., Hedrick, D. B., Gale, P. M., & White, D. C. (2001). Soil microbial community responses to dairy manure or ammonium nitrate applications. *Soil Biology and Biochemistry, 33*, 1011–1019.
Piotrowska, A., & Koper, J. (2010). Soil β-glucosidase activity under winter wheat cultivated in crop rotation systems depleting and enriching the soil in organic matter. *Journal of Elementology, 15*, 593–600.
Raich, J. W., & Tufekcioglu, A. (2000). Vegetation and soil respiration: correlations and controls. *Biogeochemistry, 48*, 71–90.
Rousk, J., Baath, E., Brookes, P. C., Lauber, C. L., Lozupone, C., Caporaso, J. G., Knight, R., & Fierer, N. (2010). Soil bacterial and fungal communities across a pH gradient in arable soil. *ISME Journal, 4*, 1340–1351.
Sessitsch, A., Weilharter, A., Gerzabek, M. H., Kirchmann, H., & Kandeler, E. (2001). Microbial population structures in soil particle size fractions of a long-term fertilizer field experiment. *Applied and Environmental Microbiology, 67*, 4215–4224.
Sinsabaugh, R. L., Antabus, R. K., & Linkins, A. (1991). An enzymic approach to the analysis of microbial activity during plant litter decomposition. *Agriculture, Ecosystem and Environment, 34*, 43–54.
Stone, M. M., DeForest, J. L., & Plante, A. F. (2014). Changes in extracellular enzyme activity and microbial community structure with soil depth at the Luquillo Critical Zone Observatory. *Soil Biology and Biochemistry, 75*, 237–247.
Trasar-Cepeda, C., Leirós, C., Gil-Sotres, F., & Seoane, S. (1998). Towards a biochemical quality index for soils: an expression relating several biological and biochemical properties. *Biology and Fertility of Soils, 26*, 100–106.
Vance, E. D., Brookes, P. C., & Jenkinson, D. S. (1987). An extraction method for measuring soil microbial biomass C. *Soil Biology and Biochemistry, 19*, 703–707.
Wallenstein, M. D., Haddix, M. L., Lee, D. D., Conant, R. T., & Paul, E. A. (2012). A litter-slurry technique elucidates the key role of enzyme production and microbial dynamics in temperature sensitivity of organic matter decomposition. *Soil Biology and Biochemistry, 47*, 18–26.
Wang, X. C., & Lu, Q. (2006). Beta-glucosidase activity in paddy soils of the Taihu Lake region, China. *Pedosphere, 16*, 118–124.
WRB (World Reference Base For Soil Resource) (2002) FAO, ISRIC and ISSS

Chapter 16
Linking Ecosystem Variability and Carbon Sequestration: Estimation of Sequestered Carbon in Natural Forests and Perennial Crops

Jelena Milovanović, Uroš Radojević, and Mirjana Šijačić-Nikolić

Abstract Forests have a key role in climate change. One of the most important forest carbon stocks is forest soil. Globally it is estimated that forest soil carbon stock may vary from equal to twice that of forest vegetation. Because of that it is important to have estimates of soil carbon and soil biodiversity which plays a key role in soil carbon sequestration. This can be done by either through repeated measurements or by application of dynamic models. In this chapter up to date program solutions which can be used to model carbon sequestration under different forest management practices are reviewed. The application of modeling tools with the aim of sequestered carbon estimation and exploring its linkages to ecosystem variability are presented through literature review and a case study from Serbia. Moreover, possibilities of modeling carbon sequestration in perennial grass species are elaborated, emphasizing simplified ecosystem of perennial crop *Miscanthus x giganteus*, an attractive agro-energy crop suitable for short rotation plantation establishment, assumed as neutral for CO_2, that is retained by underground organs.

Keywords Carbon dynamics • Modeling • Natural stands • Simplified ecosystems

---

J. Milovanović (✉) • U. Radojević
Faculty of Applied Ecology Futura, Singidunum University, Belgrade, Serbia
e-mail: jelami@ptt.rs

M. Šijačić-Nikolić
Faculty of Forestry, University of Belgrade, Belgrade, Serbia

M. Lukac et al. (eds.), *Soil Biological Communities and Ecosystem Resilience*,
Sustainability in Plant and Crop Protection, DOI 10.1007/978-3-319-63336-7_16

Acronyms

Agro-energy crops	crops used for production of biofuels
ALMANAC	model that simulates crop growth, competition, light interception, biomass accumulation, water use, nutrient uptake
CAI	current annual increment
CO2FIX	model for quantifying carbon sequestration in forest ecosystems and wood products
Degree days above zero	measure of heat accumulation by plants at temperatures above zero
EPIC	generalized crop model that simulates daily crop growth on a hectare scale
MISCANMOD	model for estimating biomass production from *Miscanthus* throughout Europe
Rotation	growth period required to derive maximum value from a stand of timber
Yasso	soil carbon model
Yield table	tabular statement which summarizes per unit area basis of all essential data relating to the development of managed forests
WIMOVAC	model of the carbon balance of vegetation and related systems which allows prediction of responses to climate change

16.1 Introduction

Terrestrial part of the biosphere has an important role in the global carbon cycle, and due to the actual issues related to climate change there is an increased interest in this process, specially related to sequestration of carbon in the biosphere. During the last decade of the past century the average carbon sequestration in the biosphere was estimated at 2.3 billion tons per year, which is 36% of the total carbon emission coming from combustion of fossil fuels (Chapin et al. 2002). Forest ecosystems play an important role in the global carbon cycle since they have a large potential for carbon sequestration within forest biomass and forest soils. However, they can also represent a source of carbon due to natural processes such as ageing, seasonal variations in carbon dynamics, and anthropogenic impacts related to forest management or forest degradation.

Soils also represent an important reservoir of carbon since they can store it even after plants which have been grown on them have been removed (forestry, agriculture, etc.). Soils are especially important because they are able to store carbon for longer periods of time then organic biomass. It is estimated that soils contain more than 4.5 times carbon than living biomass (Lal 2004). Importance of soil's role as a carbon reservoir has also been recognized within Articles 3.3 and 3.4 of the Kyoto Protocol which state that reporting of changes in soil organic carbon which result from land-use change is necessary within annual GHG inventories. This has contributed to an increase in cultivation of agro-energy crops which are used for biofuel production. This is thought to have an offsetting effect on anthropogenic CO_2 emissions through soil carbon sequestration, organic biomass carbon sequestration as well as fossil fuel substitution.

So far both food crops (for example corn) and lingo-cellulosic biomass crops have been used to produce biofuels. However, this has led to a food vs. fuel debate since there is limited availability of land for growing crops. Currently, many countries and regions in the world already feel pressure in land available for critical socioeconomic activities, converting the existing cropland or developing new land for biofuel production raises immediate concerns, including the food versus fuel debate, effects on the livelihoods of small-scale farmers, pastoralists and indigenous people, threat to nature conservation, and possible increase of carbon emissions, also, land use change usually causes changes in water use, and consequently, biofuel production may aggravate water stress, which is already a growing worldwide issue (Service 2007). Possible solutions for this problem include using marginal, degraded and/or abandoned agricultural land as a source of biomass for biofuel production. According to Campbell et al. (2008) abandoned land globally available for the production of bioenergy crops varies between 385 and 472 Mha.

Degraded land, defined as "areas where human activities have induced soil and/or vegetable degradation" (Hoogwijk et al. 2003), is assumed to have a potential between 430 and 580 Mha. Plants which could be suitable for agro-energy crops should then satisfy several criteria which include suitability for growth on marginal/degraded lands, fast growth of biomass, and none or low negative environmental impact. Possible species include fast growing woody species like poplars, willows or eucalyptus, as well as high biomass yield perennial grasses like *Miscanthus x giganteus*.

The final aim of a model-based approach is to determine the applicability of available soil and forest biomass models in determining the links between forest soil biodiversity and forest soil carbon cycle. In this chapter up to date program solutions which can be used to model carbon sequestration under different forest management practices is reviewed. The application of modeling tools with the aim of sequestered carbon estimation and exploring its linkages to ecosystem variability is presented through latest literature review and a case study from Serbia.

16.2 Modeling Forest Ecosystems Carbon Dynamics

In general, modeling of forest ecosystems, and of carbon cycle within them, has to encompass all relevant factors and processes such as environmental conditions, photosynthesis, autotrophic and heterotrophic respiration, carbon allocation and differences associated with different tree species. Different models use different approaches for representing these parameters.

Modeling of forest ecosystems has from its starting applications been primarily oriented towards forestry activities. Traditional forestry revolves around the cultivation of woody species for obtaining wood products. It includes vegetation, soil and people management as the total resources which jointly have to achieve multiple objectives. The key to responsible management of forest ecosystems is twofold and it includes an understanding of how forest ecosystems function, and secondly, that the understanding and knowledge be successfully used to meet social needs without or with minimal impact to the environment (Šijačić-Nikolić et al. 2014).

Models used in forestry can be grouped into several types based on the process which they try to simulate. The first and most common are models of growth and yield of timber that predict increment during a certain period. Initially, these models were focused on trees of individual species which are of same age and are located within nurseries. As the application of modeling became more sophisticated, different models had started to include the mixing of tree species and trees of different ages into assessments growth and yield.

As importance of forests and their environmental values, which are beyond the timber production, started to become more apparent the focus of research shifted, and further efforts were made towards modeling these "new" processes. Management of the forest ecosystems needed to recognize other values that forests carry, such as: that forests are a habitat of wild animals; that they provide boundless opportunities for recreation; that they have the function of holding and purification of water; and that these are only some of the benefits from forests aside from the classical timber production.

In forest management, it is important to anticipate the effects of the implementation of various alternatives. Although there is a vast scientific knowledge about the relationship between the existing form of forest structure and characteristics of the ecosystem, relevant information which should be derived from this is often difficult to interpret and apply. In order to use models that incorporate best available knowledge, it is necessary to incorporate the best available knowledge on the biological systems that is accessible, and that the results are presented in a usable way. The development, evaluation and adaptation of new decision-making processes and their supporting modern software, based on adequate models is a very important endeavor in forest management.

Based on the approach which models use to describe and simulate different processes they can be differentiated between empirical and mechanistic models. While empirical models are more rooted in statistics since they base their projections on a large sample of observed datasets, mechanistic model aim to imitate natural processes through the use of different equations.

Today models in forestry are used for different purposes which include growth and yield predictions, forest regeneration, tree mortality, habitat modeling, harvest planning as well as carbon dynamics and sequestration.

One of the models which has been used in various number of published studies to estimate carbon sequestration is CO2FIX model (Schelhaas et al. 2004; Masera et al. 2003) developed within the CASFOR I and CASFOR II ("Carbon sequestration in afforestation and sustainable forest management") projects. This model can be used for temporal analysis of forest carbon cycle and also for carbon calculation related to projects aimed at climate change mitigation within forestry sector, such as CDM A/R (Clean Development Mechanism Afforestation and Reforestation) or REDD (Reducing Emissions from Deforestation and Forest Degradation) projects.

The latest version of this program CO2FIX 3.2 conducts simulations at ecosystem level in order to quantify carbon reserves and fluxes in forests by calculating the change in all relevant spheres and stocks. The model is divided in six modules, four of them are designed to follow the flow of carbon to and from forest ecosystems and include: Biomass module, Litter module, Forest products module and Bioenergy module. The remaining two modules are for additional options related to carbon credit calculation (Carbon credits calculation module) and finance (Financial module).

Stocks and flows of carbon within the living forest biomass (both above and below ground) are estimated by separation on specific cohorts (Reed 1980). Cohort consists of trees of same or similar species and age which have similar characteristics so they can be treated as a same entity for the process of modeling. Carbon stored in a forest ecosystem can then be calculated as the sum of carbon stored within the biomass of each cohort. Changes in biomass that influence the carbon content during one-time step (one year) are balance of the original biomass from previous time step, increases due to biomass growth and reductions due to turnover of branches, foliage and roots, tree mortality related to senescence, harvest and logging associated mortality.

CO2FIX model differentiates four types of tree biomass which include stem, foliage, branches and roots. Biomass growth is simulated from growth rate of stem volume (gross annual increment) which can be obtained from yield tables for different tree species. Based on the growth rate of tree volume and allocation coefficients other biomass compartments growth (branches, foliage and roots) is then calculated (Nabuurs and Mohren 1993).

Tree mortality due to senescence can be modeled in two ways in CO2FIX, as a function of tree age or as a function of relative biomass (current biomass related to the maximal biomass). Besides tree mortality, in order to precisely asses the carbon cycle dynamics it is important to model the annual turnover of foliage, branches and roots. Turnover of biomass from these compartments influences the carbon content in forest litter and soil. Biomass turnover is calculated by multiplying the current biomass content in one compartment and turnover constant which has values between 0 and 1 depending on the particular compartment and tree species (Schelhaas et al. 2004).

If the modeled forest ecosystem is managed it is very likely that tree biomass will be removed from it through some form of thinning and logging. Biomass harvested in this way is always subtracted from the total biomass and is then analyzed within Forest products module and Litter module depending on its type and usage. Forest logging can lead to higher tree mortality of remaining trees due to damages that they suffer during the logging process (Pinard and Putz 1997). The extent of this effect depends on the tree species, type of logging and type of used equipment.

Modeling of litter and soil carbon in CO2FIX is based on Yasso model (Liski et al. 2005), which consists of 3 sub-modules for litter and 5 sub-modules for decomposition. Litter is divided in non-woody litter (foliage and fine roots), fine woody litter (branches and coarse roots) and coarse woody litter (stems and stumps). Each of these sub-modules has a specific rate of decay which determines the amount of degradation during one-time step. During the decomposition process carbon within biomass in these sub-models is partially lost form the system, mostly as CO_2 to the atmosphere, and also transferred to decomposition sub-modules which include extractives, cellulose, lignin compound, humus 1 and humus 2 sub-modules. Modeling of biomass decomposition within forest litter is influenced by temperature and humidity, which is why it is important to have relevant meteorological data.

Calculation of carbon credits is based on CDM A/R methodology introduced at the CoP9 in 2003 which establishes two types of certified emissions reductions (CER), these are temporary tCER and long-term lCER. One CER is equal to 1 ton of CO_2 equivalent. Difference between tCER and lCER is in their duration. tCER lasts till the end of the following commitment period, while a lCER will last till the end of the whole CDM project for which it has been issued.

Forest carbon cycle dynamics modeling based on CO2FIX model was used in the research for several case studies in regularly managed forest units in Serbia with different dominant species such as European black pine, beech and Norway spruce (Radojević et al. 2012; Zlatar et al. 2014).

16.2.1 Case Study of Carbon Dynamics Modeling in CO2FIX

The methodology of modeling forest carbon dynamics with the use of CO2FIX model will be shown for the forestry unit Dajićke Mountains. It is located in Serbia between eastern longitude of 17°54′ and 18°00′ and northern latitude of 43°23′ and 43° 29′. This is a managed forested area with a total area of 2873.93 ha. Most of the area is covered by forest ecosystems while a smaller area is characterized by grasslands. Most of the area is mountainous, located between 800 and 1200 m above sea level. Soils within this area are mostly shallow and very susceptible to erosion. Climate characteristics of the area are continental with shorter summer and cold and long winters. Average monthly temperatures and rainfall are shown in Table 16.1, this data was used for inputs of CO2FIX model.

Table 16.1 Climate characteristics of the Dajićke Mountains forestry management unit

Months	I	II	III	IV	V	VI	VII	VIII	IX	X	XI	XII
Average monthly temperature (°C)	−3.6	−2.3	1.8	7.0	11.7	15.2	17.7	17.4	14.2	8.5	3.9	−1.7
Average monthly rainfall (mm)	47	56	57	63	93	110	80	90	61	91	79	65

In this forestry management unit, the most widespread tree species are beech (*Fagus moesiaca*), which accounts for 48.6% (439,562 m^3) of total volume in the forestry management unit, followed by spruce (*Picea excelsa*), which participates in the total volume with 44.9% (406,288.5 m^3). Because of their overwhelming presence within the forest ecosystem these two species were modeled in CO2FIX. In both cases the modeling was done with a single cohort, even-aged forests with the start of the simulation being their planting time. CO2FIX requires a variety of biomass parameters which are necessary for the calculation of its growth, the most important being current annual increment (CAI) which was taken from yield tables for trees at this location (*** 2011; Simeunović 1957). CO2FIX converts the values of stemwood volume to biomass and carbon using wood density and carbon content coefficients. Wood densities were derived from CO2FIX manual (Nabuurs et al. 2002) and carbon content for all biomass was assumed to be 50%. Allocation coefficients which are necessary for the growth of different biomass compartments which include foliage, branches and root were obtained by using different biomass equations (Simeunović 1957; Ter-Mikaelian and Korzukhin 1997). Rotation and thinning regimes were obtained from the current management plan for this forestry unit (*** 2011), for beech the rotation is 100 years while spruce has a slightly shorter rotation of 80 years. Initial soil carbon stocks were calculated and added to the soil module on the basis of preparatory simulations, which were done to determine the mean annual carbon input to forest soil with each rotation length. Climate parameters where obtained from the yearly report of the Republic Hydrometeorological Service of Serbia (*** 2015). Apart from current climate data we also used predicted changes due to climate change to calculate the climate input data for the CO2FIX model. Predictions for this location which were used estimate an increase of temperature around +4.5 °C and a decrease in precipitation by 10%, this is based on IPCC highest emission scenario (IPCC 2013). Comparison of calculated climate input data between current climate and predicted climate changes is given in Table 16.2.

The total length of the simulation was set at 200 years. Modules for wood products and bioenergy were not used because there is no reliable data on the lifecycles of different wood products.

Results show that there are no changes in biomass growth with different climate inputs. This is to be expected since CO2FIX model is setup in such a way that climate data primarily influence processes of organic matter decomposition in soil. Changes in climate parameters would obviously affect the growth of biomass,

Table 16.2 Climate input data for CO2FIX model

	Current climate	Climate change
Degree days above zero (°C)	2542,8	3366,3
Potential evapotranspiration in growing season (mm)	431,372	525,807
Precipitation in growing season (mm)	497	447

however in CO2FIX these influences could only be detected and modeled if different CAI and yield tables are used. All biomass growth that is simulated in the model is dependent on these two parameters which already encompass within them how the climate has affected the growth of biomass. In order to include the effects of different climate conditions within the model it would be necessary to measure tree growth and CAI of specimens from same species (which are possibly planted from same seed source and have similar genetic characteristics) which have grown in different climate conditions.

Slight changes in soil carbon were detected with the change of climate inputs which reflect different conditions due to climate change. With the increase of temperature and the reduction of rainfall the average amount of carbon stored in forest soil for the duration of the simulation has increased by almost 1 Mg C/ha for spruce and by 1.6 Mg C/ha for beech. If just temperature input was increased by 4.5 °C there was a slight decrease in soil carbon, 0.08 Mg C/ha for spruce and 0.12 Mg C/ha for beech. On the other hand, reducing the amount of precipitation in the growing season by 10% and using the current temperature has resulted in the highest soil carbon content. Average increase in soil carbon for this case was slightly over 1 Mg C/ha for spruce and 1.8 MgC/ha for beech. Reduction of rainfall by 10% has produced a greater impact on carbon soil then an increase of temperature by 4.5 °C. In the case of just reduced rainfall we can expect a decrease in soil organic matter decomposition since less water is available. Higher temperatures speed up decomposition and mineralization of soil organic matter which caused more carbon to be released from forest soil and led to lowest soil carbon content associated with the scenario in which just the temperature was changed.

16.3 Modeling Carbon Sequestration in Perennial Grass Species

Modeling biomass of perennial grass species has so far mostly been done in two ways which use different methodological approaches. The first method involves making a descriptive model, while the second is based on the development of mechanistic models. Descriptive, statistical, regression or empirical models, as they are often called, are based on the application of statistical and correlation method based on known data obtained by measurements. Consequently, these models often have a short duration of the simulation, a smaller number of variables and their parameters are easier to assess. Although their prediction accuracy can be very high, considering that they include both known and unknown effects on the growth of biomass, descriptive models have significant limitations. Extrapolation of these models to other locations and/or species is generally impossible, because they primarily rely on empirical data. In addition, input of new variables means re-creating models from scratch based on a wider set of input data. To calculate the biomass in these models different functions are used to connect biomass yield with environmental factors.

Mechanistic models are models that are based on the "imitation" of different, especially physiological, processes. To create such a model, instead of large amounts of data obtained by observation (as is the case with descriptive models), knowledge about specific processes is used in order to try to represent a certain process with a mathematical function. Mechanistic models are usually consisting of several sub-models, which are at least one hierarchical level below the parameter model should provide. For example, when it comes to biomass yield, sub-models can describe the process of photosynthesis and the increase of leaf surface. It is important to emphasize that it is not possible to transfer all of the physiological, cellular and other processes that are now known to one model, because this would require a lot of time, and the question is whether such a model could be successfully applied (Marcelis et al. 1998). For this reason, it is always important to select key necessary processes, such as photosynthesis, respiration, growth, leaves, absorption of CO_2, water and other nutrients, which will be represented in one of the sub-models of the mechanistic model.

There are already several models which have been developed during the past years which can be used to calculate biomass growth of perennial grass species. EPIC – Erosion Productivity Impact Calculator (Williams et al. 1984) is a model that was originally developed to determine the relation between erosion and land productivity in the United States. Since then, its use has been expanded allowing it become a generalized model for crops that simulates the daily crop growth. Prediction of biomass growth is based on the simulation of carbon fixation through photosynthesis and respiration. EPIC is a comprehensive model that can be used both for grass and woody species. In addition, it is able to simulate a large number of ecosystem processes, such as plant growth, yield, balance of water and nutrients, as well as soil erosion.

ALMANAC Model – Agricultural Land Management Alternatives with Numerical Assessment Criteria (Kiniry et al. 1992) has emerged as a model that simulates the competition of crops and weeds for different plant species and different locations. Based on the way this model is set, it is quite similar to the EPIC model. The ecological processes which are included in this model are crop growth, absorption of solar radiation by the leaves of the plant, production of dry matter, allocation of biomass growth between plant organs and the balance of water in the soil.

Both EPIC and ALMANAC models offer great flexibility and are currently, with certain modifications, used for the simulation of growth and development of over a hundred different plant species, which include all the major agricultural crops, grasses, legumes, as well as agro-energy crops such as Miscanthus. Although these species are not capable of sequestering a large quantity of carbon in their biomass for a longer period of time like woody species, they can transfer significant quantities of carbon to soil through litter, harvest residues and roots. This is especially the case for fast growing energy crops like Miscanthus. *Miscanthus*×*giganteus* Greef et Deu is an attractive agro-energy crop suitable for short rotation plantation establishment and it is assumed as CO_2 neutral with underground organs which retain CO_2. Currently estimates of carbon sequestration in soil under these crops range from 0.6 to 3.0 Mg C ha^{-1} $year^{-1}$ (Lemus and Lal 2005; Sartori et al. 2007). Application of biomass modeling for Miscanthus species as a bioenergy crop has been done by few specific models which have been applied for this purpose with greater or lesser success rate. Interest in this species is not surprising considering its favorable physiological characteristics (Babović et al. 2012), and the possibility of its cultivation on degraded land. Models vary in terms of complexity, in their approach to modeling plant growth and other ecological and eco-physiological processes. Most models have a time step of one day, while some models perform simulations on the level of hours or less for specific processes.

MISCANMOD is a model that originated in Ireland as a result of several years of monitoring the growth of *Miscanthus* on multiple locations throughout Europe, and is based on the principles established by the Monteith (1977). Calculation of biomass growth in this model is based on its dependency on air temperature and solar radiation, provided that water is not a limiting factor. The growth begins in the spring when the air temperature rises above 10 °C, while length of frost-free period is used to calculate the sum of active temperature. Validation of this model was done by comparing the calculated and actual yield on 15 locations throughout the Europe, which were observed to be closely related (Clifton-Brown et al. 2004). By upgrading this model MISCANFOR model was created. This newer version represents the FORTRAN version of MISCANMOD model with additional functions for specific plant growth depending on the genotype, the efficiency of utilization of solar radiation depending on temperature, mortality due to drought or frost, moisture content at the time of harvest and transport of nutrients into rhizome. In addition, processes of absorption of solar radiation and the physiological effects of temperature and lack of water were improved. By comparing the simulated results of both models with experimental data, it was observed that MISCANFOR can predict the yields of biomass with even greater precision.

Miguez et al. (2009) applied the mechanistic eco-physiological model WIMOVAC – Windows Intuitive Model of Vegetation response to Atmosphere and Climate Change, which is specially adapted to the modeling of biomass for Miscanthus species. WIMOVAC is a software package for the Windows platform, which allows the modeling of various aspects of photosynthesis of plants with special emphasis on the effects of climate change. Model enables the control and simulation of photosynthesis through the Windows user interface and automatically gives the results formatted as a table or a flexible graph, which is significantly easier to use and provides opportunities for use not only in science but also in practical and educational purposes. This model was applied in the mentioned study for several locations in Europe. The results show that the model predicts a good assimilation of CO_2 and growth patterns of biomass with a slight tendency to have higher values than actual measurements obtained in the field. The authors suggest that further parameterized models, particularly in relation to leaf area index, can achieve more precise results. These results provide a framework for predicting the growth of *Miscanthus* in conditions that are not its natural habitat and allow assessment of the impact of climate change, which is of great importance for the commencement of the process of commercial cultivation of this species in order to produce biofuels. In addition, it is important to emphasize that the program WIMOVAC can be adapted for other agro-energy crops as well.

16.4 Conclusions

Application and testing of soil carbon models typically require information about the amount and quality of litter input to the soil, the amount of carbon in different soils, environmental factors (like climate) and the changes in the amount of soil carbon over time. Although models need to be validated, their application provides quite a few benefits, especially since many countries have comprehensive and representative national forest inventories which can be used as input data for biomass turnover. It is well known that environmental factors like temperature and humidity influence the forest soil microbial activity and in turn the forest soil carbon cycle. However, other factors, like forest management, can also have a significant impact on forest soil carbon cycle. To determine the extent of these types of impacts on microbial activity and associated forest soil carbon cycle an application of different types of models in combination with measurements and laboratory analysis can be used.

References

***. (2011). Osnova gazdovanja šumama za gazdinsku jedinicu "Dajićke planine" (2011–2020). Biro za planiranje i projektovanje u šumarstvu, Beograd (Forest management plan for management unit "Dajićke planine" (2011–2020). Birou for planning and projects in forestry, Belgrade).

***. (2015). Meteorološki godišnjak. 1 – klimatološki podaci 2014. Republički hidrometeorološki zavod, Beograd. (Meteorology annual report 1 – climatology data 2014. Republic Hydrometeorological Service of Serbia).

Babović, N., Dražić, G., & Đorđević, A. (2012). Mogućnosti korišćenja biomase poreklom od brzorastuće trske *Miscanthus* x *giganteus*. *Chemical Industry, 66*, 223–233.

Campbell, J. E., Lobell, D. B., Genova, R. C., & Field, C. B. (2008). The global potential of bioenergy on abandoned agriculture lands. *Environmental Science and Technology, 42*, 5791–5794.

Chapin, F. S., Matson, P. A., & Moony, H. A. (2002). *Principles of terrestrial ecosystem ecology*. New York: Springer Science.

Clifton-Brown, J. C., Stampfl, P., & Jones, M. B. (2004). *Miscanthus* biomass production for energy in Europe and its potential contribution to decreasing fossil fuel carbon emissions. *Global Change Biology, 10*, 509–518.

Hoogwijk, M., Faaij, A., Broek, R., Berndes, G., Gielen, D., & Turkenburg, W. (2003). Exploration of the ranges of the global potential of biomass for energy. *Biomass Bioenergy, 25*, 119–133.

IPCC. (2013). *Climate Change 2013: The physical science basis*. Working Group I Contribution to the Fifth Assessment Report of the Intergovernmental Panel on Climate Change (T. F. Stocker, D. Qin, G.-K. Plattner, et al., Eds.). Cambridge/New York: Cambridge University Press.

Kiniry, J. R., Williams, J. R., Gassman, P. W., & Debaeke, P. (1992). A general, process oriented model for two competing plant species. *Transactions of the American Society of Agricultural Engineers, 35*, 801–810.
Lal, R. (2004). Soil carbon sequestration impacts on global climate change and food security. *Science, 304*, 1623–1627.
Lemus, R., & Lal, R. (2005). Bioenergy crops and carbon sequestration. *Critical Reviews in Plant Science, 24*, 1–21.
Liski, J., Palosuo, T., Peltoniemi, M., & Sievänen, R. (2005). Carbon and decomposition model Yasso for forest soils. *Ecological Modelling, 189*, 168–182.
Marcelis, L. F. M., Heuvelink, E., & Goudriaan, J. (1998). Modelling biomass production and yield of horticultural crops: A review. *Scientia Horticulturae, 74*(1–2), 83–111.
Masera, O., Garza-Caligaris, J. F., Kanninen, M., Karjalainen, T., Liski, J., et al. (2003). Modelling carbon sequestration in afforestation, agroforestry and forest management projects: The CO2FIX V.2 approach. *Ecological Modelling, 164*, 177–199.
Miguez, F. E., Zhu, X., Humphries, S., Bollero, G. A., & Long, S. P. (2009). A semimechanistic model predicting the growth and production of the bioenergy crop *Miscanthus* x *giganteus*: Description, parameterization and validation. *Global Change Biology Bioenergy, 1*, 282–296.
Monteith, J. L. (1977). Climate and the efficiency of crop production in Britain. *Philosophical Transactions of the Royal Society of London, B281*, 277–297.
Nabuurs, G. J., & Mohren, G. M. J. (1993). *Carbon fixation through forestation activities a study of the carbon sequestration potential of selected types, Commissioned by the foundation FACE* (IBN Research Report 93/4). Wageningen: Institute for Forestry and Nature Research.
Nabuurs, G. J., Garza-Caligaris, J. F., Kanninen, M., Karjalainen, T., Lapvetelainen, T., et al. (2002). *CO2FIX V2.0—Manual of a model for quantifying carbon sequestration in forest ecosystems and wood products* (ALTERRA report 445, 45). Wageningen.
Pinard, M., & Putz, F. (1997). Monitoring carbon sequestration benefits associated with a reduced impact logging in Malaysia. *Mitigation and Adaptptation Strategies Global Change, 2*, 203.
Radojević, U., Milovanović, J., & Šijačić-Nikolić, M. (2012, October 4–5). Application of carbon cycle modeling in sustainable management of forest ecosystems. In *Proceedings of the international scientific conference forests in future: Sustainable use, risks and challenges*, Belgrade, pp 957–964.
Reed, K. L. (1980). An ecological approach to modeling the growth of forest trees. *Forest Science, 26*, 33–50.
Sartori, F., Lal, R., Ebinger, M. H., & Eaton, J. A. (2007). Changes in soil carbon and nutrient pools along a chronosequence of poplar plantations in the Columbia Plateau, Oregon, USA. *Agriculture, Ecosystem and Environment, 122*(3), 325–339.
Schelhaas, M. J., van Esch, P.W., Groen, T.A., de Jong, B.J., Kanninen, M., et al. (2004). *CO2FIX V 3.1 – Description of a model for quantifying carbon sequestration in forest ecosystems and wood products* (ALTERRA Report 1068). Wageningen.
Service, R. F. (2007). Biofuel Researchers Prepare to reap a New Harvest. *Science, 315*, 1488–1491.
Šijačić-Nikolić, M., Milovanović, J., & Nonić, M. (2014). Conservation of forest genetic resources. In: M. R. Ahuja, & K. G. Ramawat (Eds.), *Biotechnology and biodiversity*. (pp. 103–128). New York: Springer.
Simeunović, D. (1957). *Priručnik za šumarske inženjere*. Beograd: Naučna knjiga.
Ter-Mikaelian, M. T., & Korzukhin, M. D. (1997). Biomass equations for sixty-five North American tree species. *Forest Ecology and Management, 97*, 1–24.
Williams, J. R., Jones, C. A., & Dyke, P. T. (1984). A modeling approach to determining the relationship between erosion and soil productivity. *Transactions of the American Society of Agricultural Engineers, 27*, 129–144.
Zlatar, J., Milovanović, J., Radojević, U., Aranđelović, M., & Aleksić, S. (2014, September 29–30). Estimating sequestered carbon in forest ecosystems in Serbia. In *Proceedings of the international conference: Ecological improvement of devastated locations for sustainable development* (pp. 37–43). Belgrade.

Chapter 17
Bioactive Peptaibols of Forest-Derived *Trichoderma* Isolates from Section *Longibrachiatum*

Tamás Marik, András Szekeres, Maria A. Andersson, Mirja Salkinoja-Salonen, Chetna Tyagi, Balázs Leitgeb, Csaba Vágvölgyi, Irina S. Druzhinina, and László Kredics

Abstract Filamentous fungi are producers of a large number of secondary metabolites with wide spectra of biological effects. Among them, peptaibols represent a group of compounds produced mainly by members of the mycotrophic filamentous fungal genus *Trichoderma*. A simple peptaibol characterization strategy including purification and structural elucidation steps was applied to examine the peptaibol production of three strains from the *Longibrachiatum* section of genus *Trichoderma*, *T. aethiopicum* TUCIM 1817, *T. novae-zelandiae* TUCIM 4158 and *T. pseudokoningii* TUCIM 1277, all deriving from natural forest habitats (disturbed semiforest, native *Notophagus* forest and the bark of *Beilschmiedia tawa*, respectively). After the solid phase clean-up of culture extracts, mass spectrometric analysis of peptaibols produced by the examined strains was performed by on-line reversed-phase

T. Marik • A. Szekeres • C. Vágvölgyi • L. Kredics (✉)
Department of Microbiology, Faculty of Science and Informatics, University of Szeged, Közép fasor 52, H-6726 Szeged, Hungary
e-mail: kredics@bio.u-szeged.hu

M.A. Andersson • M. Salkinoja-Salonen
Department of Food and Environmental Science, University of Helsinki, P.O. Box 56, FI-00014 Helsinki, Finland

C. Tyagi
Department of Microbiology, Faculty of Science and Informatics, University of Szeged, Közép fasor 52, H-6726 Szeged, Hungary

Institute of Biophysics, Biological Research Centre, Hungarian Academy of Sciences, Temesvári krt. 62, H-6726 Szeged, Hungary

B. Leitgeb
Institute of Biophysics, Biological Research Centre, Hungarian Academy of Sciences, Temesvári krt. 62, H-6726 Szeged, Hungary

I.S. Druzhinina
Area Gene Technology and Applied Biochemistry, Institute of Chemical Engineering, Vienna University of Technology, Gumpendorfer Straße 1a/1665, 1060 Wien, Austria

M. Lukac et al. (eds.), *Soil Biological Communities and Ecosystem Resilience*, Sustainability in Plant and Crop Protection, DOI 10.1007/978-3-319-63336-7_17

high performance liquid chromatography coupled to electrospray ionization ion trap mass spectrometry. All three examined species produced 20-residue trichobrachin-like compounds, some of which are known from the literature, while others proved to be different from any peptaibols reported so far. The spectra of the peptaibols produced by these isolates were entirely different from each other. The largest amount of peptaibols consisting of four yet unknown compounds was produced by *T. pseudokoningii* TUCIM 1277, while ten and eight new, trichobrachin-like compounds were detected from *T. aethiopicum* TUCIM 1817 and *T. novae-zelandiae* TUCIM 4158, respectively. Feline fetal lung cell proliferation inhibition tests and membrane damage bio-assay with boar sperm cells revealed that although *T. novae-zelandiae* TUCIM 4158 produced the least amount of peptaibols, its compounds were the most inhibitory to mammalian cells.

Keywords Electrospray ionization ion trap mass-spectrometry (ESI-IT-MS) • Forest-derived fungi • Mammalian cell toxicity • Peptaibol • Reversed-phase high performance liquid chromatography (HPLC) • Section *Longibrachiatum* • Sperm motility inhibition • Trichobrachin • *Trichoderma* • *Trichoderma aethiopicum* • *Trichoderma novae-zelandiae* • *Trichoderma pseudokoningii*

17.1 Introduction

The *Longibrachiatum* section – one of the evolutionarily youngest clades within the filamentous fungal genus *Trichoderma,* belonging to the *Hypocreales* order of Ascomycota – includes *T. reesei,* the most extensively studied *Trichoderma* species known for its applicability in the food and feed industries, textile manufacturing and biofuel technology due to its production of cellulolytic and hemicellulolytic enzymes (Harman and Kubicek 1998; Kubicek et al. 2009). Other representatives of the section were studied as producers of numerous secondary metabolites including antibiotics and extracellular enzymes, as plant growth promoters, as degraders of xenobiotics and as commercial biofungicides (Sivasithamparam and Ghisalberti 1998). On the other hand, certain species of the section like *T. longibrachiatum*, *T. orientale* (formerly known as *Hypocrea orientalis*) and *T. citrinoviride* are opportunistic pathogens of mainly immunocompromised humans (Kredics et al. 2003; Hatvani et al. 2013). Moreover, *T. citrinoviride* and *T. longibrachiatum* are frequently isolated as indoor molds with high allergenic potential to humans (Thrane et al. 2001). The significance of other members of the section – like *T. aethiopicum* (Samuels et al. 2012) or *T. novae-zelandiae* (Samuels et al. 1998) has not yet been studied in detail.

Peptaibiotics are secondary metabolites belonging to the family of fungal peptide antibiotics which are mostly produced by *Trichoderma*. They are amphipathic, linear oligopeptides with special features: the N-terminal amino-acids of the peptides are usually acetylated or rarely acylated; the C-terminals of the sequences contain an amino alcohol, mostly phenylalaninol, or in certain cases valinol, leucinol, isoleucinol or tryptophanol is linked by a peptide bond at the C-terminal end (Szekeres et al. 2005). The inner sequences always contain special residues like isovaline (Iva) and α-aminoisobutyric acid (Aib), which are α,α-dialkylated amino acids occurring in high proportions and representing characteristic building blocks of the structure.

In our recent work, three species selected from the *Longibrachiatum* section of the genus *Trichoderma* were investigated for the production of peptaibols and screened for their bioactivities by studying the motility inhibition and membrane integrity disruption in boar sperm bioassays, and the inhibition of proliferation in the case of feline fetal lung cells. Several studies applied these tests for screening bioactive secondary metabolites, e.g. *Bipolaris oryzae*, the causal agent of the brown spot disease in rice produces substances – identified later as ophiobolins – that can inhibit these cells in bioassays (Bencsik et al. 2014).

17.2 Materials and Methods

17.2.1 Strains, Culture Conditions and Extraction Procedures

Strains *T. aethiopicum* TUCIM 1817 isolated from a disturbed semiforest, *T. novae-zelandiae* TUCIM 4158 from a native *Notophagus* forest and *T. pseudokoningii* TUCIM 1277 from the bark of *Beilschmiedia tawa* derived from the TU Collection of Industrially Important Microorganisms (Vienna University of Technology, Austria) and were cultivated on malt extract agar medium (MEA: 30 g l^{-1} malt-extract, 5 g l^{-1} soy peptone, 15 g l^{-1} agar in distilled water) for 7 days at 25 °C. After the cultivation period, the biomass of each strain was harvested from a full Petri plate and the produced peptaibols were extracted by chloroform. After evaporation, the extracts were diluted in 200 μl MeOH. These samples were used for the identification of peptaibols and for the bioassays. Results are given for the concentration of the toxic compounds from the extraction of the biomass of 1 Petri plate.

17.2.2 Analytical Procedures

The crude extracts were investigated by on-line reversed-phase high performance liquid chromatography (HPLC) coupled to electrospray ionization ion trap mass spectrometry (ESI-IT-MS) (Marik et al. 2013). Separations were performed by an Agilent 1100 modular HPLC system (Agilent, USA) consisting of a degasser (G1379A), a binary pump (G1376A), a micro-well plate autosampler (G1229A), and using a Phenomenex Gemini NX-C18 (Gen-Lab Ltd., Hungary) (150 mm × 2.0 mm, 3 μm) HPLC column with mobile phase at a flow rate of 0.2 ml/min. The solvent A was water with 0.05% (v/v) trifluoroacetic acid (TFA), while solvent B was MeCN/MeOH 1/1 (v/v) with 0.05% (v/v) TFA. The column temperature was maintained at 40 °C using a Jones Model 7990 Space Column Heater (Jones Chromatography Ltd., UK). The gradient elution started with 65% B held for 5 min and increased linearly to 80% B at 45 min, and then to 100% B at 70 min, which was decreased to the initial percentage until the pressure stabilized. The injection volume was 5 μl. The mass spectrometric examinations were made with a Varian 500MS Ion Trap mass spectrometer (Agilent, USA) equipped with an atmospheric-pressure ESI source operated in positive mode at normal scan speed. The ESI spray chamber temperature and drying gas (N_2) pressure/temperature were 50 °C and 30 psi/350 °C, respectively. The nebulizer gas (N_2) pressure, the needle voltage and the spray shield voltage were 50 psi, 5704 V and 600 V, respectively. The general parameters were the following: maximum scan times – 2.78; μScans averaged – 2 μScans; data rate – 0.36 Hz; multiplier offset – 0. Ionization control parameters were: target TIC – 100%; max ion time – 250000 μsec. Full scan measurements were carried out within the mass range of *m/z* 100 – 2000, with 66 V capillary voltage and 147% RF loading. Chemstation B.02.01 software and MS Workstation 6.6 software were applied to control the instruments and to collect and analyze the results. For the *de novo* sequencing of the C-terminal part of peptaibols, the y ions were used as precursor ions in the MS^2 measurements, which were originated from the split of the labile Aib-Pro bond in MS^1 mode. The nomenclature for fragment ions observed on the MS^1 and MS^2 spectra followed the terminology published by Roepstorff and Fohlman (1984) as well as by Biemann (1990).

17.2.3 Mammalian Somatic Cell Toxicity Assays

Feline fetal lung cells were pre-cultivated in RPMI 1640 medium (BioWhittaker) supplemented with L-glutamine at 37 °C in an atmosphere of 95% air and 5% CO_2 in a cell culture cabinet (Heracell 150i; Thermo Fisher Scientific, Vantaa, Finland). The bioassays were carried out in 96 well microtiter plates with the separated and detached cells. Initially, the first columns were filled with 180 μl from the media and the other wells were filled with 100 μl of media. Into the first column, 20 μl of the samples were added, mixed and then in a half dilution 100 μl amounts were pipetted to the next columns. Next, 100 μl of the media containing 5000–10000 lung cells were added to each microplate well except from the control positions which contained only the media without cells. After 2 days of incubation at 37 °C, 10 μl resazurin (Sigma Chemical Co., St. Louis, MO, USA) (400 μg ml^{-1} in normal saline) was added to each well and incubated by shaking for 2 hours under the same conditions. After incubation, the plates were investigated by a microtiter plate reader (Fluoroskan Ascent, Thermo Scientific, Vantaa, Finland) at the excitation wavelength of 544 nm and the emission wavelength of 590 nm. The endpoints of toxicity were specified by the differences from the OD of the control wells that means the living cells were less than 50%. Typically, peptaibols have very sharp endpoints between two dilution steps. After 2 more days of incubation, the $NADH^+$ production of the living cells causes a color changing to red for visible examination. Purified ophiobolins were used as references for our peptaibol inhibitions in the bioassays.

17.2.4 Sperm Motility Inhibition Assay

For the sperm motility bioassays, boar semen was obtained from Figen Oy Finland and extended in MR-A (Kubus, S.A., Madrid, Spain) containing 27×10^6 spermatozoa ml^{-1}. Four different amounts, 10 μl, 5 μl, 2 μl and 1 μl of the samples were added to 2 ml of 100× diluted MR-A boar semen stock solution, while 20 μl methanol was added to the blank assay. At room temperature two different incubation times, 30 min and 24 hours were used for the short and long term bio-assays, respectively. After the incubation, optimal conditions (shaking at 37 °C with providing oxygen) were ensured for 5 minutes to the sperm cells as a pre-incubation step before the examination of their motility. For the detection of the living, motile sperm cells, a phase contrast microscope (Olympus CKX31 with Software CellSens standard version 11.0.06, Olympus Soft Imaging Solution GmbH, Münster, Germany, 2009–2012) was used to check the tail beating of the sperm cells proving that they are alive. A living normal cell seems as if it has two tails by looking at the pictures of the microscope and the frequency of these "two-tailed" cells were compared to the blank samples in order to define the effect of the peptaibol samples.

17.2.5 Sperm Membrane Integrity Disruption Assay

Disruption of the cell membrane integrity of motile sperm cells was assayed by staining with propidium iodide (PI) (Molecular Probes, Eugene, OR, USA) in sperm cells induced by shaking to swim at 37 °C as described by Bencsik et al. (2014) with minor modifications. The test was performed as follows: 50 μl aliquots of Dulbecco's phosphate buffered saline were pipetted into 96 wells of a microtiter plate. Fifty μl of the test suspension (methanol extracts) was added to the first vertical row of the wells and serially diluted (twofold steps) until the 11th row of wells. The wells of the 12th row representing reagent blank were filled with 200 μl Dulbecco's phosphate buffered saline. One hundred and fifty μl aliquots of commercially extended boar semen (27×10^6 sperms ml^{-1}) were pipetted into the microtiter plate wells and pre-incubated for 2 h on an orbital shaker (Innova 5000, New Brunswick Scientific, Enfield, CT, USA) at 160 rpm and 37 °C. Three parallel dilutions were performed for each sample. After incubation, PI solution (stock solution stored as frozen 1 mg ml^{-1} in Dulbecco's phosphate buffered saline) was diluted to a concentration of 10 μg ml^{-1}. One hundred μl of this stain mixture was added to the exposed boar semen. This suspension was incubated at 37 °C in darkness for 15 min and measured with a Fluoroskan Ascent microplate reader (Thermo Scientific, Vantaa, Finland) at excitation and emission wavelengths of 544 nm and 590–600 nm, respectively. During the observation of plasma membrane damage by staining with PI, the positive control was frozen-thawed semen (for mortality of 100%) representing the maximal fluorescence emitted by the cells permeable to PI. Dulbecco's phosphate buffered saline was used as the background control (blank). The frozen semen was represented by the wells in the last horizontal row. Loss of viability in the sample was calculated as described by Alm et al. (2001). The frozen sample represented 100% loss of viability. The assay was calibrated with triclosan (Sigma Chemical Co., St. Louis, MO, USA) in 5 parallel tests, the EC_{50} was 2 μg ml^{-1} (SD ± 0.6). The possible autofluorescence of the tested petaibols was measured against alamethicin (50 μl of alamethicin in 150 μl of PBS) The cell membrane permeability assay on static sperm cells at room temperature was performed by viability staining with PI and calcein-AM (Molecular Probes, Eugene, OR, USA) as described by Hoornstra et al. (2003).

17.3 Results and Discussion

17.3.1 Peptaibols Detected in Trichoderma *Isolates from Section* Longibrachiatum

Different peptaibols belonging to the long-sequence group of trichobrachins were produced by *T. aethiopicum, T. pseudokoningii* and *T. novae-zelandiae* (Table 17.1). Nomenclature of the compounds were generated from 3 parts: i) the first number is the molecular mass of the compound, ii) the second alphabetic character refers to the variety of the y-part, and iii) the third number refers to the elution order of the compounds. Most of the molecules were produced only by single strains, although one sequence, Pept-1936-a-1 was present in the peptaibol extracts of all three strains. The strains *T. pseudokoningii* TUCIM 1277 and *T. novae-zelandiae* TUCIM 4158 showed higher similarity in terms of their peptaibol profiles. Three blocks of the peptaibol sequences proved to be conserved. One of the conserved parts is the R14-R20 region, which could be recorded on the MS spectra as the y-ion, where the compounds showed almost exact match with each other. The difference appeared only in the 17th amino acid residue, except from the sequences of Pept-1937-b-1, Pept-1951-b-1, Pept-1951-b-2, Pept-1966-d-1 and Pept-1966-d-2 where Glu was observed instead of Gln in position R19, indicated with the distinguishing marks (a, b, c and d) in their names. The first four amino acid residues (R1-R4) at the N terminal region showed also strong homology containing alternated Aib and Ala residues, which is also characteristic to the previously described peptaibols produced by members of section *Longibrachiatum*. In the spectra of *T. aethiopicum* TUCIM 1817 the 3rd residue (R3) was observed as Vxx instead of Aib in the sequences of Pept-1936-a-2, Pept-1950-a-3, Pept-1965-a-1, Pept-1965-a-2, Pept-1978-c-1, Pept-1978-c-2, Pept-1978-c-3 and Pept-1978-c-4. Furthermore, in the R7-R8 positions a very stable Gln-Aib bond was always observed. The sequences starting from the R5 residue to the labile Aib-Pro motif (R13-R14) were highly variable and showed

Table 17.1 Sequences and characteristic ions of the peptaibol compounds produced by forest-derived *Trichoderma* strains from section *Longibrachiatum*

Strains	Comp. name	M	M+Na$^+$	b-ion	y-ion	Rt (min)	Area size (*10^6)	R	R1	R2	R3	R4	R5	R6	R7	R8	R9	R10	R11	R12	R13	R14	R15	R16	R17	R18	R19	R20	
Trichoderma pseudokoningii (TUCIM 1277)	Pept-1936-c-1	1936	1959	1149	788	40	16.3	Ac	Aib	Ala	Aib	Ala	Aib	Ala	Gln	Aib	Vxx	Aib	Gly	Vxx	Aib	Pro	Vxx	Aib	Vxx	Gln	Gln	Pheol	new
	Pept-1936-a-1	1936	1959	1163	774	40	219.6	Ac	Aib	Ala	Aib	Ala	Aib	Ala	Gln	Aib	Vxx	Aib	Gly	Lxx	Aib	Pro	Vxx	Aib	Aib	Gln	Gln	Pheol	
	Pept-1950-a-1	1950	1974	1177	774	41	69.7	Ac	Aib	Ala	Aib	Ala	Aib	Aib	Gln	Aib	Vxx	Aib	Gly	Lxx	Aib	Pro	Vxx	Aib	Aib	Gln	Gln	Pheol	
	Pept-1950-c-1	1950	1974	1163	788	43	397.8	Ac	Aib	Ala	Aib	Ala	Aib	Ala	Gln	Aib	Vxx	Aib	Gly	Lxx	Aib	Pro	Vxx	Aib	Vxx	Gln	Gln	Pheol	
	Pept-1950-a-2	1950	1974	1177	774	44	18.9	Ac	Aib	Ala	Aib	Ala	Aib	Aib	Gln	Aib	Vxx	Aib	Gly	Lxx	Aib	Pro	Vxx	Aib	Aib	Gln	Gln	Pheol	new
	Pept-1950-c-2	1950	1974	1163	788	45	35.2	Ac	Aib	Ala	Aib	Ala	Aib	Ala	Gln	Aib	Vxx	Aib	Gly	Lxx	Aib	Pro	Vxx	Aib	Vxx	Gln	Gln	Pheol	
	Pept-1950-a-4	1950	1974	1177	774	45	107.6	Ac	Aib	Ala	Aib	Ala	Aib	Aib	Gln	Aib	Vxx	Aib	Gly	Lxx	Aib	Pro	Vxx	Aib	Aib	Gln	Gln	Pheol	
	Pept-1951-b-1	1951	1975	1177	775	41	28.1	Ac	Aib	Ala	Aib	Ala	Aib	Aib	Gln	Aib	Vxx	Aib	Gly	Lxx	Aib	Pro	Vxx	Aib	Aib	Glu	Gln	Pheol	
	Pept-1951-b-2	1951	1975	1177	775	47	8	Ac	Aib	Ala	Aib	Ala	Aib	Aib	Gln	Aib	Vxx	Aib	Gly	Lxx	Aib	Pro	Vxx	Aib	Aib	Glu	Gln	Pheol	
	Pept-1965-c-1	1965	1988	1177	788	43	114.2	Ac	Aib	Ala	Aib	Ala	Aib	Aib	Gln	Aib	Vxx	Aib	Gly	Lxx	Aib	Pro	Vxx	Aib	Vxx	Gln	Gln	Pheol	
	Pept-1965-c-2	1965	1988	1177	788	44	37.1	Ac	Aib	Ala	Aib	Ala	Aib	Aib	Gln	Aib	Vxx	Aib	Gly	Lxx	Aib	Pro	Vxx	Aib	Vxx	Gln	Gln	Pheol	new
	Pept-1965-c-3	1965	1988	1177	788	46	67	Ac	Aib	Ala	Aib	Ala	Aib	Aib	Gln	Aib	Vxx	Aib	Gly	Lxx	Aib	Pro	Vxx	Aib	Vxx	Gln	Gln	Pheol	new
	Pept-1965-c-4	1965	1988	1177	788	48	194.5	Ac	Aib	Ala	Aib	Ala	Aib	Aib	Gln	Aib	Vxx	Aib	Gly	Lxx	Aib	Pro	Vxx	Aib	Vxx	Gln	Gln	Pheol	
Trichoderma aethiopicum (TUCIM 1817)	*Pept-1936-a-1*	1936	1959	1163	774	40	249	Ac	Aib	Ala	Aib	Ala	Aib	Ala	Gln	Aib	Vxx	Aib	Gly	Lxx	Aib	Pro	Vxx	Aib	Aib	Gln	Gln	Pheol	
	Pept-1936-a-2	1936	1959	1163	774	43	34.7	Ac	Aib	Ala	Vxx	Ala	Aib	Ala	Gln	Aib	Vxx	Ala	Gly	Lxx	Aib	Pro	Vxx	Aib	Aib	Gln	Gln	Pheol	new
	Pept-1937-b-1	1937	1960	1163	775	42	63.3	Ac	Aib	Ala	Aib	Ala	Aib	Ala	Gln	Aib	Lxx	Ala	Gly	Lxx	Aib	Pro	Vxx	Aib	Aib	Glu	Gln	Pheol	new
	Pept-1950-c-1	1950	1974	1163	788	43	28	Ac	Aib	Ala	Aib	Ala	Aib	Ala	Gln	Aib	Vxx	Aib	Gly	Lxx	Aib	Pro	Vxx	Aib	Vxx	Gln	Gln	Pheol	
	Pept-1950-a-3	1950	1974	1177	774	44	18.8	Ac	Aib	Ala	Vxx	Ala	Aib	Ala	Gln	Aib	Vxx	Aib	Gly	Lxx	Aib	Pro	Vxx	Aib	Aib	Gln	Gln	Pheol	new
	Pept-1950-a-4	1950	1974	1177	774	45	335.3	Ac	Aib	Ala	Aib	Ala	Aib	Aib	Gln	Aib	Vxx	Aib	Gly	Lxx	Aib	Pro	Vxx	Aib	Aib	Gln	Gln	Pheol	
	Pept-1951-b-2	1951	1975	1177	775	47	62	Ac	Aib	Ala	Aib	Ala	Aib	Aib	Gln	Aib	Vxx	Aib	Gly	Lxx	Aib	Pro	Vxx	Aib	Aib	Glu	Gln	Pheol	
	Pept-1965-a-1	1965	1988	1191	774	48	44.6	Ac	Aib	Ala	Vxx	Ala	Aib	Aib	Gln	Aib	Vxx	Aib	Gly	Lxx	Aib	Pro	Vxx	Aib	Aib	Gln	Gln	Pheol	new
	Pept-1965-c-4	1965	1988	1177	788	48	32.2	Ac	Aib	Ala	Aib	Ala	Aib	Aib	Gln	Aib	Vxx	Aib	Gly	Lxx	Aib	Pro	Vxx	Aib	Vxx	Gln	Gln	Pheol	
	Pept-1965-a-2	1965	1988	1191	774	49	27.6	Ac	Aib	Ala	Vxx	Ala	Aib	Aib	Gln	Aib	Vxx	Aib	Gly	Lxx	Aib	Pro	Vxx	Aib	Aib	Gln	Gln	Pheol	new
	Pept-1966-d-1	1966	1989	1177	789	44	4.4	Ac	Aib	Ala	Aib	Ala	Aib	Aib	Gln	Aib	Lxx	Ala	Gly	Lxx	Aib	Pro	Vxx	Aib	Vxx	Glu	Gln	Pheol	new
	Pept-1966-d-2	1966	1989	1177	789	49	2.8	Ac	Aib	Ala	Aib	Ala	Aib	Aib	Gln	Aib	Vxx	Aib	Gly	Lxx	Aib	Pro	Vxx	Aib	Vxx	Glu	Gln	Pheol	
	Pept-1978-c-1	1978	2001	1191	788	45	2.7	Ac	Aib	Ala	Vxx	Ala	Aib	Aib	Gln	Aib	Vxx	Aib	Gly	Lxx	Aib	Pro	Vxx	Aib	Vxx	Gln	Gln	Pheol	new
	Pept-1978-c-2	1978	2001	1191	788	46	2.9	Ac	Aib	Ala	Vxx	Ala	Aib	Aib	Gln	Aib	Vxx	Aib	Gly	Lxx	Aib	Pro	Vxx	Aib	Vxx	Gln	Gln	Pheol	new
	Pept-1978-c-3	1978	2001	1191	788	50	6.9	Ac	Aib	Ala	Vxx	Ala	Aib	Aib	Gln	Aib	Vxx	Aib	Gly	Lxx	Aib	Pro	Vxx	Aib	Vxx	Gln	Gln	Pheol	new
	Pept-1978-c-4	1978	2001	1191	788	51	3.6	Ac	Aib	Ala	Vxx	Ala	Aib	Aib	Gln	Aib	Vxx	Aib	Gly	Lxx	Aib	Pro	Vxx	Aib	Vxx	Gln	Gln	Pheol	new
Trichoderma novae-zelandiae (TUCIM 4158)	Pept-1922-a-1	1922	1945	1149	774	37	17.7	Ac	Aib	Ala	Aib	Ala	Aib	Aib	Gln	Aib	Vxx	Ala	Gly	Vxx	Aib	Pro	Vxx	Aib	Aib	Gln	Gln	Pheol	new
	Pept-1936-c-2	1936	1959	1149	788	40	11	Ac	Aib	Ala	Aib	Ala	Aib	Aib	Gln	Aib	Vxx	Ala	Gly	Vxx	Aib	Pro	Vxx	Aib	Vxx	Gln	Gln	Pheol	new
	Pept-1936-a-1	1936	1959	1163	774	40	146.7	Ac	Aib	Ala	Aib	Ala	Aib	Ala	Gln	Aib	Vxx	Aib	Gly	Lxx	Aib	Pro	Vxx	Aib	Aib	Gln	Gln	Pheol	
	Pept-1937-b-1	1937	1960	1163	775	43	3	Ac	Aib	Ala	Aib	Ala	Aib	Ala	Gln	Aib	Lxx	Ala	Gly	Lxx	Aib	Pro	Vxx	Aib	Aib	Glu	Gln	Pheol	new
	Pept-1950-a-5	1950	1974	1177	774	41	20.3	Ac	Aib	Ala	Aib	Ala	Aib	Ala	Gln	Aib	Lxx	Aib	Gly	Lxx	Aib	Pro	Vxx	Aib	Aib	Gln	Gln	Pheol	new
	Pept-1950-c-3	1950	1974	1163	788	43	260.9	Ac	Aib	Ala	Aib	Ala	Aib	Ala	Gln	Aib	Lxx	Ala	Gly	Lxx	Aib	Pro	Vxx	Aib	Vxx	Gln	Gln	Pheol	new
	Pept-1950-c-4	1950	1974	1163	788	45	26.8	Ac	Aib	Ala	Aib	Ala	Aib	Ala	Gln	Aib	Lxx	Ala	Gly	Lxx	Aib	Pro	Vxx	Aib	Vxx	Gln	Gln	Pheol	new
	Pept-1950-a-6	1950	1974	1177	774	45	16.6	Ac	Aib	Ala	Aib	Ala	Aib	Aib	Gln	Aib	Lxx	Ala	Gly	Lxx	Aib	Pro	Vxx	Aib	Aib	Gln	Gln	Pheol	new
	Pept-1965-c-5	1965	1988	1177	788	44	19.3	Ac	Aib	Ala	Aib	Ala	Aib	Ala	Gln	Aib	Lxx	Aib	Gly	Lxx	Aib	Pro	Vxx	Aib	Vxx	Gln	Gln	Pheol	
	Pept-1965-c-6	1965	1988	1177	788	48	32.7	Ac	Aib	Ala	Aib	Ala	Aib	Aib	Gln	Aib	Lxx	Ala	Gly	Lxx	Aib	Pro	Vxx	Aib	Vxx	Gln	Gln	Pheol	new

remarkable differences. The produced peptaibols can be divided into four groups based on their molecular masses (1936, 1950, 1965 and 1978 Da). The 1978 Da peptaibols were produced only by *T. aethiopicum* TUCIM 1817 and attached to the c type (*m/z* 788) y-ion part, while the 1937, 1950 and 1965 Da peptaibols were produced by all of the three strains and attached to any of the four types of the y-ions, where the differences were in the R17 position with an Aib to Vxx (valine/isovaline) change, or in the R19 position with a Gln to Glu change.

Seventeen detected compounds showed sequences corresponding with known peptaibols, while the remaining 22 compounds proved to be new sequences and showed high similarities to the sequences of the compounds described in the literature (Table 17.2). The same characteristic stable blocks of the sequences can also be nicely observed in the sequences. Trilongins (Mikkola et al. 2012) and longibrachins (Leclerc et al. 1998) are described as trichobrachins with high similarities to the new sequences presented in Table 17.1, as well as with suzukacillins (Krause et al. 2006) or trichoaureocins (Brückner et al. 2002). The *Trichoderma citrinoviride* sequences (Maddau et al. 2009) are different from trichobrachins in the R12 position with a change from Lxx to Ala or Vxx, respectively, which can be detected also in the case of Pept-1936-c-1, Pept-1936-c-2 and Pept-1922-a-1 compounds.

Table 17.2 Similar peptaibol compounds described in the literature (sequences also detected on the spectra of *T. pseudokoningii* TUCIM 1277, *T. aethiopicum* TUCIM 1817 and *T. novae-zelandiae* TUCIM 4158 are underlined)

Strains	Compound name	Mol. Mass	R	R1	R2	R3	R4	R5	R6	R7	R8	R9	R10	R11	R12	R13	R14	R15	R16	R17	R18	R19	R20	References
Trichoderma parceramosum Bissett (*Trichoderma longibrochiotum* Rifai) CBS 936.69	Trichobrachin II 03	1922	Ac	Aib	Ala	Ala	Ala	Aib	Ala	Gln	Aib	Vxx	Aib	Gly	Lxx	Aib	Pro	Vxx	Aib	Aib	Gln	Gln	Pheol	Krause et. al. (2007)
	Trichobrachin II 04	1922	Ac	Aib	Ala	Aib	Ala	Aib	Ala	Gln	Aib	Vxx	Aib	Gly	Lxx	Aib	Pro	Vxx	Aib	Ala	Gln	Gln	Pheol	
	Trichobrachin II 05	1936	Ac	Aib	Ala	Aib	Ala	Aib	Ala	Gln	Aib	Vxx	Aib	Gly	Lxx	Aib	Pro	Vxx	Aib	Aib	Gln	Gln	Pheol	
	Trichobrachin II 06	1936	Ac	Aib	Ala	Aib	Ala	Aib	Ala	Gln	Aib	Vxx	Aib	Gly	Lxx	Aib	Pro	Vxx	Aib	Aib	Gln	Gln	Pheol	
	Trichobrachin IIb A	1936	Ac	Aib	Ala	Aib	Ala	Aib	Ala	Gln	Aib	Val	Aib	Gly	Leu	Aib	Pro	Val	Aib	Aib	Gln	Gln	Pheol	
	<u>Trichobrachin II 07</u>	1950	Ac	Aib	Ala	Aib	Aia	Aib	Ala	Gln	Aib	Vxx	Aib	Gly	Lxx	Aib	Pro	Vxx	Aib	Vxx	Gln	Gln	Pheol	
	<u>Trichobrachin II 08</u>	1950	Ac	Aib	Ala	Aib	Ala	Aib	Ala	Gln	Aib	Vxx	Aib	Gly	Lxx	Aib	Pro	Vxx	Aib	Vxx	Gln	Gln	Pheol	
	<u>Trichobrachin II 09</u>	1950	Ac	Aib	Ala	Aib	Ala	Aib	Ala	Gln	Aib	Vxx	Aib	Gly	Lxx	Aib	Pro	Vxx	Aib	Vxx	Gln	Gln	Pheol	
	<u>Trichobrachin IIb B</u>	1950	Ac	Aib	Ala	Aib	Ala	Aib	Ala	Gln	Aib	Val	Aib	Gly	Leu	Aib	Pro	Val	Aib	Iva	Gln	Gln	Pheol	
	<u>Trichobrachin IIb C</u>	1950	Ac	Aib	Ala	Aib	Ala	Aib	Aib	Gln	Aib	Val	Aib	Gly	Leu	Aib	Pro	Val	Aib	Aib	Gln	Gln	Pheol	
	<u>Trichobrachin II 10</u>	1964	Ac	Aib	Ala	Aib	Ala	Aib	Aib	Gln	Aib	Vxx	Aib	Gly	Lxx	Aib	Pro	Vxx	Aib	Vxx	Gln	Gln	Pheol	
	<u>Trichobrachin IIb D</u>	1964	Ac	Aib	Ala	Aib	Ala	Aib	Aib	Gln	Aib	Val	Aib	Gly	Leu	Aib	Pro	Val	Aib	Iva	Gln	Gln	Pheol	

Trichoderma longibrachiatum (Thb, Thd, SzMCThg)	Trilongin BI	1936	Ac	Aib	Ala	Aib	Ala	Aib	Ala	Gln	Aib	Vxx	Aib	Gly	Lxx	Aib	Pro	Vxx	Aib	Aib	Gln	Gln	Pheol	Mikkola et al. (2012)
	Trilongin CI	1937	Ac	Aib	Ala	Aib	Ala	Aib	Ala	Gln	Aib	Vxx	Aib	Gly	Lxx	Aib	Pro	Vxx	Aib	Aib	Glu	Gln	Pheol	
	Trilongin BII	1950	Ac	Aib	Ala	Aib	Ala	Aib	Ala	Gln	Aib	Vxx	Aib	Gly	Lxx	Aib	Pro	Vxx	Aib	Vxx	Gln	Gln	Pheol	
	Trilongin BIII	1950	Ac	Aib	Ala	Aib	Ala	Aib	Aib	Gln	Aib	Vxx	Aib	Gly	Lxx	Aib	Pro	Vxx	Aib	Aib	Gln	Gln	Pheol	
	Trilongin CIII	1951	Ac	Aib	Ala	Aib	Ala	Aib	Aib	Gln	Aib	Vxx	Aib	Gly	Lxx	Aib	Pro	Vxx	Aib	Aib	Glu	Gln	Pheol	
	Trilongin CIV	1965	Ac	Aib	Ala	Aib	Ala	Aib	Aib	Gln	Aib	Vxx	Aib	Gly	Lxx	Aib	Pro	Vxx	Aib	Vxx	Glu	Gln	Pheol	
	Trilongin BIV	1964	Ac	Aib	Ala	Aib	Ala	Aib	Aib	Gln	Aib	Vxx	Aib	Gly	Lxx	Aib	Pro	Vxx	Aib	Vxx	Gln	Gln	Pheol	
Trichoderma viride (63 C-1)	Suzukacillin A 09	1950	Ac	Aib	Ala	Aib	Ala	Aib	Aia	Gln	Aib	Lxx	Aib	Gly	Lxx	Aib	Pro	Vxx	Aib	Aib	Gln	Gln	Pheol	Krause et al. (2006)
	Suzukacillin A 10a	1950	Ac	Aib	Ala	Aib	Ala	Aib	Ala	Gln	Aib	Vxx	Aib	Gly	Lxx	Aib	Pro	Vxx	Aib	Vxx	Gln	Gln	Pheol	
	Suzukacillin A 11a	1950	Ac	Aib	Ala	Aib	Aia	Aib	Ala	Gln	Aib	Lxx	Aib	Gly	Lxx	Aib	Pro	Vxx	Aib	Aib	Gln	Gln	Pheol	
	Suzukacillin A 10b	1964	Ac	Aib	Ala	Aib	Ala	Aib	Ala	Gln	Aib	Lxx	Aib	Gly	Lxx	Aib	Pro	Vxx	Aib	Vxx	Gln	Gln	Pheol	
	Suzukacillin A 11b	1964	Ac	Aib	Ala	Aib	Ala	Aib	Ala	Gln	Aib	Lxx	Aib	Gly	Lxx	Aib	Pro	Vxx	Aib	Vxx	Gln	Gln	Pheol	
	Suzukacillin A 13	1964	Ac	Aib	Ala	Aib	Ala	Aib	Ala	Gln	Aib	Lxx	Aib	Gly	Lxx	Aib	Pro	Vxx	Aib	Vxx	Gln	Gln	Pheol	
	Suzukacillin A 12	1978	Ac	Aib	Ala	Aib	Ala	Aib	Aib	Gln	Aib	Lxx	Aib	Gly	Lxx	Aib	Pro	Vxx	Aib	Vxx	Gln	Gln	Pheol	

(continued)

Table 17.2 (continued)

Strains	Compound name	Mol. Mass	R	R1	R2	R3	R4	R5	R6	R7	R8	R9	R10	R11	R12	R13	R14	R15	R16	R17	R18	R19	R20	References
Trichoderma citrinoviride (S25)	Trichoderma citrinoviride seq. 1	1922	Ac	Aib	Ala	Aib	Ala	Aib	Ala	Gln	Aib	Lxx	Aib	Gly	Ala	Aib	Pro	Vxx	Aib	Vxx	Gln	Gln	Pheol	Maddau et al. (2009)
	Trichoderma citrinoviride seq. 2	1936	Ac	Aib	Ala	Aib	Aia	Aib	Ala	Gln	Aib	Lxx	Aib	Gly	Vxx	Aib	Pro	Vxx	Aib	Aib	Gln	Gln	Pheol	
	Trichoderma citrinoviride seq. 3	1950	Ac	Aib	Ala	Aib	Ala	Aib	Aib	Gln	Aib	Vxx	Aib	Gly	Vxx	Aib	Pro	Vxx	Aib	Vxx	Gln	Gln	Pheol	
	Trichoderma citrinoviride seq. 4	1950	Ac	Aib	Ala	Aib	Ala	Aib	Aib	Gln	Aib	Lxx	Aib	Gly	Vxx	Aib	Pro	Vxx	Aib	Aib	Gln	Gln	Pheol	
	Trichoderma citrinoviride seq. 5	1964	Ac	Aib	Ala	Aib	Ala	Aib	Aib	Gln	Aib	Lxx	Aib	Gly	Vxx	Aib	Pro	Vxx	Aib	Vxx	Gln	Gln	Pheol	
	Trichoderma citrinoviride seq. 6	1964	Ac	Aib	Ala	Aib	Ala	Aib	Aib	Gln	Aib	Lxx	Aib	Gly	Vxx	Aib	Pro	Vxx	Aib	Vxx	Gln	Gln	Pheol	
	Trichoderma citrinoviride seq. 7	1964	Ac	Aib	Ala	Aib	Ala	Aib	Aib	Gln	Aib	Lxx	Aib	Gly	Lxx	Aib	Pro	Vxx	Aib	Aib	Gln	Gln	Pheol	
	Trichoderma citrinoviride seq. 8	1964	Ac	Aib	Ala	Aib	Ala	Aib	Aib	Gln	Aib	Lxx	Aib	Gly	Vxx	Aib	Pro	Vxx	Aib	Vxx	Gln	Gln	Pheol	

Trichoderma longibrachiatum (M-853431)	Longibrachin A I	1936	Ac	Aib	Ala	Aib	Ala	Aib	Ala	Gln	Aib	Val	Aib	Gly	Leu	Aib	Pro	Val	Aib	Aib	Gln	Gln	Pheol	Leclerc et al. (1998)
	Longibrachin A II	1950	Ac	Aib	Ala	Aib	Aia	Aib	Ala	Gln	Aib	Vaf	Aib	Gly	Leu	Aib	Pro	Val	Aib	Iva	Gln	Gln	Pheol	
	Longibrachin A III	1950	Ac	Aib	Ala	Aib	Ala	Aib	Aib	Gln	Aib	Val	Aib	Gly	Leu	Aib	Pro	Val	Aib	Aib	Gln	Gln	Pheol	
	Longibrachin A IV	1964	Ac	Aib	Ala	Aib	Aia	Aib	Aib	Gln	Aib	Val	Aib	Gly	Leu	Aib	Pro	Vat	Aib	Iva	Gln	Gln	Pheol	
Trichoderma aureoviride (IMI 91968)	Trichoaureocin 1e	1908	Ac	Aib	Ala	Aib	Ala	Aib	Ala	Gln	Ala	Val	Ala	Gly	Leu	Aib	Pro	Val	Aib	Aib	Gln	Gln	Pheol	Brückner et al. (2002)
	Trichoaureocin 7	1908	Ac	Aib	Ala	Aib	Ala	Aib	Ala	Gln	Ala	Val	Ala	Gly	Lys	Aib	Pro	Val	Aib	Aib	Gln	Gln	Pheol	
	Trichoaureocin 1b	1922	Ac	Aib	Ala	Aib	Ala	Ala	Ala	Gln	Aib	Val	Aib	Gly	Leu	Aib	Pro	Val	Aib	Aib	Gln	Gln	Pheol	
	Trichoaureocin 1c	1922	Ac	Aib	Ala	Aib	Ala	Aib	Ala	Gln	Aib	Val	Aib	Gly	Leu	Aib	Pro	Val	Aib	Ala	Gln	Gln	Pheol	
	Trichoaureocin 1d	1922	Ac	Aib	Ala	Aib	Ala	Aib	Ala	Gln	Aib	Val	Ala	Gly	Leu	Aib	Pro	Val	Aib	Aib	Gln	Gln	Pheol	
	Trichoaureocin 1a	1936	Ac	Aib	Ala	Aib	Ala	Aib	Ala	Gln	Aib	Val	Aib	Gly	Leu	Aib	Pro	VaT	Aib	Val	Gln	Gln	Pheol	
	Trichoaureocin 3	1936	Ac	Aib	Ala	Aib	Ala	Aib	Ala	Gln	Aib	Val	Aib	Gly	Leu	Aib	Pro	Val	Aib	Aib	Gln	Gln	Pheol	
	Trichoaureocin 4	1950	Ac	Aib	Ala	Aib	Ala	Aib	Ala	Gln	Aib	Vxx	Aib	Gly	Leu	Aib	Pro	Vxx	Aib	Vxx	Gln	Gln	Pheol	
	Trichoaureocin 5	1950	Ac	Aib	Ala	Aib	Ala	Aib	Aib	Gln	Aib	Val	Aib	Gly	Leu	Aib	Pro	Val	Aib	Aib	Gln	Gln	Pheol	
	Trichoaureocin 6	1964	Ac	Aib	Ala	Aib	Ala	Aib	Aib	Gln	Aib	Vxx	Aib	Gly	Leu	Aib	Pro	Vxx	Aib	Vxx	Gln	Gln	Pheol	

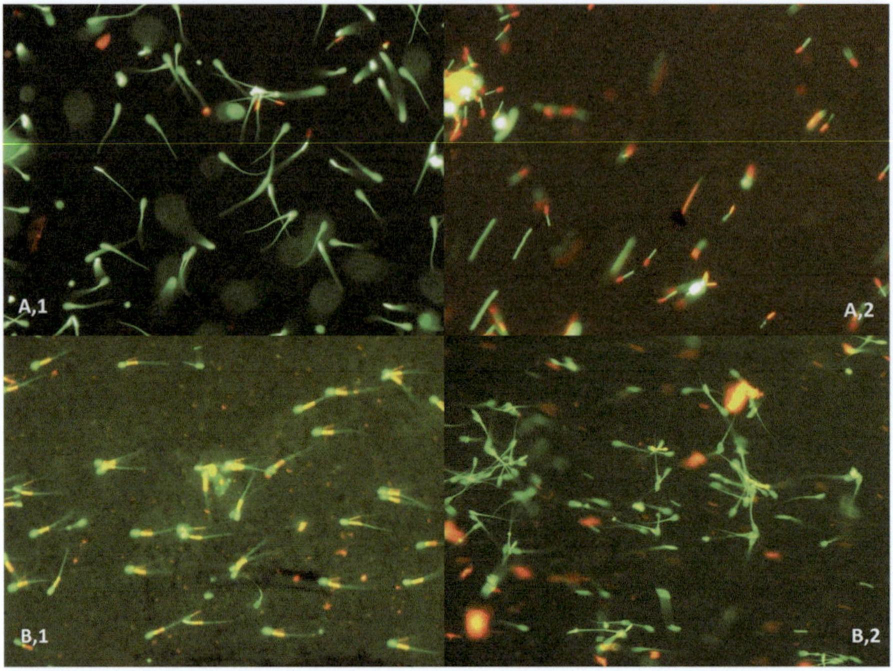

Fig. 17.1 Membrane damage in boar semen cells by crude peptaibol extracts from forest-derived *Trichoderma* strains of section *Longibrachiatum*. (**A**,1) living cells (*green*) after staining with propidium-iodide; (**A**,2, dead cells (*red*) after staining with propidium-iodide; (**B**,1) living cells (*orange neck part*) after staining with JC-1; (**B**,2) dead cells (*green neck part*) after staining with JC-1

17.3.2 Mammalian Somatic Cell Toxicity Assays

During toxicity examinations of the peptaibol extracts, similar results were obtained in both the feline fetal lung cell proliferation inhibition test and the sperm membrane integrity disruption assay, with the latter being more sensitive to peptaibols.

Peptaibols are capable of causing membrane damage by forming ion channels in lipid membranes (Szekeres et al. 2005). After the boar semen cell membrane damage bioassay, the membrane damage was detected by staining the cells with propidium-iodide (PI) which is permeable through the cell membrane. The living cells with a healthy membrane are impermeable to the dye, while in the dead, membrane-damaged cells PI – when bound to nucleic acids – causes an emission wavelength color shift from green to red (Fig. 17.1A,1, A,2). After incubation, the fluorescence microscopic detection of the minimum quantity of the toxic peptaibol sample added to the sperm cell mixture (end-point) represented both alive and dead

Table 17.3 Mammalian cell toxicity of crude peptaibol extracts from forest-derived *Trichoderma* strains of section *Longibrachiatum* (OD ex: 544 nm, em: 590–600 nm)

Type of bio-assay	Collection number	Production (Area sizes	No. of peptides	Dilution rate 1/8	1/16	1/32	1/64	1/128	1/256	1/512	1/1024	1/2048	1/4096	1/8192	media
Feline fetal lung cell toxicity assay	TUCIM 1277	13.1	13	69.43	66.97	139.9	241.4	222.6	204.5	199.9	222.8	n.m.	n.m.	n.m.	0.846
	TUCIM 1817	9.19	16	72.49	68.42	64.49	158.8	195.7	196.1	238	253.5	n.m.	n.m.	n.m.	0.8493
	TUCIM 4158	5.55	10	67.47	65.52	59.51	72.5	108.2	121.2	151.3	56.7	n.m.	n.m.	n.m.	0.8804
Membrane integrity disruption assay	TUCIM 1277	13.1	13	10.89	10.23	10.56	10.61	10.57	11.09	10.61	10.53	5.234	2.437	2.323	1.538
	TUCIM 1817	9.19	16	10.84	10.34	10.34	10.49	10.95	11.26	10.36	8.482	3.186	2.343	2.273	1.544
	TUCIM 4158	5.55	10	8.885	7.756	8.427	8.643	8.646	8.694	8.821	8.741	8.855	9.155	9.158	1.829

The highlighted parts are where the toxic compounds were effective and giving an end-point of the toxicity. *n.m.*: not measured

cells in the samples. The membrane damage of the mitochondria was also observed by using 5,5′,6,6′-tetrachloro-1,1′,3,3′-tetraethylbenzimidazolyl-carbocyanine iodide (JC-1), which can exhibit the potential-dependent accumulation in the mitochondria by showing the emission color as orange (Fig. 17.1B,1, B,2). Therefore, in the samples of the end-points, the mitochondrial membrane was depolarized in the case of dead cells represented with green color.

The toxicity of extracts from *T. aethiopicum* TUCIM 1817 and *T. pseudokoningii* TUCIM 1277 occurred at the same dilution levels in both tests but the amount of the peptaibols produced was different. Furthermore, despite much lower peptaibol production than for the other two isolates, strain *T. novae-zelandiae* TUCIM 4158 was more effective against mammalian cells in both tests (Table 17.3).

Based on our results we can presume that the peptaibol products of strain *T. novae-zelandiae* TUCIM 4158 possess higher toxicity towards the examined mammalian cells even at lower concentration than the compounds produced by the other two examined strains. Probably the presence of the Pept-1950-c-3 peptaibol compound in *T. novae-zelandiae* TUCIM 4158 – which was the only compound not detected in the other strains and produced in the highest amount by *T. novae-zelandiae* – results in the higher toxicity towards mammalian cells.

Acknowledgements This study was supported by grants NKFI K-105972 (National Research, Development and Innovation Office, Hungary), GINOP-2.3.2-15-2016-00052 (Széchenyi 2020 Programme, Hungary), TSR 112134 (Finnish Work Environment Fund, Finland), SA 289161 (Academy of Finland) and 95öu4 (Austrian-Hungarian Action Fund). The technical assistance of L. Atanasova is highly acknowledged.

References

Alm, K., Taponen, J., Dahlbom, M., Tuunainen, E., Koskinen, E., & Andersson, M. (2001). A novel automated fluorometric assay to evaluate sperm viability and fertility in dairy bulls. *Theriogenology, 56*, 677–684.

Bencsik, O., Papp, T., Berta, M., Zana, A., Forgó, P., Dombi, G., Andersson, M. A., Salkinoja-Salonen, M., Vágvölgyi, C., & Szekeres, A. (2014). Ophiobolin A from *Bipolaris oryzae* perturbs motility and membrane integrities of porcine sperm and induces cell death on mammalian somatic cell lines. *Toxins, 23*, 2857–2871.

Biemann, K. (1990). Sequencing of peptides by tandem mass spectrometry and high-energy collision-induced dissociation. *Methods in Enzymology, 193*, 455–479.
Brückner, H., Kirschbaum, J., & Jaworski, A. (2002). Sequences of peptaibol antibiotics trichoaureocins from *Trichoderma aureoviride*. In E. Benedetti, & C. Pedone (Eds.), *Peptides 2002 – Proceedings of the 27th European Peptide Symposium*, Edizioni Ziino, Napoli, pp. 362–363.
Harman, G. E., & Kubicek, C. P. (1998). *Trichoderma and Gliocladium, enzymes, biological control and commercial applications* (Vol. 2, p. 393). London: Taylor & Francis.
Hatvani, L., Manczinger, L., Vágvölgyi, C., & Kredics, L. (2013). *Trichoderma* as a human pathogen. In P. K. Mukherjee, B. A. Horwitz, U. S. Singh, M. Mukherjee, & M. Schmoll (Eds.), *Trichoderma – Biology and applications* (pp. 292–313). Wallingford: CAB International.
Hoornstra, D., Andersson, M. A., Mikkola, R., & Salkinoja-Salonen, M. S. (2003). A new method for in vitro detection of microbially produced mitochondrial toxins. *Toxicology in Vitro, 17*, 745–751.
Krause, C., Kirschbaum, J., Jung, G., & Brückner, H. (2006). Sequence diversity of the peptaibol antibiotic suzukacillin-A from the mold *Trichoderma viride*. *Journal of Peptide Science, 12*, 321–327.
Krause, C., Kirschbaum, J., & Brückner, H. (2007). Peptaibiomics: microheterogeneity, dynamics, and sequences of trichobrachins, peptaibiotics from *Trichoderma parceramosum* BISSETT (*T. longibrachiatum* RIFAI). *Chemistry & Biodiversity, 4*, 1083–1102.
Kredics, L., Antal, Z., Dóczi, I., Manczinger, L., Kevei, F., & Nagy, E. (2003). Clinical importance of the genus *Trichoderma*. A review. *Acta Microbiologica et Immunologica Hungarica, 50*, 105–117.
Kubicek, C. P., Mikus, M., Schuster, A., Schmoll, M., & Seiboth, B. (2009). Metabolic engineering strategies for the improvement of cellulase production by *Hypocrea jecorina*. *Biotechnology for Biofuels, 2*, 19.
Leclerc, G., Rebuffat, S., Goulard, C., & Bodo, B. (1998). Direct biosynthesis of peptaibol antibiotics in two *Trichoderma* strains. *Journal of Antibiotics, 51*, 170–177.
Maddau, L., Cabras, A., Franceschini, A., Linaldeddu, B. T., Crobu, S., Roggio, T., & Pagnozzi, D. (2009). Occurrence and characterization of peptaibols from *Trichoderma citrinoviride*, an endophytic fungus of cork oak, using electrospray ionization quadrupole time-of-flight mass spectrometry. *Microbiology (UK), 155*, 3371–3381.
Marik, T., Várszegi, C., Kredics, L., Vágvölgyi, C., & Szekeres, A. (2013). Mass spectrometric investigation of alamethicin. *Acta Biologica Szegediensis, 57*, 109–112.
Mikkola, R., Andersson, M. A., Kredics, L., Grigoriev, P. A., Sundell, N., & Salkinoja-Salonen, M. S. (2012). 20-residue and 11-residue peptaibols from the fungus *Trichoderma longibrachiatum* are synergistic in forming Na+/K+ permeable channels and adverse action towards mammalian cells. *FEBS Journal, 279*, 4172–4190.
Roepstorff, P., & Fohlman, J. (1984). Proposal for a common nomenclature for sequence ions in mass spectra of peptides. *Biomedical Mass Spectrometry, 11*, 601.
Samuels, G. J., Petrini, O., Kuhls, K., Lieckfeldt, E., & Kubicek, C. P. (1998). The *Hypocrea schweinitzii* complex and *Trichoderma* sect. *Longibrachiatum*. *Studies in Mycology, 41*, 1–54.
Samuels, G. J., Ismaiel, A., Mulaw, T. B., Szakacs, G., Druzhinina, I. S., Kubicek, C. P., & Jaklitsch, W. M. (2012). The *Longibrachiatum* clade of *Trichoderma*: A revision with new species. *Fungal Diversity, 55*, 77–108.
Sivasithamparam, K. & Ghisalberti, E. L. (1998). Secondary metabolism in *Trichoderma* and *Gliocladium*. In C. P. Kubicek, & G. E. Harman (Eds.), *Trichoderma* and *Gliocladium*. Vol. 1. Basic biology, taxonomy and genetics (pp. 139–191). London: Taylor and Francis Ltd, London.
Szekeres, A., Leitgeb, B., Kredics, L., Antal, Z., Hatvani, L., Manczinger, L., & Vágvölgyi, C. (2005). Peptaibols and related peptaibiotics of *Trichoderma*. A review. *Acta Microbiologica et Immunologica Hungarica, 52*, 137–168.
Thrane, U., Poulsen, S. B., Nirenberg, H. I., & Lieckfeldt, E. (2001). Identification of *Trichoderma* strains by image analysis of HPLC chromatograms. *FEMS Microbiology Letters, 203*, 249–255.

Chapter 18
Plant-Assisted Bioremediation: An Ecological Approach for Recovering Multi-contaminated Areas

Valeria Ancona, Paola Grenni, Anna Barra Caracciolo, Claudia Campanale, Martina Di Lenola, Ida Rascio, Vito Felice Uricchio, and Angelo Massacci

Abstract Plant-based clean up technologies are gaining popularity as a sustainable solution to contaminated soil remediation. In particular, plant-assisted bioremediation or phyto-assisted bioremediation exploits the synergistic action between plant root systems and natural microorganisms (bacteria and fungi) to remove, convert or contain toxic substances in soils, sediments or water. It can be applied successfully to contaminated areas. It relies on the use of a selected appropriate plant species for stimulating the biodegradation activity of natural soil microorganisms in the rhizosphere (e.g. through root exudates production or oxygen transport). Plant species can also produce extracellular enzymes that directly transform contaminants and/or make them more bioavailable. Moreover, they can also phyto-contain them. In selecting the plant species, the specific contaminant/s to be removed, and the local geopedological and climatic conditions need to be considered. Beyond the contaminant removal, there are additional benefits such as soil quality improvement, soil carbon sequestration and biomass production for energy purposes. The difficulties in remediating areas characterized by multiple pollutant occurrence (e.g. organic and inorganic toxic compounds) make the study of plant-microbial interactions important if sustainable soil recovery strategies are to be achieved. Consequently, in recent years, several plant species have been tested for stimulating natural microbial communities and supporting the remediation of contaminated soils. Among these, the poplar tree can be considered suitable for plant-assisted bioremediation purposes.

V. Ancona • C. Campanale • I. Rascio • V.F. Uricchio
CNR-IRSA, National Research Council, Water Research Institute, Bari, Italy

P. Grenni • A. Barra Caracciolo (✉) • M. Di Lenola
National Research Council, Water Research Institute, CNR-IRSA,
Monterotondo, Rome, Italy
e-mail: barracaracciolo@irsa.cnr.it

A. Massacci
CNR-IBAF, National Research Council, Institute of Agroenvironmental and Forestry Biology,
Rome, Italy

M. Lukac et al. (eds.), *Soil Biological Communities and Ecosystem Resilience*,
Sustainability in Plant and Crop Protection, DOI 10.1007/978-3-319-63336-7_18

In this chapter an example of the methodological approach used for its application to an area multi-contaminated (by polychlorinated biphenyls and heavy metals) is illustrated.

Keywords Polychlorinated biphenyls (PCBs) • Heavy metals • Poplar

18.1 Introduction

Phytoremediation is an environmentally sustainable, solar-powered, and cost-effective technology which relies on the capability of plants to intercept, take up, accumulate, sequestrate, stabilize or translocate contaminants (Thijs et al. 2016; Pilon-Smits 2005). In particular, plant-assisted bioremediation or phyto-assisted bioremediation is an *in situ* treatment of contaminated soils, based on the complex rhizosphere interactions between plant roots and naturally occurring microorganisms (Wenzel 2009). Soil properties are modified by a range of processes occurring during plant growth, which in turn affect the rhizosphere microbiota. Roots release low-molecular-mass compounds (e.g. sugars, amino acids and organic acids), polymerized sugar, root border cells and dead root cap cells. These rhizodeposits are used as carbon sources by soil microorganisms and can also contain secondary metabolites, such as antimicrobial compounds, nematicides and flavonoids (Bais et al. 2006), which are involved in establishing symbiosis or in warding off pathogens and pests.

Soil pH, another important driver of microbial communities (Fierer and Jackson 2006; Bru et al. 2011), can increase or decrease by up to two units in the rhizosphere owing to the ions release and uptake by roots (Hinsinger et al. 2009). Water uptake and root respiration affect soil oxygen pressure, thereby influencing microbial respiration. Finally, soil nutrient availability is modified in the rhizosphere by the plant uptake and by the secretion of chelators, such as phytosiderophores, to sequester metallic micronutrients (Tsednee et al. 2012; Philippot et al. 2013). In the rhizosphere, contaminants can therefore change their bioavailability and/or be sequestered by root systems and partially transformed by extracellular enzymes. However, the complete degradation of organic contaminants can occur only if there are microbial populations able to metabolize them through co-metabolic and metabolic pathways, until their mineralization.

Microbial metabolism is potentially capable of degrading several persistent organic pollutants, including polychlorinated biphenyls (PCBs) (Pieper and Seeger 2008). Laboratory studies identified several bacterial strains able to aerobically transform lower chlorinated congeners of PCBs through metabolic and co-metabolic pathways and, anaerobically, highly chlorinated molecules by using them as electron acceptors (Furukawa 2000; Ohtsubo et al. 2004; Pieper 2005; Field and Sierra-Alvarez 2008). The role of specific microbial species in promoting PCB de-halogenation has been evaluated in several studies (Wiegel and Wu 2000; Zanaroli et al.

2012; Matturro et al. 2016). However, the complete degradative pathways and the environmental parameters involved in PCB transformation need to be better clarified (Sylvestre and Toussaint 2011). Consequently, the next major focus of research into PCBs aims at maximizing the potential of these natural degraders in order to accelerate environmental degradation processes (Meggo and Schnoor 2013).

Plant-microorganism associations can improve PCB degradation in the rhizosphere, thanks to synergic interactions between roots and soil microbial communities. Some root exudates may contain plant secondary metabolites, which can act as chemical signals promoting or inducing bacterial enzymes, involved in PCB degradation (Sylvestre and Toussaint 2011; Toussaint et al. 2012; Meggo and Schnoor 2013; Qin et al. 2014; Pham et al. 2015; Musilova et al. 2016). In return, degrading bacteria can produce plant growth stimulators or can suppress pathogens, through competition and antibiotic production.

A wide range of plants has been shown to enhance the dissipation of PCBs in soil, ranging from trees genera such as *Populus* and *Salix* to different forages, both grasses and legumes (Dzantor et al. 2000; Chekol et al. 2004; Xu et al. 2010; Ding et al. 2011). In particular poplar, owing to its fast growth rate and its deep and widespreading root system, which adds to its ability to grow in nutrient-poor soil and to resist high metal concentrations (Di Baccio et al. 2003; Sebastiani et al. 2004; Soudek et al. 2004; Baldantoni et al. 2011).

Poplar has been used successfully to stimulate the biodegradation of xenobiotic compounds in both field (Liu and Schnoor 2008; Zhai et al. 2011; Massacci et al. 2012) and microcosm studies (Meggo and Schnoor 2013; Meggo et al. 2013). It was also effective in heavy metal phytoremediation experiments (Gamalero et al. 2012). However, only a few investigations with large-scale trials have been attempted and often they were not completed (Xu et al. 2010; Teng et al. 2010). No evidence of PCB degradation in the field has been found so far in historically and multi-contaminated soils, where PCB molecules are strongly bound to soil particles and therefore less available for biodegradation processes. Recently, Ancona et al. (2017) successfully carried out a plant-assisted bioremediation of a historically PCB and heavy metal-contaminated area in Southern Italy. To provide a case study example, their experimental design will be shown and discussed below.

18.2 Before Starting: Preparatory and Preliminary Steps

18.2.1 Characterization of the Contaminated Area

An initial and complete chemical and microbiological characterization of the polluted area where the plant-assisted bioremediation strategy is to be carried out is required. Physico-chemical analyses include soil texture, pH, moisture, electrical conductivity and nutrients (such as organic carbon, nitrogen and available phosphorous content). At the same time there needs to be an evaluation of different contaminants (e.g. PCBs and heavy metals) using analytical instruments (e.g. GC-MS for organic contaminants and ICP-MS for inorganic ones such as heavy metals) to assess if their concentrations exceed legal limits and to identify tolerant and suitable plant species for phytoremediation purposes.

Moreover, microbiological analyses for evaluating the structure (biodiversity) and activity of the soil autochthonous microbial community are very important. In fact, the presence of an abundant and varied natural microbial community is a prerequisite for an effective homeostatic response to the various chemicals that can contaminate an ecosystem (Artigas et al. 2012). The direct soil extraction and transesterification of total ester-linked fatty acids (ELFAs) is a reliable method for microbial profiling (Schutter and Dick 2000; Hinojosa et al. 2005, 2016; Laudicina et al. 2012) and phenotypic fingerprinting of the main microbial groups (Gram-positive and Gram-negative bacteria, Fungi), occurring in soil.

A measurement of dehydrogenase activity can provide useful information on the status and functioning of a soil microbial community as it reflects the respiration rate, giving information on the active community portion (Barra Caracciolo et al. 2015; Ancona et al. 2017). Moreover, the methods successfully used in these studies include those which adopt epifluorescence microscopy, such as the direct measurement of total microbial abundance by DAPI counts (Barra Caracciolo et al. 2015), cell viability (% live cells/live + dead, Grenni et al. 2012) and the Fluorescence *in situ* Hybridization (FISH) technique, which makes it possible to determine the phylogenetic characterization of active *Bacteria* and *Archaea* populations (Barra Caracciolo et al. 2015). Moreover, it is also advisable to perform a complete modelling of ground- and surface-water, including any runoff and drainage of contaminants in the studied area.

18.2.2 Evaluation of Local Climatic Conditions and Plant Species Selection

Each phytoremediation strategy is site-dependent and has to be planned in detail, based on soil and climatic conditions, and on the type and history of the contamination event. The choice of the plant species and varieties is of crucial importance to optimize each remediation process. The search for a plant able to degrade/transform a specific contaminant is performed in appropriate databases available on line. As an example, the Italian Institute of Agro-environmental and Forest Biology (IBAF-CNR) has published a detailed online database (see http://www.ibaf.cnr.it/phytoremediation/fitorimedio.pdf). The results of several phytoremediation projects carried out in the USA have been included in the Guideline "*Phytotechnology Technical and Regulatory Guidance and Decision Trees*" – Interstate Technology & Regulatory Council (ITRC 2009), available online.

In the case of the historically contaminated area located in Southern Italy described herein, the *Monviso* poplar clone was selected for its ability to promote hexachlorocyclohexane degradation, tested in previous bioremediation studies (Massacci et al. 2012; Bianconi et al. 2010; Pietrini et al. 2010).

18.3 Assessment of the Natural Bioremediation Capacity in Microcosm Experiments

Prior to the field phytoremediation application, the bioremediation capability of the autochthonous soil microbial community must be tested, using microcosm experiments. Laboratory microcosms are ecosystem models in which a portion of the natural environment (soil or water) is circumscribed and then studied. They contain natural biotic communities which are maintained under controlled environmental conditions (e.g. temperature, light, humidity and so on) corresponding to natural ones. This approach makes it possible to firmly secure the causal relationship between a toxicant and its effects on the microbial communities and/or plant. The presence of a contaminant can act as a selective force, affecting the microbial community. Some populations, which are not capable of resisting its toxic effects, either die or enter a static metabolic phase, while others can evolve resistance mechanisms and utilize the excess chemical as an energy source or in their co-metabolism. In this case they can proliferate and become dominant members of the ecosystem (Barra Caracciolo et al. 2013). Consequently, the natural capability of organic pollutant degradation by microbial community could be evaluated.

A set of experimental microcosms, using soil from the contaminated area, are performed for evaluating PCB degradation (Fig. 18.1). Half of the microcosms are planted with *Populus* cuttings. Other experimental conditions could be also simulated, such as soil amended with compost in absence/presence of plants, in order to assess if an additional source of carbon can increase PCB degradation, and/or pre-sterilized soil in order to underline the effectiveness of plants in degrading chemicals, without considering the microbial population.

For each condition (soil, soil + poplar, sterilized soil + poplar, soil + compost, soil + poplar + compost) at least three replicated microcosms are necessary. Aliquots of planted and unplanted soils are sampled at fixed times from the microcosms, for a chemical and microbiological evaluation of the decontamination process (PCB degradation and heavy metal decrease in the soil). Plant materials (roots, stems,

Fig. 18.1 PCB contaminated soil planted with poplar and maintained in a greenhouse for 6 months under different conditions

leaves) are also sampled to identify any PCB and heavy metal bioaccumulation. The microcosms are maintained in a greenhouse for 6 months.

The chemical analysis of the contaminants, together with the evaluation of the microbial structure (diversity) and functioning, provide very useful information such as:

1. the occurrence in soil of autochthonous microbial populations able to degrade the contaminants studied (natural bioremediation)
2. a possible increase in the chemicals degradation in presence of the selected plant species (plant-assisted bioremediation)
3. a further bio-stimulation of biodegradation, achieved adding external organic sources (i.e. compost) or nutrients (water, oxygen, etc.).

18.4 Field Scale Application

18.4.1 Experimental Design and Set Up of Planting Area

The remediation strategy start has to be planned based on the ecology (e.g. suitable season for plant growth) of the species selected and the local climatic conditions. In the case of a poplar-bioremediation strategy performed in Southern Italy (Ancona et al. 2017), the selected *Monviso* clone was planted in spring (April), while in another application in Northern Italy, it was planted by the end of February (Bianconi et al. 2010).

The planting scheme (e.g. the number of plants and distance between them) must take into account the plant ecology and the root and shoot development. Poplar tree display a fast growth rate and a deep and wide-spreading root system (Baldantoni et al. 2011), promoting root anastomoses and diffusion in soil, and rhizodegradation processes. For this purpose, cuttings of poplars can be placed in a number of rows (each 2 m from the other) that cover the planting area. Inside each row the cuttings should be at a distance of 0.5 m from each other (Ancona et al. 2017). Moreover, specific interventions to prepare soil for planting are also performed (e.g. stone removal, ploughing, scarifying, hand weeding and soil milling). At the same time of planting operations, a suitable irrigation system must be set up (e.g. a tube for each row) to provide the necessary water supply. Mulching is also necessary, for each planting row, to control weeds in the early growth stages (Fig. 18.2).

18.4.2 Monitoring Early Stages of Vegetative Growth and Optimized Agronomic Management of Treated Area

The experimental area has to be supervised weekly to assess plant growth and health. In particular, it is important to monitor the early vegetative growth stages within a few days of planting and, if necessary, to replace plants if rooting failure is

Fig. 18.2 Mulching set up before planting poplar cuttings in a PCB historically contaminated area located in Southern Italy (Ancona et al. 2017)

observed. A continuous monitoring of plant health during the application of the remediation technology is necessary in order to evaluate, in time, any plant requirement/need. In this regard, various strategies need to be planned and implemented during the different seasons to control weeds and/or pests, to identify water consumption for managing irrigation and to carry out pruning.

18.4.3 Sampling of Soils and Plant Materials at Different Times

In order to evaluate the efficiency of the phytoremediation strategy in the experimental area, samplings of soil and plant materials (roots, stems, leaves) should to be carried out at different times (e.g. 12, 24, 36, 48 months after planting), both in the treated area and outside the planting area. Soil samples are collected from target plants in accordance with a sampling procedure designed for the site-specific treatment, taking into account the plantation scheme (Fig. 18.3). A number of soil/subsoil samples have to be collected at different distances (for example 1 m, 25 cm) and depths (for example 0–20 cm, 20–40 cm), from each target plant (Fig. 18.4).

Correct sampling is essential for evaluating the possible existence of a contamination gradient in the area below the plant (ranging from rhizosphere to bulk soil) and the presence of synergistic interactions between the plants and the microorganisms present. Plant material has also to be sampled, i.e. roots from the rhizosphere (to a

Fig. 18.3 Soil sampling from a PCB historically contaminated area, located in Southern Italy and planted with poplar (Ancona et al. 2017)

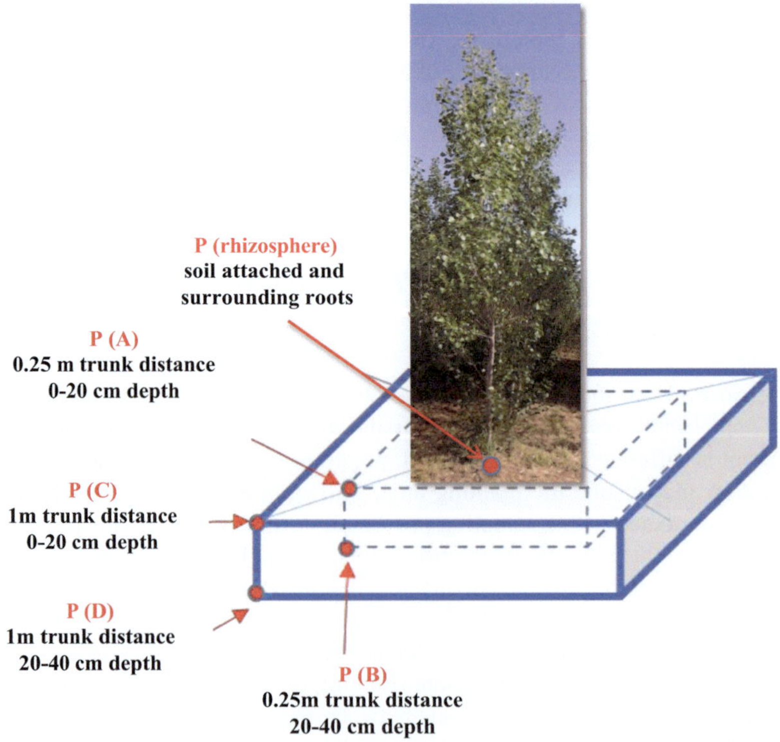

Fig. 18.4 Soil sampling design for each poplar in the studied area (from Ancona et al. 2017)

maximum depth of 30 cm) and leaves from the outer canopy located in the upper third of each tree, the area in which leaves tend to accumulate more mineral elements than inner ones, due to their higher transpiration rates (Madejón et al. 2004).

18.4.4 Chemical Analyses

Physico-chemical analyses (soil moisture, pH, conductivity, organic carbon, nitrogen content, etc.) performed at different times after the plant-treatment make it possible to evaluate the improvement in soil quality and fertility. Such analyses are useful for measuring, at different sampling times, any nutritional deficiency that might affect development and/or make plant more susceptible to pathogens attacks. Such data are necessary to identify the right agronomic practices (e.g. fertilization), supporting plant growth.

The analysis of the different PCBs and heavy metal congeners makes it possible to assess their reduction in soil as compared to their initial values and legal limits. In the case of PCBs, it is very important to assess if the congeners with a higher chlorine content (*octa*, *hepta*, *hexa*-chlorinated) decrease in favor of those with a lower content (*penta*, *tetra*, *tri*-chlorinated). In our experiment, chemical analyses performed at day 420 in selected plots at different distances from the trunk (0.25–1 m) and at different soil depths (0–20 and 20–40 cm) showed a significant decrease in PCB congeners and an overall reduction in all heavy metals (phyto-containment) when the poplar trees were present. Moreover, root and leaf PCB analyses showed that the poplar plant was not substantially able to phyto-extract PCBs (Ancona et al. 2017).

18.4.5 Microbiological Analyses

The organic contaminants decrease in the rhizosphere can be associated with an increase in the microbial activity and the number of microbial populations connected with PCB degradation. Measuring dehydrogenase activity can be informative about the active portion of the soil microbial community (Barra Caracciolo et al. 2015; Ancona et al. 2017) and the microbial abundance and diversity (phylogenetic characterization by ELFA and FISH analyses) of the interactions established between plant roots, autochthonous microbial populations and contaminant transformation. In the poplar-assisted bioremediation of a historically PCB and heavy metal-contaminated area in Southern Italy, the overall analyses of the autochthonous microbial community showed an improvement in soil quality. In particular, the activity of microorganisms generally increased in the poplar-rhizosphere. A positive effect was observed, in some cases at up to 1 m distance from the trunk and at up to 40 cm in depth. The *Monviso* clone was effective in promoting a general decrease in contaminant occurrence (plant-assisted PCB bioremediation and phyto-containment of heavy metals), with an increase in microbial activity and diversity in a chronically polluted area, just over one year after planting.

18.4.6 Quality Assessment of Plant Biomass

Plant assisted bioremediation of contaminated areas also has the potential to provide valuable sources of renewable biomass, which can be used to produce bioenergy, in line with the sustainability criteria of the Renewable Energy Directive 2009/28/EC (EC 2009). In this context, several studies have demonstrated that biomass cultivated for plant based remediation purposes in metal polluted soils can be thermally treated (combustion, gasification, torrefaction and pyrolysis) and used as potential fuel (biofuel, bioliquids) to obtain bioenergy (Bert et al. 2017). In fact, these plants may enter valorization pathways if the trace elements (metals) do not disturb the functioning and performance of the processes, if the metal transfer is controlled and impacts on the environment are prevented and, finally, if such plants use complies with current regulation (Bert et al. 2017). In this context, a chemical investigation of the biomass produced after 3-5 years from the planting is required to determine its quality and possible contaminant concentrations. It would be currently very useful to develop best practice guidelines at a European level to propose biomass properties, required for best "energy crops", to be used for bioremediation purposes.

18.5 Conclusions

The application of phyto-technologies requires interdisciplinary knowledge (chemistry, microbial ecology, plant physiology, botany, agronomy, soil science, hydrogeology, etc.) of site-specific ecosystem processes. The success of plant-assisted bioremediation and other phyto-technologies for recovering contaminated soils (phyto-containment/phyto-stabilization, phyto-extraction) relies on a holistic approach ensuring a full recovery of the contaminated area, including in terms of soil quality and fertility, and a possible use of the biomass produced. The potential of microorganisms to degrade contaminants and of their synergic interactions with plant are unlimited and can be exploited in the future as an economical, environmentally sustainable and landscape enhancing technology. Plant-assisted bioremediation complies with green remediation and has a low impact on climate change and on the environment, in line with environmental sustainability (US EPA 2010).

Acknowledgements The authors acknowledge CISA S.p.A. (Massafra, Italy), which partially funded the Research Project "*Applicazione di tecniche di fitorimedio a basso costo in località ex campo Cimino-Manganecchia a Taranto*", Prot. IRSA-CNR N. 0005159, 04/12/2012.

Authors thank contribution by COST Action FP1305 "BioLink-Linking belowground biodiversity and ecosystem function in European forests".

References

Ancona, V., Barra Caracciolo, A., Grenni, P., Di Lenola, M., Campanale, C., Calabrese, A., Uricchio, V. F., Mascolo, G., & Massacci, A. (2017). Plant-assisted bioremediation of a historically PCB and heavy metal-contaminated area in Southern Italy. *New Biotechnology: Part B, 38*, 65–73.

Artigas, J., Arts, G., Babut, B., Barra Caracciolo, A., Charles, S., Chaumot, A., Combourieu, B., Dahllöf, I., Despréaux, D., Ferrari, B., Friberg, N., Garric, J., Geffard, O., Gourlay-Francé, C., Hein, M., Hjorth, M., Krauss, M., De Lange, H. J., Lahr, J., Lehtonen, K. K., Lettieri, T., Liess, M., Lofts, S., Mayer, P., Morin, S., Paschke, A., Svendsen, C., Usseglio-Polatera, P., van den Brink, N., Vindimian, E., & Williams, R. (2012). Towards a renewed research agenda in eco toxicology. *Environmental Pollution, 160*, 201–206.

Bais, H., Weir, T., Perry, L., Gilroy, S., & Vivanco, J. (2006). The role of root exudates in rhizosphere interactions with plants and other organisms. *Annual Review of Plant Biology, 57*, 233–266.

Barra Caracciolo, A., Bottoni, P., & Grenni, P. (2013). Microcosm studies to evaluate microbial potential to degrade pollutants in soil and water ecosystems. *Microchemical Journal, 107*, 126–130.

Barra Caracciolo, A., Bustamante, M. A., Nogues, I., Di Lenola, M., Luprano, M. L., & Grenni, P. (2015). Changes in microbial community structure and functioning of a semiarid soil due to the use of anaerobic digestate derived composts and rosemary plants. *Geoderma, 245–246*, 89–97.

Baldantoni, D., Bellino, A., Cicatelli, A., & Castiglione, S. (2011). Artificial mycorrhization does not influence the effects of iron availability on Fe, Zn, Cu, Pb and Cd accumulation in leaves of a heavy metal tolerant white poplar clone. *Plant Biosystems, 145*, 236–240.

Bert, V., Allemon, J., Sajet, P., Dieu, S., Papin, A., Collet, S., Gaucher, R., Chalot, M., Michiels, B., & Raventos, C. (2017). Torrefaction and pyrolysis of metal-enriched poplars from phytotechnologies: Effect of temperature and biomass chlorine content on metal distribution in endproducts and valorization options. *Biomass and Bioenergy, 96*, 1–11.

Bianconi, D., De Paolis, M. R., Agnello, M. C., Lippi, D., Pietrini, F., Zacchini, M., Polcaro, C., Donati, E., Paris, P., Spina, S., & Massacci, A. (2010). Field-scale rhyzoremediation of a contaminated soil with hexachlorocyclohexane (HCH) isomers: The potential of poplars for environmental restoration. In I. A. Golubev (Ed.), *Phytoremediation: Processes, characteristics, and applications* (pp. 783–794). Hauppauge: Nova Science Publisher.

Bru, D., Ramette, A., Saby, N. P., Dequiedt, S., Ranjard, L., Jolivet, C., Arrouays, D., & Philippot, L. (2011). Determinants of the distribution of nitrogen-cycling microbial communities at the landscape scale. *The ISME Journal, 5*, 532–542.

Chekol, T., Vough, L. R., & Chaney, R. L. (2004). Phytoremediation of polychlorinated biphenyl-contaminated soils: The rhizosphere effect. *Environment International Journal, 30*, 799–804.

Di Baccio, D., Tognetti, R., Sebastiani, L., & Vitagliano, C. (2003). Responses of *Populus deltoides* x *Populus nigra* (*Populus x euramericana*) clone I-214 to high zinc concentrations. *New Phytologist, 159*, 443–452.

Ding, N., Hayat, T., Wang, J. E., Wang, H. Z., Liu, X. M., & Xu, J. M. (2011). Responses of microbial community in rhizosphere soils when ryegrass was subjected to stress from PCBs. *Journal of Soils Sediments, 11*, 1355–1362.

Dzantor, E. K., Chekol, T., & Vough, L. R. (2000). Feasibility of using forage grasses and legumes for phytoremediation of organic pollutants. *Journal of Environmental Science and Health, 35*, 1645–1661.

EC (European Commission). (2009). Directive 2009/28/EC of the European Parliament and of the Council of 23 April 2009 on the promotion of the use of energy from renewable sources and amending and subsequently repealing Directives 2001/77/EC and 2003/30/EC. *Official Journal of the European Union L, 140*, 16–62.

Field, J. A., & Sierra-Alvarez, R. (2008). Microbial transformation and degradation of polychlorinated biphenyls. *Environmental Pollution, 155*, 1–12.

Fierer, N., & Jackson, R. B. (2006). The diversity and biogeography of soil bacterial communities. *Proceedings of the National Academy of Sciences, USA, 103*, 626–631.
Furukawa, K. (2000). Biochemical and genetic bases of microbial degradation of polychlorinated biphenyls (PCBs). *Journal of General and Applied Microbiology, 46*, 283–296.
Gamalero, E., Cesaro, P., Cicatelli, A., Todeschini, V., Musso, C., et al. (2012). Poplar clones of different sizes, grown on a heavy metal polluted site, are associated with microbial populations of varying composition. *Science of Total Environment, 425*, 262–270.
Grenni, P., Rodriguez-Cruz, M. S., Herrero-Hernandez, E., Marin-Benito, J. M., Sanchez-Martin, M. J., & Barra Caracciolo, A. (2012). Effects of wood amendments on the degradation of terbuthylazine and on soil microbial community activity in a clay loam soil. *Water, Air and Soil Pollution, 223*, 5401–5412.
Hinojosa, M. B., Carreira, J. A., García-Ruíz, R., & Dick, R. P. (2005). Microbial response to heavy metal-polluted soils: Community analysis from phospholipid-linked fatty acids and ester-linked fatty acids extracts. *Journal of Environmental Quality, 34*, 1789–1800.
Hinojosa, M. B., Parra, A., Laudicina, V. A., & Moreno, J. M. (2016). Post-fire soil functionality and microbial community structure in a Mediterranean shrubland subjected to experimental drought. *Science of the Total Environment, 573*, 1178–1189.
Hinsinger, P., Bengough, A. G., Vetterlein, D., & Young, I. M. (2009). Rhizosphere: Biophysics, biogeochemistry and ecological relevance. *Plant and Soil, 321*, 117–152.
ITRC. (2009). *Phytotechnology. Technical and regulatory guidance and decision trees*, revised. Available at http://www.scribd.com/doc/67998102/Phytotechnology-Technical-and-Regulatory-Guidance-and-Decision-Trees-Revised
Laudicina, V. A., Dennis, P. G., Palazzolo, E., & Badalucco, L. (2012). Key biochemical attributes to assess soil ecosystem sustainability. In A. Malik & E. Grohmann (Eds.), *Environmental protection strategies for sustainable development* (pp. 193–227). Cham: Springer.
Liu, J., & Schnoor, J. L. (2008). Uptake and translocation of lesser chlorinated polychlorinated biphenyls (PCBs) in whole hybrid of poplar plants after hydroponic exposure. *Chemosphere, 73*, 1608–1616.
Madejón, P., Marañón, T., Murillo, J. M., & Robinson, B. (2004). White poplar (*Populus alba*) as a biomonitor of trace elements in contaminated riparian forests. *Environmental Pollution, 132*, 145–155.
Matturro, B., Ubaldi, C., Grenni, P., Barra Caracciolo, A., & Rossetti, S. (2016). Polychlorinated biphenyl (PCB) anaerobic degradation in marine sediments: Microcosm study and role of autochtonous microbial communities. *Environmental Science and Pollution Research, 23*, 12613–12623.
Massacci, A., Bianconi, D., & Paris, P. (2012). Pioppicoltura a turno di taglio breve per bioenergia e fitorimedio. *SILVÆ, 7*, 125–144.
Meggo, R. E., & Schnoor, J. L. (2013). Rhizosphere redox cycling and implications for rhizosphere biotransformation of selected polychlorinated biphenyl (PCB) congeners. *Ecological Engineering, 57*, 285–292.
Meggo, R. E., Schnoor, J. L., & Hu, D. (2013). Dechlorination of PCBs in the rhizosphere of switchgrass and poplar. *Environmental Pollution, 178*, 312–321.
Musilova, L., Ridl, J., Polivkova, M., Macek, T., & Uhlik, O. (2016). Effects of secondary plant metabolites on microbial populations: Changes in community structure and metabolic activity in contaminated environments. *International Journal of Molecular Science, 17*, E1205.
Ohtsubo, Y., Kudo, T., Tsuda, M., & Nagata, Y. (2004). Strategies for bioremediation of polychlorinated biphenyls. *Applied Microbiology and Biotechnology, 65*, 250–258.
Pham, T. T., Pino Rodriguez, N. J., Hijri, M., & Sylvestre, M. (2015). Optimizing polychlorinated biphenyl degradation by flavonoid-induced cells of the rhizobacterium *Rhodococcus erythropolis* U23A. *PLoS One, 10*, e0126033.
Philippot, L., Raaijmakers, J. M., Lemanceau, P., & van der Putten, W. M. (2013). Going back to the roots: the microbial ecology of the rhizosphere. *Nature Review Microbiology, 11*, 789–799.

Pieper, D. H. (2005). Aerobic degradation of polychlorinated biphenyls. *Applied Microbiology and Biotechnology, 67*, 170–191.

Pieper, D., & Seeger, M. (2008). Bacterial metabolism of polychlorinated biphenyls. *Journal of Molecular Microbiology and Biotechnology, 15*, 121–133.

Pietrini, F., Zacchini, M., Iori, V., Pietrosanti, L., Bianconi, D., & Massacci, A. (2010). Screening of poplar clones for cadmium phytoremediation using photosynthesis, biomass and cadmium content analyses. *International Journal of Phytoremediation, 12*, 1–16.

Pilon-Smits, E. (2005). Phytoremediation. *Annual Review of Plant Biology, 56*, 15–39.

Qin, H., Brookes, P. C., & Xu, J. (2014). *Cucurbita* spp. and *Cucumis sativus* enhance the dissipation of polychlorinated biphenyl congeners by stimulating soil microbial community development. *Environmental Pollution, 184*, 306–312.

Schutter, M. E., & Dick, R. P. (2000). Comparison of fatty acid methyl ester (FAME) methods for characterizing microbial communities. *Soil Science Society of America Journal, 64*, 1659–1668.

Sebastiani, L., Scebba, F., & Tognetti, R. (2004). Heavy metal accumulation and growth responses in poplar clones Eridano (*Populus deltoides* × *maximowiczii*) and I-214 (*P.* × *euramericana*) exposed to industrial waste. *Environmental and Experimental Botany, 52*, 79–88.

Soudek, P., Tykva, R., & Vanek, T. (2004). Laboratory analyses of ^{137}Cs uptake by sunflower, reed and poplar. *Chemosphere, 55*, 1081–1087.

Sylvestre, M., & Toussaint, J. P. (2011). Engineering microbial enzymes and plants to promote PCB degradation in soil: Current state of knowledge. In A. I. Koukkou (Ed.), *Microbial bioremediation of nonmetals – Current research* (pp. 177–196). Norfolk: Caister Academic.

Teng, Y., Luo, Y., Sun, X., Tu, C., Xu, L., Liu, W., Li, Z., & Christie, P. (2010). Influence of arbuscular mycorrhiza and Rhizobium on phytoremediation by alfalfa of an agricultural soil contaminated with weathered PCBs: A field study. *International Journal of Phytoremediation, 12*, 516–533.

Thijs, S., Sillen, W., Rineau, F., Weyens, N., & Vangronsveld, J. (2016). Towards an enhanced understanding of plant-microbiome interactions to improve phytoremediation: Engineering the metaorganism. *Frontiers in Microbiology, 7*, 341.

Tsednee, M., Mak, Y. W., Chen, Y. R., & Yeh, K. C. (2012). A sensitive LC-ESI-Q-TOF-MS method reveals novel phytosiderophores and phytosiderophore–iron complexes in barley. *New Phytologist, 195*, 951–961.

Toussaint, J. P., Pham, T. T., Barriault, D., & Sylvestre, M. (2012). Plant exudates promote PCB degradation by a rhodococcal rhizobacteria. *Applied Microbiology and Biotechnology, 95*, 1589–1603.

US EPA. (2010). *Superfund green remediation strategy*. U.S. Environmental Protection Agency, Office of Solid Waste and Emergency Response, Office of Superfund Remediation and Technology Innovation. Available at https://www.epa.gov/sites/production/files/2016-01/documents/175857.pdf

Wenzel, W. W. (2009). Rhizosphere processes and management in plant-assisted bioremediation (phytoremediation) of soils. *Plant and Soil, 321*, 385–408.

Wiegel, J., & Wu, Q. (2000). Microbial reductive dehalogenation of polychlorinated biphenyls. *FEMS Microbiology and Ecology, 32*, 1–15.

Xu, L., Teng, Y., Li, Z. G., Norton, J. M., & Luo, Y. M. (2010). Enhanced removal of polychlorinated biphenyls from alfalfa rhizosphere soil in a field study: The impact of a rhizobial inoculum. *Science of Total Environment, 408*, 1007–1013.

Zanaroli, G., Balloi, A., Negroni, A., Borruso, L., Daffonchio, D., & Fava, F. (2012). Chloroflexi bacterium dechlorinates polychlorinated biphenyls in marine sediments under in sit-like biogeochemical conditions. *Journal of Hazardous Materials, 209*, 449–457.

Zhai, G., Hu, D., Lehmler, H. J., & Schnoor, J. L. (2011). Enantioselective biotransformation on chiral PCBs in whole poplar plants. *Environmental Science and Technology, 45*, 2308–2316.

Chapter 19
Bioavailability of Polycyclic Aromatic Hydrocarbons in Soil as Affected by Microorganisms and Plants

Jose Julio Ortega-Calvo, Rosa Posada-Baquero, José Luis Garcia, and Manuel Cantos

Abstract The bioavailability of polycyclic aromatic hydrocarbons (PAHs) in soil can be enhanced through a variety of microbial and plant functions, that can be incorporated into optimized bioremediation technologies. In this review, we examine the potential of (bio)surfactants, the chemotactic mobilization of pollutant-degrading bacteria, and the role of bacterial attachment, to enhance biodegradation of PAHs. Plants can also play an active role in enhancing bioavailability of PAHs through rhizosphere-related mechanisms associated to specific exudate components that affect bacterial chemotaxis, pollutant mobilization, and intra-aggregate bacterial growth.

Keywords Biodegradation • Bioremediation • Bioavailability • Bioaccessibility • PAHs • Roots • Biosurfactant • Chemotaxis • Attachment • Desorption • Exudates • Bacteria • Transport

19.1 Introduction

Polycyclic aromatic hydrocarbons (PAHs) are ubiquitous soil pollutants. They are typically produced in the incomplete combustion of wood and petroleum products, and they are transferred to soils through dry or wet deposition. Accidental spills of petroleum or the disposal of industrial wastes cause point source pollution. Unacceptably high PAH concentrations in soils are the main reason for the remediation of many contaminated sites. In this context, PAHs are among the soil organic contaminants that have significant potential for bioremediation. However, they are the best representatives of chemicals for which specific limitations in bioremediation

---

J.J. Ortega-Calvo (✉) • R. Posada-Baquero • J.L. Garcia • M. Cantos
Instituto de Recursos Naturales y Agrobiologia de Sevilla, CSIC, Sevilla, Spain
e-mail: jjortega@irnase.csic.es

M. Lukac et al. (eds.), *Soil Biological Communities and Ecosystem Resilience*, Sustainability in Plant and Crop Protection, DOI 10.1007/978-3-319-63336-7_19

exist due to low bioavailability. The residual concentrations of the PAHs after bioremediation are crucial because they may limit the use of the area after treatment, or land use might not even be possible if the residual concentrations do not meet the legal requirements.

Of the 16 PAHs listed by the US-EPA as priority pollutants, those with a molecular weight of up to 202 g/mol can be degraded microbially through growth-linked aerobic reactions, including the high-molecular-weight PAHs pyrene and fluoranthene, while the rest of the PAHs in that list, such as benzo[a]pyrene, are susceptible to co-metabolic removal (Niqui-Arroyo et al. 2011). These reactions are the basis of a variety of approaches that have been used and validated to biologically treat soils contaminated with PAHs, which include landfarming, composting, bioreactor treatments and phytoremediation. In this review, we examine the potential of microorganisms and plants to act as bioavailability-promoting agents. The focus will be on mechanisms that eventually allow the integration of a bioavailability-efficient technology into current remediation practices based on biodegradation of PAHs.

19.2 The State-of-the-Art in Bioavailability Science

19.2.1 The Bioavailability Concept

For bioavailability to be incorporated into soil (bio)remediation, three questions must be addressed: (1) what is meant by "bioavailability", (2) how should it be measured? and (3) is it possible to increase bioavailability but not environmental risk of the pollutants? Over the last 30 years, numerous publications have discussed the concepts and definitions of bioavailability of organic chemicals. These have been summarized recently in the context of risk assessment and regulation (Ortega-Calvo et al. 2015), and are illustrated in Fig. 19.1.

Main schools of thought consider bioavailability (focusing on the aqueous or dissolved contaminant), bioaccessibility (incorporating the rapidly desorbing contaminant in the exposure), and chemical activity (determining the potential of the dissolved contaminant for biological effects). Using the same framework, the figure places these different schools (Ehlers and Luthy 2003; Semple et al. 2004; Reichenberg and Mayer 2006) that have dissected bioavailability into the different processes that are involved (A to E), the dissimilar endpoints (bioaccessibility and chemical activity), and the different methodologies (desorption extraction, passive sampling and biological tests).

Each of these processes, endpoints and methods has been considered differently in a wide variety of bioavailability scenarios. Depending on the processes investigated, bioavailability can be examined through chemical activity, the potential of the contaminant for direct transport and interaction with the cell membrane (processes B, C and D), or bioaccessibility measurements, which incorporate the time-dependent phase exchange of the contaminant between the soil/sediment and the

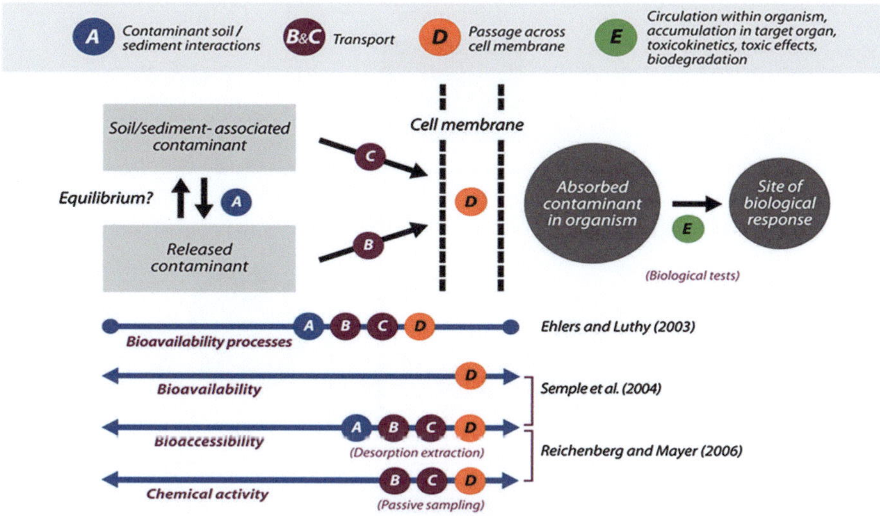

Fig. 19.1 Overview of scientific concepts of the bioavailability of organic chemicals (Reproduced with permission from Ortega-Calvo et al. 2015)

water phase (process A). Depending on biological complexity, the passage of the contaminant molecule across the cell membrane (process D) may represent multiple stages within a given organism before the site of biological response is reached (process E).

Chemical and biological approaches can be used to measure the bioavailability of organic chemicals. The results of infinite sink methods using Tenax and cyclodextrin extraction during approximately 20 hours are currently used to predict toxicity and biodegradation, and are in the process of being standardized. The results of these methods represent and define what is referred to as the rapidly desorbing fraction. The second complementary approach is the use of passive sampling to determine the freely dissolved concentration as a measure of the chemical activity of organic chemicals in soils and sediments. This approach proposes that chemical activity drives bioavailability (Fig. 19.1). Finally, several (mostly standardized by ISO and OECD) ecotoxicological test methods are available to determine bioavailability in the soil and sediment compartments to invertebrates, plants and microorganisms (Ortega-Calvo et al. 2015).

19.2.2 Biodegradability vs Bioavailability of PAHs

The majority of contaminated areas show significant levels of PAHs pollution. During the execution of research projects at the Spanish National Research Council (CSIC) over the last 15 years we have prospected bioremediation strategies that minimize risks, such as *(i)* the modulation of pollutant release relative to the actual biodegradation

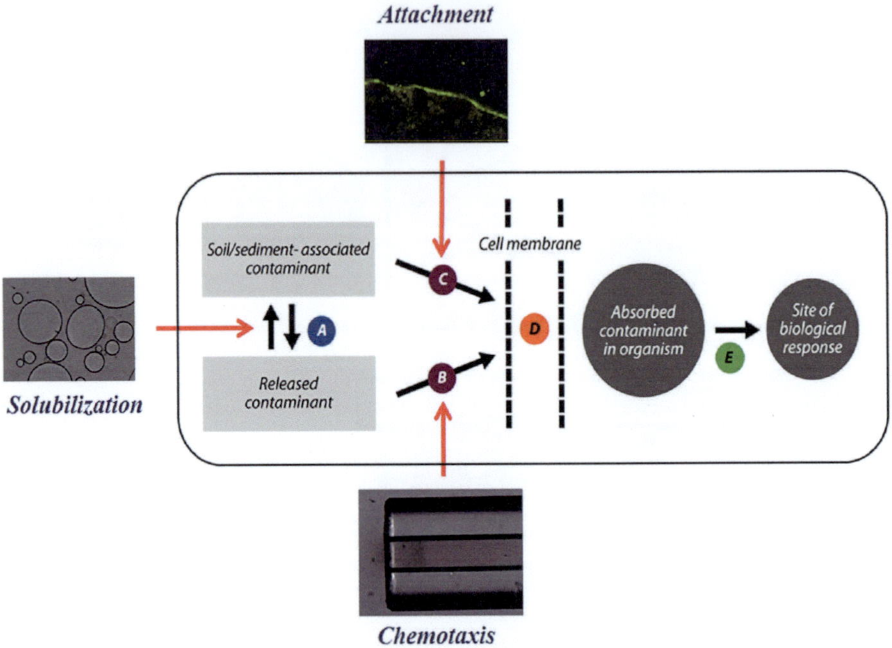

Fig. 19.2 Strategies to influence bioavailability mechanisms (*A* to *E*, as described in Fig. 19.1), in connection with the microbial degradation of PAHs. Biodegradation can be enhanced through solubilization (representing the surfactant action on phase exchange-process *A*), tactically- driven microbial mobilization (represented in the figure by the band of chemotactic bacteria attracted by a chemical gradient inside a capillary-process *B*) and attachment to interfaces, what allows the direct acquisition of the soil/sediment-associated contaminant (process *C*). In this context, *D* represents the pollutant uptake by the microbial cells, necessary for biodegradation processing (*E*)

potential, *(ii)* the use of environmentally acceptable agents, such as surfactants, microorganisms and plants and *(iii)* the application of treatment methods requiring minimal handling to reduce the risk of the contaminants within and from the soil or sediment (Ortega-Calvo et al. 2013). These strategies can operate, at different levels, on bioavailability processes to enhance the microbial degradation of PAHs (Fig. 19.2).

Surfactants can increase the biodegradation rates of PAHs that desorb slowly from soils (Fig. 19.2, process A). In general, nonionic surfactants are the most frequently used types of surfactants in biodegradation studies, mainly because they are electrically neutral, which minimizes any eventual toxic effects (Martín et al. 2014). We applied a nonionic surfactant, Brij 35, to a soil originating directly from a site polluted by creosote and a soil from a manufactured gas plant (MGP) that had been treated by bioremediation (Bueno-Montes et al. 2011). In the creosote-polluted soil, biodegradation was inhibited by the surfactant, but biodegradation in the bioremediated MGP soil was enhanced. The different outcomes were likely a consequence of the balance of two effects, namely an increase in the bioaccessibility of the chemicals and an enhancement of the consumption rate of other PAHs present in the soil at the same time, yielding subsequent competition effects. Our results indicate that electrokinetics can also be successfully employed to enhance the bioavailability of PAHs, particularly for the surfactant-assisted bioremediation of soils that are rich in clay fractions and/or aged contaminants (Niqui-Arroyo et al. 2006; Niqui-Arroyo and Ortega-Calvo 2007, 2010).

It is conceivable, therefore, that microbial or plant biosurfactants, if properly managed, are able to improve PAH-bioremediation performance. Indeed, we have shown that rhamnolipid biosurfactants can dissolve pure, solid PAHs, such as phenanthrene, thus increasing their rate of biodegradation (Garcia-Junco et al. 2001, 2003; Resina-Pelfort et al. 2003), and enhance the desorption and biodegradation of slowly-desorbing and aged pyrene present in soil (Congiu and Ortega-Calvo 2014; Congiu et al. 2015). A similar effect was observed with dissolved humic acids, which enhanced the phase exchange and release of PAHs and thus promoted their biodegradation (Tejeda-Agredano et al. 2014).

The bioavailability of PAHs can be increased not only by mobilizing the pollutants, but also by promoting the dispersal of microorganisms throughout the polluted matrix, to have a better access to the released contaminant (Fig. 19.2, process B). To this end, chemotactic mobilization of flagellated bacteria has gained attention in bioremediation (Krell et al. 2013). A technological innovation based on this concept rests on the mobilization potential of pollutant-degrading microbes from rhizospheres, with chemotactic responses demonstrated toward moderately hydrophobic PAHs, such as phenanthrene, anthracene and pyrene, and are able to move chemotactically at speeds of approximately 1 mm/min (Ortega-Calvo et al. 2003). It has also been shown in later research that motility and transport of pollutant-degrading bacteria can be controlled through a suitable choice of chemical effectors, including carbon sources and nanomaterials (Velasco-Casal et al. 2008; Ortega-Calvo et al. 2011; Jimenez-Sanchez et al. 2012, 2015).

Microorganisms can also increase the bioavailability of PAHs when in direct contact with the pollutant through attachment (Fig. 19.2, process C), thereby enabling biodegradation to proceed more rapidly (Ortega-Calvo and Alexander 1994; Garcia-Junco et al. 2003). The main goal of a recent study (Tejeda-Agredano et al. 2011) was to target the potential nutritional limitations of microorganisms to enhance the biodegradation of PAHs at the interface between the nonaqueous-phase liquid (NAPL) and the water phase. The results indicated that the biodegradation of PAHs by bacterial cells attached to NAPLs can be limited by nutrient availability as a result of the simultaneous consumption of PAHs within the NAPLs, but this limitation can be overcome by interface fertilization. Another approach was based on facilitated microbial (bacterial/oomycete) interactions for the formation of biofilms at pollutant interfaces (Sungthong et al. 2015; Sungthong et al. 2016).

19.3 Rhizoremediation

19.3.1 PAHs Degradation in the Rhizosphere

The remediation method based on plants to extract, sequester and detoxify soil pollutants, phytoremediation, has in the rhizosphere one of the most important agents. Rhizoremediation has recently gained attention with regard to the remediation of organic pollutants such as PAHs, because, among other advantages, it increases the bioavailability and biodegradation of PAHs but does not necessarily increase the risk to the environment. Plants release rhizodeposits, such as exudates in the rhizosphere thereby providing nutrients to the microflora and influencing their activity and diversity. In consequence, the rhizosphere is rich in nutrients and therefore its microbial density is higher than that of bulk soil (Yang et al. 2013). Furthermore, translocation of dissolved contaminants in the rhizosphere as a result of interactions with dissolved organic matter (DOM) and the microbial utilization of root exudates as co-substrates for the co-metabolism of PAHs and chemotaxis, have been proposed as mechanisms through which plants contribute to biodegradation and bioavailability of these pollutants (Newman and Reynolds 2004).

Plants utilized for PAHs rhizoremediation must have developed an extensive root system and a strong positive geotropism to achieve the maximum colonization in the contaminated area. It is important for the plant to be able to tolerate the presence of these contaminants. The aerial and root parts are influencing each other. As a consequence, the transpiration process at leaves is important in the water flux regulation in the soil. Losses of soil water through transpiration might reduce microbial mineralization of organic matter (Bottner et al. 1999). As the soil dries, its hydraulic potential together with the surface tension of exudates decreases, increasing the exudate viscosity. As viscosity increases, the resistance to movement of soil particles in contact with exudates will increase (Walker et al. 2003), influencing e.g. on chemotactic movements through the porous matrix of soils and, in consequence, on PAHs biodegradation process. For all these reasons, an optimal plant selection to stimulating the biodegradation of PAHs by rhizoremediation is critical.

19.3.2 Optimal Plant Species for Rhizoremediation Purposes

Many studies have shown the potential of herbaceous species, mainly grasses (Aprill and Sims 1990), such as tall fescue (*Festuca arundinacea* Scherb.) (Parrish et al. 2004, 2005) and ryegrass (*Lolium perenne* L.), (Chigbo and Batty 2013; D'Orazio et al. 2013), to remediate soils polluted with PAHs. Other family used for PAHs rhizorremedation is *Fabaceae* with species such as alfalfa (*Medicago sativa* L.) with a dominant effect on the structure of rhizospherical microbial communities, supporting 100-fold more PAH degraders than other species and stimulating relative increases in specific *Bacteroidetes* and Proteobacteria populations (Phillips et al.

2006). The high capability of *M. sativa* promoting the degradation of pyrene in pyrene-contaminated soils allowed that after 90 days, pyrene concentration declined 32%, whereas it decreased only by 18% in the control soil without plants (D'Orazio et al. 2013). Rhizoremediation process can be enhanced both for soil PAH degraders and plants adding soil fertilizers. In the case of biodegradation of petroleum polluted soil, the fertilizers optimal balance for *F. arundinacea* was 100C/ 2N/ 0.2P (Reichenauer and Germida 2008).

On the contrary, the studies evaluating the efficiency of woody species on PAHs dissipation in polluted soils are scarcer and with limited results. Pyrene disappearance from unplanted soil was more rapid than in the soil planted to the pine species, jack pinus (*Pinus banksiana* Lamb), white pinus (*Pinus strobus* L.) and red pinus (*Pinus resinosa* Aiton) (Liste and Alexander 2000). The dynamics of PAHs in the fine roots in four typical woody species belonging to a Chinese National Nature Reserve was also reported (Chen et al. 2015), but the potential degradation of PAHs in the rhizosphere of these plants was not considered. Species in the genus *Paulownia*, a deciduous tree native to Central and Western China, are extremely adaptive to wide variations in edaphic and climatic factors, and grow well on lands deemed marginal. studied the bioremediation ability of *Paulownia tomentosa,* (Thunb.) Steud on an Italian soil located in an industrial area used for about 10 years as dump for a range of wastes, including fuel oils and solvents, has also been studied (Macci et al. 2013, 2016). After two years of bioremediation, the highest reduction (about 60%) in total hydrocarbons occurred in presence of this species at both soil depths tested (0–30 cm and 30–60 cm).

However, in both cases (herbaceous and woody plants), fast growth and deep roots have been considered desirable features in potential candidates for the development of feasible phytoremediation.

19.3.3 The Sunflower Case

The sunflower (*Helianthus annuus* L.), including in the *Asteraceae* family, has also been used by several authors as a model plant for PAH dissipation. The sunflower rhizosphere removed a greater quantity of various PAHs from contaminated soil than rhizospheres from other species (Maliszewska-Kordybach and Smreczak 2000; Kummerova et al. 2001; Olson et al. 2007), mainly due to its characteristics such as very good response to seed germination and root architecture and density. In a recent study, we employed a soil that was polluted with aged PAHs at concentrations considered realistic for soils that have undergone extensive bioremediation to test the hypothesis that the germination and development of sunflower plants would enhance the bioavailability of PAHs in the soil. Total PAHs concentration decreased by 93% in 90 days when the polluted soil was cultivated with sunflowers, representing an improvement of 16% compared to unplanted polluted soil (Tejeda-Agredano et al. 2013). We observed that the dissipation of PAHs from soil promoted by sunflower plants after the greenhouse experiment had been run for 90 d could be

Table 19.1 Effect of planting with sunflowers and slurry-phase treatment with exudates on residual contents of polycyclic aromatic hydrocarbons (PAHs) in polluted soil

	Greenhouse			Slurry		
PAH[a, b]	Initial	Control	Planted	Initial	Control	Exudates
Phenanthrene	4.0	0.5	0.1	4.6	1.6	1.0
Pyrene	2.0	0.3	0.1	1.0	0.5	0.2
Chrysene	1.3	1.0	0.5	1.5	0.8	0.5
∑ PAH [c]	21.8	5.0	1.5	19.3	4.8	2.5

Source: Modified with permission from Tejeda-Agredano et al. (2013)

[a]Data of native PAHs in mg kg^{-1} obtained from greenhouse and bioaccessibility experiments. As biodegradation was tested successively, initial and control data are reported for each experiment separately

[b]Greenhouse and slurry indicate, respectively, whether the polluted soil was planted with sunflower and maintained during 90 d under greenhouse conditions, or shaken in a slurry under laboratory conditions during 10 d in the presence of sunflower exudates produced previously *in vitro*

[c]Sum of six PAHs: fluorene, phenanthrene, anthracene, fluoranthene, pyrene and chrysene

reproduced over a shorter period of 10 d by incubating the soil in shacked slurries and adding root exudates (Table 19.1).

Sunflower root exudates have the potential to increase the degradation of xenobiotics due to its influence on the soil microorganisms, including the growth of PAH-degrading populations such as *Sphingomonas* (α-*Proteobacteria*), *Commamonas* and *Oxalobacteria* (β-*Proteobacteria*), and *Xhanthomonas* (γ-*Proteobacteria*), where these exudates act improving the bioavailability of the contaminant for degradation.

Root exudation and the identification and role of compounds resulting from of this process under natural conditions are difficult to evaluate due to the interference with different soil components and microbial metabolites (Grayston et al. 1997). These difficulties can be overcome using appropriate *in vitro* methodologies to obtain root exudates allowing, later analysis. A method to produce *in vitro* sunflower root exudates which were chemically characterized and tested for their effects on the biodegradation of PAHs have been developed (Tejeda-Agredano et al. 2013). These authors indicate that these exudates present the same superficial tension that the basal medium used, indicating the absence of surfactant capacity, and that the level of total organic carbon (TOC) in exudates obtained in vitro from *H. annuus* (130 mg/l) was much higher than in exudates from *Lolium perenne* and *Festuca arundinaceae*, cultivated in the same conditions.

The TOC content observed in the sunflower root extracts, was in agreement with TOC contents from root extracts of hybrid willow (*Salix alba x matsudana*), kou (*Cordia subcordata*) and milo (*Thespesia populnea*) (Rentz et al. 2005) and for slender oat root exudates (Miya and Firestone 2001), in both cases with promoting effects of root extracts on PAH-degrading microorganisms. The bioavailability of PAHs can be enhanced by the capacity of dissolved organic carbon to mobilize PAHs (Tejeda-Agredano et al. 2013). This enhancement can occur through a variety of process such as: (i) enhanced desorption (Haderlein et al. 2001), (ii) direct access to DOM-sorbet PAHs due to the physical association of bacteria and DOM (Ortega-Calvo and Saiz-Jimenez 1998) and (iii) increased diffusional flux through unstirred boundary layers around bacterial cells (Haftka et al. 2008), similarly to the enhanced uptake of cadmium by spinach plantlets originated from seed germination in the presence of labile metal complexes (Degryse et al. 2006).

19.3.4 Exudation Process

The estimation of total allocation of photosynthetic carbon to roots is in a range from 20–30% for cereals to 30–50% for pasture plants (Kuzyakov and Domanski 2000). Whereas 50% of this carbon remains in the roots, 11% of net fixed carbon and 27% of carbon allocated to roots is released as rhizodeposits and, for this proportion, only 1–10% are exudates (Paterson 2003). According to these authors the highest intensity of root exudation is located at the apices of roots and where lateral roots emerge. However, other reports specifically indicate that the zone immediately behind these apices, the root collar and root hair are major sites of exudation in comparison to root distal parts such as tips (Compant et al. 2010). The older parts of the roots may also exude organic compounds (Badri and Vivanco 2009). In any case, the exudation process by roots is strongly depended of the plant species (Bertin et al. 2003), root age (Nguyen 2003; Micallef et al. 2009) and environmental conditions like temperature, light and soil moisture (Badri and Vivanco 2009).

Phosphorous or potassium deficiencies or nitrogen availability (as nitrate) in the soil provoke an increase of carbon compounds excreted by roots (Badri and Vivanco 2009). Exogenous addition of elicitors also changes the composition of exudates (Schachtman and Shin 2007). Alder (*Alnus glutinosa*, L.) roots exposed to continuous white light for 5 days excrete a higher quantity of flavonoids (Hughes et al. 1999). Consequently, it is possible to modify either the *in vivo* or *in vitro* conditions for improving the release of organic compounds to increase the dissipation of PAHs in soil.

19.4 Bioavailability Processes in the Rhizosphere

19.4.1 Influence of Root Exudates on Soil Microbiota

The rhizosphere microbiota has continuous access to a flow of organic compounds derived from roots. Various factors hinder the relationship studies between roots and microbiota in the rhizosphere space, especially those taking place at specific microsites (Nannipieri 2006): (i) space explored by main roots is different from that explored by the lateral roots; (ii) nutrients immobilized by microbial assimilation are likely to be recycled following the death and degradation of microorganisms; (iii) influence of microorganism activity on the ratio efflux/influx of dissolved organic compounds released from root exudation; (iv) the uptake of any nutrient by plants and microorganisms depends on the form of this nutrient in the rhizosphere soil.

These issues can be dealt with by applying specific methods, currently available, such as, DNA or RNA extraction and PCR amplification to study the presence or expression of microbial and plant genes; biosensors, microorganisms with specific genetically incorporated genes with bioluminescence properties that act as indicators related with nutrient movements; isotope additions to aerial part of plant or to soil to follow their pathway in rhizosphere and in microbial progression, this microbial growth rates can be followed by leucine/thymidine incorporation (in the case of bacteria) or acetate-in-ergosterol incorporation (for fungi) (Marschner et al. 2011); fluorescent in-situ hybridization (FISH), in this technique oligonucleotide probes or antibodies are labelled with a fluorescent marker to detect microorganisms in situ by confocal microscopy (Briones et al. 2003); root-microbe interactions between individual root cells and individual microbes in situ using secondary ion mass spectrometry (SIMS) (Clode et al. 2009). These techniques have always been used to study the rhizosphere effect on soil microorganisms, but their application on pollutant bioavailability studies as related to plants remains highly unexplored.

19.4.2 Chemical Composition of Exudates and Role in Bioavailability

The microbial biostimulation exerted by the roots is mainly based on the chemical composition of the exudates. Flavonoids, organic acids, phenolic compounds and carbohydrates are among the relevant biostimulant compounds included in exudates, often metabolized by rhizobacteria, with bioremediation potential (Hegde and Fletcher 1996; Ryan et al. 2001). Exudates produced *in vitro* from sunflower roots (Tejeda-Agredano et al. 2013) had a concentration of total organic carbon (TOC) of 129.73 mg L^{-1}, and contained specific substances with the potential to directly or indirectly increase the bioavailability of the PAHs.

These substances include chemicals that are able to induce chemotaxis as several sugars as fructose and galactose, dominant root exudate sugars both in C3 and C4 plants (Vranova et al. 2013). Fructose and galactose cause positive chemotaxis in *Phytophthora sojae* zoospores (Suo et al. 2016), and provide a favorable environment for the growth of rhizosphere microorganisms (Grayston et al. 1997; Bertin et al. 2003). Sunflower root exudates also include amino acids which are powerful chemoattractants for different microorganisms (Zheng and Sinclair 1996; Pandya et al. 1999) such as asparagine and glutamine both in high concentration in the sunflower root exudates are chemoattractants for the *Bacillus megaterium* strain B153-2-2 (Zheng and Sinclair 1996).

Fatty acids were detected and quantified in sunflower root exudates. These compounds have a known potential to enhance the bioavailability of PAHs in soil by acting as surfactants (Yi and Crowley 2007). It is possible that the preferential development of rhizosphere microorganisms often observed on the exudate components at specific sites inside soil aggregates causes colony growth in the vicinity of pollutant sources and may modify the structure of the soil aggregates, promoting bioaccessibility through the excretion of extracellular polymeric substances and biosurfactants. These substances are able to induce chemotaxis in bacteria such as the soil bacterium *Pseudomonas putida* G7 (Jimenez-Sanchez et al. 2015) or positive tactic responses of *Pythium aphanidermatum* zoospores (Sungthong et al. 2015). It is also possible that the organic carbon in the exudates enhances the bioavailability of PAHs by a mechanism related to its ability to mobilize PAHs that are initially sorbed to the soil, as described above.

Acknowledgements This study was supported by the Spanish Ministry of Science and Innovation (CGL2013-44554-R and CGL2016-77497-R), the Andalusian Government (RNM 2337), and the European Commission (LIFE15 ENV/IT/000396).

References

Aprill, W., & Sims, R. C. (1990). Evaluation of the use of prairie grasses for stimulating polycyclic aromatic hydrocarbon treatment in soil. *Chemosphere, 20*, 253–265.

Badri, D. V., & Vivanco, J. M. (2009). Regulation and function of root exudates. *Plant Cell and Environment, 32*, 666–681.

Bertin, C., Yang, X. H., & Weston, L. A. (2003). The role of root exudates and allelochemicals in the rhizosphere. *Plant and Soil, 256*, 67–83.

Bottner, P., Pansu, M., & Sallih, Z. (1999). Modelling the effect of active roots on soil organic matter turnover. *Plant and Soil, 216*, 15–25.

Briones, A. M., Okabe, S., Umemiya, Y., Ramsing, N. B., Reichardt, W., & Okuyama, H. (2003). Ammonia-oxidizing bacteria on root biofilms and their possible contribution to N use efficiency of different rice cultivars. *Plant and Soil, 250*, 335–348.

Bueno-Montes, M., Springael, D., & Ortega-Calvo, J. J. (2011). Effect of a non-ionic surfactant on biodegradation of slowly desorbing PAHs in contaminated soils. *Environmental Science and Technology, 45*, 3019–3026.

Chen, Z. X., Ni, H. G., Jing, X., Chang, W. J., Sun, J. L., & Zeng, H. (2015). Plant uptake, translocation, and return of polycyclic aromatic hydrocarbons via fine root branch orders in a subtropical forest ecosystem. *Chemosphere, 131*, 192–200.

Chigbo, C., & Batty, L. (2013). Effect of combined pollution of chromium and benzo (a) pyrene on seed growth of *Lolium perenne*. *Chemosphere, 90*, 164–169.

Clode, P. L., Kilburn, M. R., Jones, D. L., Stockdale, E. A., Cliff, J. B., et al. (2009). In situ mapping of nutrient uptake in the rhizosphere using nanoscale secondary ion mass spectrometry. *Plant Physiology, 151*, 1751–1757.

Compant, S., Clement, C., & Sessitsch, A. (2010). Plant growth-promoting bacteria in the rhizo- and endosphere of plants: Their role, colonization, mechanisms involved and prospects for utilization. *Soil Biology and Biochemistry, 42*, 669–678.

Congiu, E., & Ortega-Calvo, J. J. (2014). Role of desorption kinetics in the rhamnolipid-enhanced biodegradation of polycyclic aromatic hydrocarbons. *Environmental Science and Technology, 48*, 10869–10877.

Congiu, E., Parsons, J. R., & Ortega-Calvo, J. J. (2015). Dual partitioning and attachment effects of rhamnolipid on pyrene biodegradation under bioavailability restrictions. *Environmental Pollution, 205*, 378–384.

Degryse, F., Smolders, E., & Merckx, R. (2006). Labile Cd complexes increase Cd availability to plants. *Environmental Science and Technology, 40*, 830–836.

D'Orazio, V., Ghanem, A., & Senesi, N. (2013). Phytoremediation of pyrene contaminated soils by different plant species. *Clean-Soil Air Water, 41*, 377–382.
Ehlers, L. J., & Luthy, R. G. (2003). Contaminant bioavailability in soil and sediment. *Environmental Science and Technology, 37*, 295A–302A.
Garcia-Junco, M., De Olmedo, E., & Ortega-Calvo, J. J. (2001). Bioavailability of solid and non-aqueous phase liquid (NAPL)-dissolved phenanthrene to the biosurfactant-producing bacterium *Pseudomonas aeruginosa* 19SJ. *Environmental Microbiology, 3*, 561–569.
Garcia-Junco, M., Gomez-Lahoz, C., Niqui-Arroyo, J. L., & Ortega-Calvo, J. J. (2003). Biodegradation- and biosurfactant-enhanced partitioning of polycyclic aromatic hydrocarbons from nonaqueous-phase liquids. *Environmental Science and Technology, 37*, 2988–2996.
Grayston, S. J., Vaughan, D., & Jones, D. (1997). Rhizosphere carbon flow in trees, in comparison with annual plants: The importance of root exudation and its impact on microbial activity and nutrient availability. *Applied Soil Ecology, 5*, 29–56.
Haderlein, A., Legros, R., & Ramsay, B. (2001). Enhancing pyrene mineralization in contaminated soil by the addition of humic acids or composted contaminated soil. *Applied Microbiology and Biotechnology, 56*, 555–559.
Haftka, J. J. H., Parsons, J. R., Govers, H. A. J., & Ortega-Calvo, J. J. (2008). Enhanced kinetics of solid-phase microextraction and biodegradation of polycyclic aromatic hydrocarbons in the presence of dissolved organic matter. *Environmental Toxicology and Chemistry, 27*, 1526–1532.
Hegde, R. S., & Fletcher, J. S. (1996). Influence of plant growth stage and season on the release of root phenolics by mulberry as related to development of phytoremediation technology. *Chemosphere, 32*, 2471–2479.
Hughes, M., Donnelly, C., Crozier, A., & Wheeler, C. T. (1999). Effects of the exposure of roots of *Alnus glutinosa* to light on flavonoids and nodulation. *Canadian Journal of Botany, 77*, 1311–1315.
Jimenez-Sanchez, C., Wick, L. Y., & Ortega-Calvo, J. J. (2012). Chemical effectors cause different motile behavior and deposition of bacteria in porous media. *Environmental Science and Technology, 46*, 6790–6797.
Jimenez-Sanchez, C., Wick, L. Y., Cantos, M., & Ortega-Calvo, J. J. (2015). Impact of dissolved organic matter on bacterial tactic motility, attachment, and transport. *Environmental Science and Technology, 49*, 4498–4505.
Krell, T., Lacal, J., Reyes-Darías, J. A., Jimenez-Sanchez, C., Sungthong, R., & Ortega-Calvo, J. J. (2013). Bioavailability of pollutants and chemotaxis. *Current Opinion in Biotechnology, 24*, 451–456.
Kummerova, M., Kmentova, E., & Koptikova, J. (2001). Effect of fluoranthene on growth and primary processes of photosynthesis in faba bean and sunflower. *Rostlinna Vyroba, 47*, 344–351.
Kuzyakov, Y., & Domanski, G. (2000). Carbon input by plants into the soil. Review. *Journal of Plant Nutrition and Soil Science-Zeitschrift Fur Pflanzenernahrung Und Bodenkunde, 163*, 421–431.
Liste, H. H., & Alexander, M. (2000). Plant-promoted pyrene degradation in soil. *Chemosphere, 40*, 7–10.
Macci, C., Doni, S., Peruzzi, E., Bardella, S., Filippis, G., et al. (2013). A real-scale soil phytoremediation. *Biodegradation, 24*, 521–538.
Macci, C., Peruzzi, E., Doni, S., Poggio, G., & Masciandaro, G. (2016). The phytoremediation of an organic and inorganic polluted soil: A real scale experience. *International Journal of Phytoremediation, 18*, 378–386.
Maliszewska-Kordybach, B., & Smreczak, B. (2000). Ecotoxicological activity of soils polluted with polycyclic aromatic hydrocarbons (PAHS) – Effect on plants. *Environmental Technology, 21*, 1099–1110.
Marschner, P., Crowley, D., & Rengel, Z. (2011). Rhizosphere interactions between microorganisms and plants govern iron and phosphorus acquisition along the root axis – Model and research methods. *Soil Biology and Biochemistry, 43*, 883–894.

Martín, V. I., de la Haba, R. R., Ventosa, A., Congiu, E., Ortega-Calvo, J. J., & Moyá, M. L. (2014). Colloidal and biological properties of cationic single-chain and dimeric surfactants. *Colloids and Surfaces B: Biointerfaces, 114*, 247–254.
Micallef, S. A., Channer, S., Shiaris, M. P., & Colon-Carmona, A. (2009). Plant age and genotype impact the progression of bacterial community succession in the *Arabidopsis* rhizosphere. *Plant Signaling and Behavior, 4*, 777–780.
Miya, R. K., & Firestone, M. K. (2001). Enhanced phenanthrene biodegradation in soil by slender oat root exudates and root debris. *Journal of Environmental Quality, 30*, 1911–1918.
Nannipieri, P. (2006). Role of stabilised enzymes in microbial ecology and enzyme extraction from soil with potential applications in soil proteomics. *Nucleic Acids and Proteins in Soil, 8*, 75–94.
Newman, L. A., & Reynolds, C. M. (2004). Phytodegradation of organic compounds. *Current Opinion in Biotechnology, 15*, 225–230.
Nguyen, C. (2003). Rhizodeposition of organic C by plants: Mechanisms and controls. *Agronomie, 23*, 375–396.
Niqui-Arroyo, J. L., & Ortega-Calvo, J. J. (2007). Integrating biodegradation and electroosmosis for the enhanced removal of polycyclic aromatic hydrocarbons from creosote-polluted soils. *Journal of Environmental Quality, 36*, 1444–1451.
Niqui-Arroyo, J. L., & Ortega-Calvo, J. J. (2010). Effect of electrokinetics on the bioaccessibility of polycyclic aromatic hydrocarbons in polluted soils. *Journal of Environmental Quality, 39*, 1993–1998.
Niqui-Arroyo, J. L., Bueno-Montes, M., Posada-Baquero, R., & Ortega-Calvo, J. J. (2006). Electrokinetic enhancement of phenanthrene biodegradation in creosote-polluted clay soil. *Environmental Pollution, 142*, 326–332.
Niqui-Arroyo, J. L., Bueno-Montes, M., & Ortega-Calvo, J. J. (2011). Biodegradation of anthropogenic organic compounds in natural environments. In B. Xing, N. Senesi, & P. M. Huang (Eds.), *Biophysico-chemical processes of anthropogenic organic compounds in environmental systems, IUPAC Series on Biophysico-Chemical Processes in Environmental Systems* (Vol. 3, pp. 483–501). Chichester: Wiley.
Olson, P. E., Castro, A., Joern, M., DuTeau, N. M., Pilon-Smits, E. A. H., & Reardon, K. F. (2007). Comparison of plant families in a greenhouse phytoremediation study on an aged polycyclic aromatic hydrocarbon-contaminated soil. *Journal of Environmental Quality, 36*, 1461–1469.
Ortega-Calvo, J. J., & Alexander, M. (1994). Roles of bacterial attachment and spontaneous partitioning in the biodegradation of naphthalene initially present in nonaqueous-phase liquids. *Applied and Environmental Microbiology, 60*, 2643–2646.
Ortega-Calvo, J. J., & Saiz-Jimenez, C. (1998). Effect of humic fractions and clay on biodegradation of phenanthrene by a *Pseudomonas fluorescens* strain isolated from soil. *Applied and Environmental Microbiology, 64*, 3123–3126.
Ortega-Calvo, J. J., Marchenko, A. I., Vorobyov, A. V., & Borovick, R. V. (2003). Chemotaxis in polycyclic aromatic hydrocarbon-degrading bacteria isolated from coal-tar- and oil-polluted rhizospheres. *FEMS Microbiology Ecology, 44*, 373–381.
Ortega-Calvo, J. J., Molina, R., Jimenez-Sanchez, C., Dobson, P. J., & Thompson, I. P. (2011). Bacterial tactic response to silver nanoparticles. *Environmental Microbiology Reports, 3*, 526–534.
Ortega-Calvo, J. J., Tejeda-Agredano, M. C., Jimenez-Sanchez, C., Congiu, E., Sungthong, R., et al. (2013). Is it possible to increase bioavailability but not environmental risk of PAHs in bioremediation? *Journal of Hazardous Materials, 261*, 733–745.
Ortega-Calvo, J. J., Harmsen, J., Parsons, J. R., Semple, K. T., Aitken, M. D., et al. (2015). From bioavailability science to regulation of organic chemicals. *Environmental Science and Technology, 49*, 10255–10264.
Pandya, S., Iyer, P., Gaitonde, V., Parekh, T., & Desai, A. (1999). Chemotaxis of *Rhizobium* SP.S2 towards *Cajanus cajan* root exudate and its major components. *Current Microbiology, 38*, 205–209.

Parrish, Z. D., Banks, M. K., & Schwab, A. P. (2004). Effectiveness of phytoremediation as a secondary treatment for polycyclic aromatic hydrocarbons (PAHs) in composted soil. *International Journal of Phytoremediation, 6*, 119–137.
Parrish, Z. D., Banks, M. K., & Schwab, A. P. (2005). Effect of root death and decay on dissipation of polycyclic aromatic hydrocarbons in the rhizosphere of yellow sweet clover and tall fescue. *Journal of Environmental Quality, 34*, 207–216.
Paterson, E. (2003). Importance of rhizodeposition in the coupling of plant and microbial productivity. *European Journal of Soil Science, 54*, 741–750.
Phillips, L. A., Greer, C. W., & Germida, J. J. (2006). Culture-based and culture-independent assessment of the impact of mixed and single plant treatments on rhizosphere microbial communities in hydrocarbon contaminated flare-pit soil. *Soil Biology and Biochemistry, 38*, 2823–2833.
Reichenauer, T. G., & Germida, J. J. (2008). Phytoremediation of organic contaminants in soil and groundwater. *ChemSusChem, 1*, 708–717.
Reichenberg, F., & Mayer, P. (2006). Two complementary sides of bioavailability: Accessibility and chemical activity of organic contaminants in sediments and soils. *Environmental Toxicology and Chemistry, 25*, 1239–1245.
Rentz, J. A., Alvarez, P. J. J., & Schnoor, J. L. (2005). Benzo a pyrene co-metabolism in the presence of plant root extracts and exudates: Implications for phytoremediation. *Environmental Pollution, 136*, 477–484.
Resina-Pelfort, O., García-Junco, M., Ortega-Calvo, J. J., Comas-Riu, J., & Vives-Rego, J. (2003). Flow cytometry discrimination between bacteria and clay humic acid particles during growth-linked biodegradation of phenanthrene by *Pseudomonas aeruginosa* 19SJ. *FEMS Microbiology Ecology, 43*, 55–61.
Ryan, P. R., Delhaize, E., & Jones, D. L. (2001). Function and mechanism of organic anion exudation from plant roots. *Annual Review of Plant Physiology and Plant Molecular Biology, 52*, 527–560.
Schachtman, D. P., & Shin, R. (2007). Nutrient sensing and signaling: NPKS. *Annual Review of Plant Biology, 58*, 47–69.
Semple, K. T., Doick, K. J., Jones, K. C., Burauel, P., Craven, A., & Harms, H. (2004). Defining bioavailability and bioaccessibility of contaminated soil and sediment is complicated. *Environmental Science and Technology, 38*, 228A–231A.
Sungthong, R., van West, P., Cantos, M., & Ortega-Calvo, J. J. (2015). Development of eukaryotic zoospores within polycyclic aromatic hydrocarbon (PAH)-polluted environments: A set of behaviors that are relevant for bioremediation. *Science of the Total Environment, 511*, 767–776.
Sungthong, R., Van West, P., Heyman, F., Jensen, D. F., & Ortega-Calvo, J. J. (2016). Mobilization of pollutant-degrading bacteria by eukaryotic zoospores. *Environmental Science and Technology, 50*, 7633–7640.
Suo, B., Chen, Q., Wu, W., Wu, D., Tian, M., et al. (2016). Chemotactic responses of *Phytophthora sojae* zoospores to amino acids and sugars in root exudates. *Journal of General Plant Pathology, 82*, 142–148.
Tejeda-Agredano, M. C., Gallego, S., Niqui-Arroyo, J. L., Vila, J., Grifoll, M., & Ortega-Calvo, J. J. (2011). Effect of interface fertilization on biodegradation of polycyclic aromatic hydrocarbons present in nonaqueous-phase liquids. *Environmental Science and Technology, 45*, 1074–1081.
Tejeda-Agredano, M. C., Gallego, S., Vila, J., Grifoll, M., Ortega-Calvo, J. J., & Cantos, M. (2013). Influence of sunflower rhizosphere on the biodegradation of PAHs in soil. *Soil Biology and Biochemistry, 57*, 830–840.
Tejeda-Agredano, M. C., Mayer, P., & Ortega-Calvo, J. J. (2014). The effect of humic acids on biodegradation of polycyclic aromatic hydrocarbons depends on the exposure regime. *Environmental Pollution, 184*, 435–442.

Velasco-Casal, P., Wick, L. Y., & Ortega-Calvo, J. J. (2008). Chemoeffectors decrease the deposition of chemotactic bacteria during transport in porous media. *Environmental Science and Technology, 42*, 1131–1137.

Vranova, V., Rejsek, K., Skene, K. R., Janous, D., & Formanek, P. (2013). Methods of collection of plant root exudates in relation to plant metabolism and purpose: A review. *Journal of Plant Nutrition and Soil Science, 176*, 175–199.

Walker, T. S., Bais, H. P., Grotewold, E., & Vivanco, J. M. (2003). Root exudation and rhizosphere biology. *Plant Physiology, 132*, 44–51.

Yang, Q., Wang, X., & Shen, Y. (2013). Comparison of soil microbial community catabolic diversity between rhizosphere and bulk soil induced by tillage or residue retention. *Journal of Soil Science and Plant Nutrition, 13*, 187–199.

Yi, H., & Crowley, D. E. (2007). Biostimulation of PAH degradation with plants containing high concentrations of linoleic acid. *Environmental Science and Technology, 41*, 4382–4388.

Zheng, X. Y., & Sinclair, J. B. (1996). Chemotactic response of *Bacillus megaterium* strain B153-2-2 to soybean root and seed exudates. *Physiological and Molecular Plant Pathology, 48*, 21–35.

Chapter 20
Soil Biodiversity and Tree Crops Resilience

Aurelio Ciancio and Mauro Gamboni

Abstract Promoting resilience capacity in agroecosystems is a very promising strategy for the sustainable management of tree crops. Belowground biological communities provide a range of different environmental services, including food webs structure stability and biodiversity conservation in aggregates, that contribute to the maintenance of crop production and effectively respond to external disturbances caused by global change. Aboveground habitats rely on a range of resilient reactions driven by factors including soil biodiversity interacting with dominant plant community species. Nevertheless, the links between soil biodiversity and aboveground resilience in time and space have been only partially studied and defined. Aboveground resilience is known to act on decadal or longer time scales while belowground communities often react on a time scale of a few years. This is the reason why coupling of both resilient reactions has been poorly studied and needs more investigations. Aim of this review is to provide a perspective on the potential of these studies in the management of tree crops, by identifying main drivers and services underpinning the implementation of sustainable strategies.

Keywords Bacteria • Ecosystem service • Invertebrates • Microorganisms • Perennial crops • Rhizosphere • Soil ecology

20.1 Introduction

The term "biodiversity" refers to the number of different species present in a given environment, and became popular after the 1992 Earth Summit in Rio de Janeiro (Brazil). In a more technical terminology, it refers to the density of species, counted per given surface or volume. The different indexes developed and applied to its measure are fundamental tools in many plant and soil ecology studies, as well as in

A. Ciancio (✉)
Istituto per la Protezione Sostenibile delle Piante, Consiglio Nazionale delle Ricerche, Via Amendola 122/D, 70126 Bari, Italy
e-mail: aurelio.ciancio@ipsp.cnr.it

M. Gamboni
Department of Biology, Agriculture and Food Sciences, CNR, Rome, Italy

M. Lukac et al. (eds.), *Soil Biological Communities and Ecosystem Resilience*, Sustainability in Plant and Crop Protection, DOI 10.1007/978-3-319-63336-7_20

related and applied management studies (Giller et al. 1997; Morris et al. 2014). The alpha diversity (observed within a sample or habitat) and the beta diversity (measured among samples or habitats) are integrated, at a higher level of analysis, in the gamma diversity (total diversity). Although of widespread use, depending on the organisms investigated and the environments sampled, the many biodiversity and species richness indexes developed thus far may be difficult to measure or derive (Kennedy and Smith 1995). This also depends on the biology of the species to count when considering, for example, organisms such as fungi, whose hyphae may be spanning over very large distances, or invertebrates such as ants or termites, exploring wide areas (Giller et al. 1997).

Soil, crops and forest ecosystems account for a large fraction of the total world telluric biodiversity (Hågvar 1998; Lavelle and Spain 2006). Different anthropic pressures have been identified worldwide, in relation to the aboveground composition found in these ecosystems and the occurring links with the belowground species. They include pressures present either in temperate and tropical ecosystems such as deforestation, intensification trends towards specialized monocultures or export-oriented tree crops productions. In the Tropics these include, apart of fruit crops, market-oriented commodities such as coffee (*Coffea* spp.), cocoa (*Theobroma cacao*), oil palm (*Elaeis guineensis*), rubber (*Hevea brasiliensis*) or tea (*Camellia sinensis*).

Negative effects on belowground biodiversity are also exacerbated by the widespread lack of efficient soil conservation and management practices, encountered worldwide in large industrial crops as well as in small farms or self-consumption agricultural systems. Soil degradation and loss of functioning or of biodiversity-based services are the major side effects of poorly managed agriculture, experienced in several tropical and temperate regions. They are mainly found in low fertility soils, where irreversible degradation thresholds are often reached in the long-term, in most marginal conditions (Oldeman et al. 1990; Antle et al. 2006).

In temperate regions, due to forest management, re-forestation (man-made or natural, as observed in the last decades in Italy due to the re-colonization of abandoned marginal farms by natural vegetation), the main threats for belowground biodiversity concern agricultural soils. Perennial tree crops are often cultivated under intensive regimes, with high external inputs such as fertilizers, herbicides and pesticides, which have significant impacts on belowground species richness and abundance.

The links between belowground biodiversity and aboveground composition reflect the structure of the production cycles involved. In the Tropics, either deforestation and intensification of perennial crops represent two major threats. Among the different types of tree cropping systems, shade coffee in Mexico showed an example of biodiversity conservation practices, compatible with agricultural productivity, which may be extended elsewhere, to other crops or world regions (Gordon et al. 2007). Experimental data on the effects of intensive tree crop plantations or monocultures, offering a certain level of canopy protection, showed possible strategies for biodiversity protection during production cycles. In the case of organic coffee, for example, protection of aboveground biodiversity is mainly due to the effect of the trees shade cover, the presence of epiphytes, and/or the canopy height.

In general, canopy protection avoids soil erosion and favours the flow of nutrients arising from primary production and decomposition, with beneficial effects on belowground food webs.

Examples of beneficial services deployed by highly biodiverse cropping systems include, apart of the carbon sequestration service in tropical forests, the maintenance and richness in pollinators, the improved plant nutrition by a diverse range of soil microorganisms (such as decomposers, endophytes, rhizobia), or an higher efficacy in control of plant pests and diseases, due to an higher richness in predators or antagonists (Gurr et al. 2003).

A basic, general concept to derive by these experiences is that problems in different areas of the world may have similar solutions. In this view, possible outcomes and practices may have a broader than local application range, with potential advantages available even for distant agroecosystems, through the application of sustainability-aimed policies that were effective elsewhere. The effects on soil biodiversity of factors such as trade and consumption by intensive economies are indeed global. They are exerted on areas already exposed to other extensive ecological stresses, ranging from the global rising of temperatures and rainfalls, to the local intensifications of farm inputs.

A possible strategy to protect biodiversity in presence of sustainable levels of production and eventual marketing is to rely on the added values of biodiversity-saving crops. To evaluate their effect, we have to look also at how agroecosystems, that are far from developed regions, are connected to these economies. The web of causes/effects linking a wide range of pressures and agro-ecological regimes may indeed act on soil biodiversity resilience over very long distances. The agricultural "space" is global, as are the factors that give shape to the food and commodity exchange at the world scale. This situation is challenging, as this century is considered as the first of the Anthropocene, an era in which many animal and plant species will probably become extinct. In this view, broader studies may be needed to underpin and promote wiser policies, embracing questions such as: *How do the social consumption levels impact soil biodiversity within distant ecosystems? At which extent do actual production regimes differ, in terms of biodiversity and stability, from traditional, self-consumption or natural undisturbed ecosystems? Finally, at which extent and how do tree crop inputs affect soil biodiversity?*

Aim of this review is to discuss some aspects and summarize achievements in the study on soil biodiversity and aboveground resilience, providing a perspective on its potential in the sustainable management of tree crops, by identifying some sustainable strategies, in the light of the above questions.

20.2 Belowground Biodiversity and Services

The services available over an agricultural or man-managed forest environment may be extended to soil, considered as an ecosystem itself (Wall et al. 2010). Bignell et al. (2005) listed the following services produced by and related to the belowground species: organic matter decomposition, cycling of nutrients, bioturbation, bio-antagonistic activity against pests and diseases, with further environmental services such as species replacement and functional biodiversity conservation, soil structuring, carbon sequestration, bioremediation.

All these services are present alone or in different combinations, depending on the peculiar traits of the environments examined and their evolutionary history. In the Tropics, biodiversity levels are often higher than those encountered at medium or high latitudes, mostly for large, aboveground species (De Deyn and Van der Putten 2005). This is likely due to higher tropical energy flows and longer favourable seasons, higher number of diversified and isolated habitats, lower rates of species and population extinctions and a longer evolutionary history (Giller et al. 1997; Botero et al. 2013). However, in these regions soil resilience and resistance levels largely differ, depending on crop type, climate regimes and biogeographic profiles, as well as on agricultural inputs applied and released outputs.

20.2.1 The Extent of Soil Biodiversity

Conservation of soil potentialities is central to any sustainable crop management. Soil biological quality can be evaluated by monitoring the changes in the structure and composition of its communities (microfauna and microorganisms). Soil is indeed one of the most important environments on earth, and all societies depend on it for a significant part of their food. However, our knowledge about the diversity of its biological constituents, as well as their role and conservation status, is still limited (Torsvik et al. 1990; Hawksworth 1991, 2001; Giller 1996; Wilson 2000). When comparing the range of microbial species present in soil with that of higher animals and plants (for which extensive inventories already provided a global knowledge about the number of species endangered and the expected extinctions), the number of soil organisms largely exceed that of vertebrates or higher plant species by several orders of magnitude (Bardgett 2002).

The biological soil components originate a complex system of food webs (Kampichler 1999; Brussaard et al. 2007). They are responsible for nutrient recycling through decomposition and transport, as well as for primary productions, in which the final outcome, including the ecosystem stability itself, does not usually correspond to the sum of each component effect (Addiscott and Mirza 1998). The rules of complexity in fact are active also at the scale of soil micro-ecosystems and ultimately affect food production and human societies on a global scale. This is an aspect often neglected in soil studies, which mainly deal with physico-chemical structural constituents, organic matter content, mineral or biological profiles, and related interactions (Kennedy and Smith 1995; Carter 2004; Havlicek and Mitchell 2014).

There is a general, increasing concern regarding the extent of the microbial biomass and diversity in soil ecosystems, and about its role. Detailed knowledge is arising thanks to the advent of DNA-based technologies. As a reference scale, it is already clear that the number of bacterial species which can be recovered from soil with traditional methods (i.e. culturing) are 100–1000 times lower than the real number of species present, that are in their majority still unclassified or undescribed. As early a quarter of century ago, Torsvik and collaborators showed, by extracting soil DNA, that around $2 \cdot 10^3$ bacterial species could be estimated to inhabit each g of soil (Torsvik et al. 1990). This order of magnitude was confirmed by later studies and analyses based on pyrosequencing or next generation sequencing (NGS) of soil total DNA (Rosello-Mora et al. 2001; Roesch et al. 2007). Moreover, also the total number of bacterial cells is usually large, as it may reach $10^7 \bullet g^{-1}$ of soil, of which only a small fraction is cultivable (Kellenberger 2001).

The mechanisms exerted by bacterial biodiversity on soil functioning and crop productivity hence represent part of a *terra incognita*, and their impact is still largely underestimated. It is now clear that cultivable species represent only a fraction of the soil microbial biodiversity on earth, since several groups, including uncultivable symbionts, parasites, endophytes and decomposers, escape any traditional census and identification, due to their specialized trophic niches, and biology (Head et al. 1998; Kellenberger 2001; Roesch et al. 2007). At a lower order of magnitude, similar considerations hold also for fungi, for which only a small fraction of total species has been described thus far.

Other organisms show a less abundant occurrence, for example nematodes, whose species richness is usually in the order of 50–100 taxa $\bullet$ 100 g^{-1} of soil. They are, however, important considering their function in soil. The functional role of soil microbial communities is as important as their composition and complexity, since they directly affect the life of aboveground plant species and the energy to biomass conversion, with direct or indirect effects on i.e. weeds, disease management or plant nutrition (Bonkowski and Roy 2005). Of particular interest are the root endophytes due to their mutualistic interactions with hosts and the beneficial effects on plant nutrition (Schulz and Boyle 2005; Schulz 2006).

The most important components of soil inhabiting microfauna are nematodes and higher invertebrates, which are not only pests but also nutrient recyclers or soil engineers. They are present in any trophic web, feeding on algae, bacteria, fungi, roots or predating other small animals. Given their abundance and together with amoebae and protozoa, nematodes increase nutrients turnover and indirectly influence soil decomposition by feeding on microbial decomposers. Moreover, some species may be very sensitive to various types of disturbances. In the last decades many studies analysed nematode community structures in relation to environmental changes or disturbances (Bongers 1990; Bongers and Ferris 1999). Data showed that they are better indicators of changes in soil quality than other organisms that are also sensitive to disturbance, since they occur everywhere, can be easily collected or counted and identified. Nematodes are also reliable indicators of the soil organic enrichment and content in microbial biomass (Ferris and Bongers 2006). Moreover, entomopathogenic species inhabiting soil contribute to the regulation of many insect pests. Their populations can be artificially increased and in most cases they are a valid alternative to chemical pest management. Also, the predatory species act as regulators of other nematode groups, including herbivores. Finally, the energy flow deriving by primary producers plays a strong effect on soil fauna complexity, as shown by the effects on nematode communities in food webs development, in primary successions of remediated soils (Hohberg 2003).

20.2.2 Tree Crops Impacts

As soil biodiversity is a natural constituent of the biosphere, contributing to its equilibrium and stability, its conservations requires, in disturbed areas, the identification and implementation of suitable actions. The Tropics are the centre for intensive productions of commodities consumed in the northern hemisphere, like coffee or palm oil, which exert strong pressures on belowground biodiversity. Coffee represents a key agricultural commodity, second only to oil for international trade value. Due to its requirements in terms of climate and environment this crop provided an interesting experimental model to study the impact of tree production systems on belowground and aboveground biodiversity, and to identify most convenient conservation actions.

Studies have been produced in countries such as Mexico (rated among the top five for endemism of plants and higher animals), which shows varied systems, ranging from organic to intensive crops. In northern Chiapas, coffee trees grown with shade cover showed higher yields at canopy coverings ranges of 23 – 38%. However, productivity decreased with shade covers higher than 50%. A reduced incidence of coffee leaf rust and weeds, with increasing shade density, increased revenues in high shade systems (Soto-Pinto et al. 2002). However, a simple change in the shading tree species composition (from natural to *Inga* sp. plantations) was observed to yield changes on aboveground biodiversity and positively affect income (Peeters et al. 2003).

In biodiversity conservation programs in Mexico, Brandon et al. (2005) showed that it is possible to protect animal species by using a small number of reserves compatible with a low agricultural productivity. By the year 2005, 11% of human population inhabited forested areas in this country, and more than 8400 rural communities owned forest resources, with a large number of communal enterprises, covering 80% of forests (Brandon et al. 2005). Over a coffee plantation intensification gradient in Vera Cruz, the highly biodiverse "*bajo monte*" plantations were among the cropping systems with the highest productivity, although no direct links were found between aboveground biodiversity richness and yields (Gordon et al. 2007).

A further, paramount case for biodiversity studies is given by Costa Rica, which is one of the 20 countries with the greatest biodiversity in the world. It has numerous and varied microclimates due its geographic position, coasts and mountain system. Such conditions explain its huge natural wealth, not only in terms of species but also of ecosystems. The more than half a million species found in the country represent nearly 4% of the total estimated worldwide, of which just over 60% are insects.

Costa Rica is also an important coffee producer, and this crop has a strong impact on the agricultural economy and export.

Currently, coffee plantations in Costa Rica are characterized by different systems, including farms with organic, conventional and sustainable management, located at different altitudes, ranging from 0 to >1200 amsl, with a biodiversity extremely rich on microorganisms (Obando 2002). It is estimated that $65 \cdot 10^3$ species of fungi and more than $25 \cdot 10^{3}$ species of bacteria and micro algae exist in this country.

Tea is a further international commodity with a production higher than 3 M tons $\cdot$ year $^{-1}$. Typically produced in the Asian and African regions for more than two centuries, tea is one of the main products of China, India, Sri Lanka and Malawi, which together cover more than 70% of the world production. This is also one of the crops with the most intensive cultivation regimes, highly destructive for surrounding biodiversity-rich areas. Tea requires several protection efforts and more than 30 different pesticides are often used in its cycle. In areas where tea was cultivated for more than 100 years, high pesticide sprays were responsible for amphibian decline, due to habitat losses and high contamination levels (Daniels 2003). Tea producers in Sri Lanka experimented several aspects of organic farming and soil protection, including non-chemical, natural and organic farming systems, with multi cropping systems that strengthen monocultural tea ecosystems. Tea crops are exposed to the effects of global warming, which induced changes in soil invertebrate biodiversity, including pests such as nematodes (especially for the nematode *Meloidogyne brevicauda*, a temperate species found only in Sri Lanka on tea plantations), and beneficial soil fauna and flora.

Oil palm production is the main cause of biodiversity losses in the Tropics (Barnes et al. 2014; Vijay et al. 2016). Deforestations due to this cropping system, and consequential losses of aboveground biodiversity, are mostly evident in South-East Asia and South America, where 45% and 30% of palm crops are grown today in previously forested areas, respectively (Vijai et al., 2016). These authors identified forests in Africa and South America as the most vulnerable to deforestation in the next decades, for conversion to oil palm production. Corporates-based conversion of either primary and secondary forests in Malaysia, one of the most important oil palm producers, and substitution of rubber and cocoa crops with oil palms, showed adverse negative effects on above- and belowground biodiversity, with indigenous and rare species replaced by invasive or generalist ones at around a 35% replacement rate, when compared to undisturbed forest habitats (Dhandapani 2015; Savilaakso et al. 2014).

20.2.3 Tree Crops Productivity and Biodiversity

Soil biodiversity profiles have a remarkable impact on the resilience and stability of tree crops productivity. This largely depends on the food webs of the soil inhabiting microbial communities and microfauna, as well as on many aspects of their interactions with the tree productions. Considering that soil is a non-renewable natural resource, any assessment about the effect of agriculture on its fertility and biology aims at identifying suitable conservative technologies, in view of a sustainable farming. Many socio-economic initiatives may be (or have been) implemented to protect soil biodiversity, including identifying the mechanisms allowing competitive advantages received by farmers, traders and consumers, once they learn how to value biodiversity (i.e. through the certification of *biodiversity-friendly* products and market/consumers' evaluation).

Actually, several DNA-based studies are advancing our knowledge on soil biodiversity status (and losses) as the result of management practices. Their global outcome should be finalised at identifying restoration or conservation actions keeping the most "biodiversity protective" agroecosystems economically competitive. Investing in biodiversity restoration and conservation plans (i.e. different types of coffee plantations in Central America producing *biodiversity-friendly* certified coffee), has often resulted in a range of financially viable and sustainable production systems, suggesting that these kinds of actions may be applied to reverse trends of conversion from shade coffee to less protective, shade-free plantations (Gobbi 2000; Perfecto et al. 2005). At this regard, and also in consideration of high-impact of commodities such as oil palm, final informed consumers may indeed play a significant role in the success of biodiversity protection actions, acting over very long distances by changing consumption trends and markets' demand.

Crossing biodiversity and resilience data with farms economic performance may lead to the identification of multiple equilibria and the selection, for eventual adoption, of most appropriate protective strategies. Antle et al. (2006) analysed soil degradation processes based on multiple equilibria and resilience thresholds, by modelling the soil degradation in terrace systems in the Andes, and the effects of the subsidy policies and economic conditions applied for restoration. A similar approach may be adopted for the paths of productivity decline often associated to biodiversity erosion and losses or extinctions, present in many areas. To verify if some returns occur in action plans applied for biorestoration (i.e. based on a range of alternative technologies for pest management, fertilization, water management etc.), their marginal-law trends, increasing to a maximum and then declining, should be quantified. Similarly, marginal economic benefits will show a growing curve up to a threshold level at which no more returns are received by the conservation action. The condition for a stable high-productivity and optimal biodiversity equilibrium to be reached and maintained as sustainable is that the biodiversity restoration plateau should occur before the economic threshold is reached. If instead it is reached after the economic threshold is achieved, the marginal gain in biodiversity conservation will no more be convenient nor sustainable, in presence of no additional economic benefits obtained by the farmer investment, with a consequential biodiversity decline and irreversible losses as the action plan is abandoned (Fig. 20.1).

Although the marginal gain view is usually applied for conservation or restoration of soil physical properties, it appears also suitable to monitor the soil biological erosion, identifying the threshold levels and checking under which conditions soil biological stability or persistent degradation may occur. It also appears suitable in

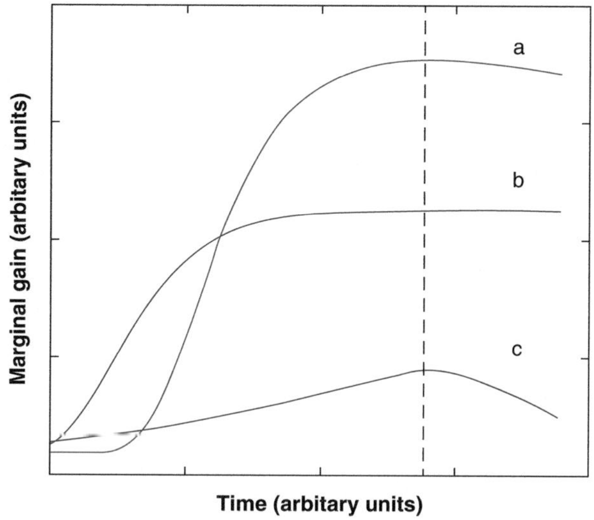

Fig. 20.1 Relationships between marginal economic benefits of farming (*a*, *red line*) and benefits of biodiversity obtained by restoration actions, in two situations with different reaction times (*b* and *c*, *blue lines*). To reach a permanent equilibrium (*b*) any restoration plan should achieve stable outcomes before the economic returns become negative (*dashed line*) and the applied restoration plan is abandoned (*c*)

tree crops, due to their persistence and the longer time during which a conservation or restoration plan may be implemented. In the case of reversing a severe degradation, the investments aiming at restoring the lost conditions (for example by abandoning costly practices such as chemical pest management, high fertilization rates or destructive soil or water management plans) could be discouraged by the higher costs of alternatives. Moreover, Fig. 20.1 shows only a part of the problem, as different rates of restoration, for different soil organisms or functional groups, may be achieved by the same initiative. Detailed information is, hence, fundamental to identify which public and/or management action or support should be implemented to promote resistance/resilience. The data should be made available at the highest possible resolution (in terms of functional groups or species). These plans should ultimately aim at restoring, in most degraded situations, at least part of the original belowground biodiversity, sustaining at the same time the economic productivity of farms.

20.2.4 Rhizosphere Services

Although the global importance of biodiversity is widely recognized for any ecosystem, the role of soil communities in maintaining the functioning and structure of the arboreal root systems and associated food webs is not fully explored. A deeper understanding of the global effects of soil microbial diversity on the aboveground level requires in fact huge research efforts, due to the large number of biogeochemical factors, soils and microbial communities that evolved on earth (Wall et al. 2010, 2015).

Two common services, plant nutrition and protection from pathogens and parasites, may be identified in the above- and belowground interactions, provided by functional groups categorized under the general terms of symbionts (such as mycorrhizae and endophytes) and antagonists, including in the latter the competitors and biological control agents. Some examples may clarify the complexity and significance of these relationships, that evolved along with the soil species and the ecosystems they occupy.

Studies on the role of arbuscular mycorrhizae (AM) and ectomycorrhizae (EM), carried out in temperate forests, showed that the type of mycorrhizal association affects the observed aboveground biodiversity of the tree species, by acting on conspecific density-dependence (Bennett et al. 2017). The effect of belowground fungal symbionts found on tree communities should then be added to the known benefits deriving by the improved nutritional exchange with roots. Higher levels of nitrogen provision and mainly of root protection, reducing the damage by pathogens or parasites, characterized the EM and root associations. These were shown to favour the development of EM-colonized seedlings, in comparison to the negative feedback observed for AM associations. EM associations thus reduced the negative density-dependency effects of conspecific trees, lowering their local biodiversity and regulating plant populations and communities, with effects on forest trees composition also visible at the regional scale (Bennett et al. 2017).

Regional scale patterns for soil invertebrates may be useful to identify soil disturbance effects, but are difficult to produce and require replicated samplings along long transects in environments that are homogeneous over long distances. This studies, however, may allow the identification of links between invertebrates such as earthworms and a number of soil properties such as pH and nitrogen contents, with potential in environment modelling and disturbance effects evaluation (Joschko et al. 2006).

The interactions among roots and symbionts provide a mechanism underpinning tree plants resilience in response to disturbances such as drought or soil pathogens. Studies on drought resilience in poplar, pines and other species, showed that the associated EMs enhanced the capacity of water absorption through an increased root specific surface. This is also the result of a longer extension provided by the hyphae and of the higher volume of soil explored. Other mechanisms include increased hydraulic conductivity and expression of cellular pathways for some aquaporin genes, improving the water transport capacity of roots (Brunner et al. 2015). Increased aquaporin gene expression was observed either for the basidiomycete *Laccaria bicolor* and the ascomycete *Tuber melanosporum* (Hacquard et al. 2013).

Nodulating endosymbiotic bacteria of the genus *Frankia* are associated to tree species, such as *Casuarina* spp. in tropical and sub-tropical regions and *Alnus* spp. in temperate regions. *Frankia* spp. are active against pathogenic fungi such as *Pythium* or *Poria* spp., through the production of anti-fungal and anti-bacterial metabolites (Akkermans et al. 1989). They are major providers of nitrogen, and significantly contribute to the success of these trees in the colonization of very poor soils.

Roots, pathogens/herbivores and microbial soil antagonists have complex links in soil, involving biochemical signals or nutritional, adaptive links. The entomopathogenic fungi of the genus *Metarhizium* provide their host plants, through their endophytic phase, with the nitrogen compounds subtracted by the root feeding insects, that they parasitize as entomopathogens (Behie et al. 2012).

Apart of being efficient decomposers or parasites, nematodes also include biocontrol agents. When attacked by an insect herbivore, the roots of some European maize varieties attracted the entomopathogenic nematode *Heterorhabditis megidis*, through the emission of a belowground signal, the sesquiterpene (E)-β-caryophyllene. The signal emission was induced by the feeding activity of larvae of the beetle *Diabrotica v. virgifera*, and was capable of significantly reducing the number of emerging adults, as shown in laboratory and field assays (Rasmann et al. 2005).

Plant nutrition and protection are not the unique services provided, as the belowground species sustain the decomposition or organic matter and nutrient recycling, as well as other services related to the soil structure and biodiversity conservation in soil aggregates.

20.2.5 Ecosystem Engineering

This term refers to the capacity of some organisms to build up and maintain stable structural constituents of soil through their biological activities, including foraging and decomposition, recycling, or direct assemblage of soil components and aggregates in elements of different dimensions and complexity (such as nests, tunnels, or termite combs). Micro- and macro-invertebrate communities play fundamental roles in the maintenance of soil fertility and other services, either in temperate and tropical soils, through the intense bioturbation that they operate and the interactions that they developed in million years of co-evolution with other soil organisms (Lavelle et al. 2006). The actions of these ecosystem engineers are usually considered as beneficial, involving nutrients and water soil adsorption. Ecosystem engineering have high impacts on the soil organic matter and structure, as well as on fertility and nutrients storage. In natural savannah soils from the Ivory Coast, for example, earthworms may ingest, mix and redistribute 800 to 1200 Mg dry soil · ha^{-1} · year $^{-1}$. In tropical environments, termites and ants are active ecosystem engineers that contribute to some general effects of great importance for tree crops productivity, including the:

- selection and redistribution of microbial communities via bioturbation of the upper 20 cm of soil and the production of energetic substrates, analogous to root exudates (Lavelle et al. 2004, 2016);
- stimulation of plant growth through accelerated nutrients release, protection from parasites (especially nematodes), stimulation of root symbionts, physical conditioning of soil and production of Plant Growth Promotors (PGP) via selective stimulation of specialized microorganisms (Brown et al. 1999);

- significant contribution to the maintenance of hydraulic services and of an important and diverse (in terms of size classes distribution) porosity;
- carbon sequestration through active macroaggregations of soil in the upper 20 cm, with positive effects on water infiltration and storage, together with sequestration in stable biogenic aggregates.

The complex of soil invertebrates also contributes to the provision of other ecosystem services by their engineering activities, including the conservation of microbial diversity in a large number of microsites (Lavelle et al. 2006).

20.3 Resilience

A resilient response is a sort of elastic reaction displayed by a system after an external disturbance or stress has been applied, allowing the return to its initial conditions. In the case of soil, it is mostly a reaction allowing the recovery of its pre-disturbance properties in a stable way, after changes have been induced by one or more external causes, often with a direct or indirect anthropic origin (Elliot and Linch 1994; Lal 1997).

Aboveground habitats rely on a range of resilient reactions towards external disturbances, driven by factors including soil and other environmental parameters, and interacting with dominant plant community species. Aboveground resilience is known to act on decadal or longer time scales. Because of its response time, resilience in forest and tree communities is often difficult to study experimentally. Data may be biased in fact by other overlapping factors, or the lack of adequate controls needed for comparison. The coupling of both above- and belowground resilient reactions has been poorly investigated as well, nor the identification of their drivers has been studied over long time scales. In one of such studies, dealing with forested lake islands of different sizes from northern Sweden, Wardle and Jonsson (2014) showed that the differences in resilience among the experimental sites were minor, and that resilience was not affected by the disturbed habitat size. In spite of being more subject to burning, due to higher frequency of lightning, the larger islands were more fertile and productive, with a lower plant biodiversity. These characteristics, however, were not sufficient to alter the resilience process, nor to reduce the recovery time over the 14 years of study, when compared to smaller islands, with the only exception of the most severely disturbed sites (Wardle and Jonsson 2014).

As soil resilience is active on a shorter time scale (1–2 years), more observational or experimental data have been made available on this subject. The most common sources of soil stress and biodiversity disturbance are belowground contamination by heavy metals and/or pesticides, the spreading of invasive pathogens and species, hydrologic stress due to climate changes, such as drought or, on the other extreme, floods.

Soil food webs respond to one or more of the cited stresses in several ways, and when a functional group is lost, its service may be partially or totally accomplished by one or more other redundant groups. Pesticide applications to soil, for example, are known to increase the likelihood of extinction of biological antagonists, due to a reduced availability of the preys or pests targeted, and to the toxic effects directly induced on these microorganisms. Although this is true for the most effective biocides such as methyl bromide, differences may be found for other pesticides. In the case of bacteria, for example, data on the effect of an organophosphorus insecticide showed an increase of some taxa in treated soil samples, in comparison to untreated controls, due to the selection of some species capable to degrade the applied pesticide and its derived compounds (Itoh et al. 2014). This is an example of a resilient reaction of soil, leading to a re-structuring among its microbial populations, depending on the richness of species present at the moment the disturbance was applied. The insurgence of one or more stress factors in a soil microhabitat may, hence, directly induce a biodiversity loss (lowering the richness of species) or reduce their abundance. However, it may also induce changes in the populations structure and in the genetic traits of the residual organisms. The reaction of a species or population to one or more disturbance factors may in fact originate novel evolutionary adaptive paths. Many soil species, including microorganisms or some invertebrates (i.e. microbiovorous nematodes) have a short (a few hours to a few days) population cycle. Genetic shifts may be then expected in a relatively short period of time, under a durable disturbance, due to structural food web alterations. This occurs mainly with changes affecting species composition or density of i.e. predators, antagonists, symbionts or parasites, altering their interactions with other components of the food web.

Moreover, the introduction of additional selective pressures or population disturbance may determine genetic bottlenecks, due to the extinction of more susceptible individuals and the selection of a few resistant ones, that eventually multiply. Similarly, the disappearance of an already occurring selective pressure (i.e. loss of a micro-predator) may induce genetic changes with the increase and accumulation of previously unfavoured traits. The final outcome for survivors is then a shift in population gene frequencies and a frequency increase of the genotypes more adapted to the newly altered conditions. This means that genetic rushes of the Red Queen type, characterizing many populations inter-relationships also in soil, are altered to follow novel and distinct paths. These processes may occur at many levels within a disturbed habitat, affecting the interactions among multiple organisms and related functional services, or include/exclude the establishment/persistence of symbiotic associations.

A similar situation concerns the functions accomplished in soil, with those relying on a low number of species (more exposed to extinction) as most sensitive to disruption. If a function is performed by one or a few species only, their extinction also produces an irreversible functional loss (Giller et al. 1997). A requirement for stability in soil functions is hence the presence of multiple groups of organisms (gilds) underpinning the same function. This will sustain soil resilience by reducing the likelihood of a functional loss due to a single species extinction. Increasing the functional redundancy must hence be regarded as a possible strategy to sustain soil services and counteract the effects of biodiversity losses. This view is, however, debated as the richness of species by itself does not warrant an increase of soil functionality, as the services that depend on a reduced number of organisms, as those encountered in agricultural soils, are the more susceptible to a disruptive loss after soil biodiversity erosion (Bender et al. 2016).

20.3.1 Soil and Tree Crops

Several benefits, directly or indirectly linked to crop biodiversity, have been identified in relation to the practices with a minimal or null environmental impact, such as organic farming (Wall et al. 2015). Some examples of above- and belowground biodiversity issues in tree crops may be used to illustrate potential benefits related to the identification and application of most convenient low-impact practices. However, general rules in agriculture do not appear frequently, and it appears more useful to search for, and then apply, local experience-based technologies, given the diversity of the world agroecosystems and of the problems that farmers have to face. These may include abandoning chemical treatments in favour of biological control methods. It is known, for example, that chemical products affect biocontrol agents too, as shown by the treatments applied to control insect pests in citrus, affecting the biodiversity of predators or other useful arthropods (Liang et al. 2007). In citrus crops several herbivores are no more considered as pests today because they are controlled by parasitoids. At the same time, due to globalization, new pests and associated transmitted diseases are dramatically emerging.

Differing from aboveground pests and predators, soil biocontrol agents do not rely on prey searching strategies, although some nematophagous fungi may attract nematodes towards their traps. Soil pests such as phytoparasitic nematodes are controlled by several fungi or bacteria, including specific endoparasites such as the endospore-forming bacteria of the genus *Pasteuria* or other antagonists of the genera *Bacillus*, *Pseudomonas* or *Streptomyces*.

The eradication of a pest or pathogen from an agricultural environment (and soil in particular) is, in many cases, difficult or impossible, due to the low likelihood of killing all individuals as the soil volume to treat increases. Many mechanisms contribute to this situation, including the nematode survival through mechanisms such as anhydrobiosis, cyst or egg mass forming, dauer arrest or multiplication on alternative hosts, as well as its dispersal through soil, water or even wind. Some properties like polyphagy allow root knot nematodes to survive on weeds or other plants, as well as on hosts present in untreated fields or parcels. Differing from chemical control, biological management methods seek the production of a stable equilibrium, which may even lead to suppressivity, a condition in which the nematode pest does not achieve high density levels, even in presence of a susceptible plant. In tree crops such as peach, suppressive activities towards plant parasitic nematodes have been associated to nematophagous fungi, such as the hyphomycetes *Hirsutella rhossiliensis* (Eayre et al. 1987) or *Brachyphoris* (syn. *Dactylella*) *oviparasitica* (Stirling and Mankau 1978).

Plants, through the energy conversion, are the most important drivers active at the agroecosystem level. Tree crops in particular act at multiple levels, as hosts for foragers and herbivores, as energy or nutrient providers, and as aboveground ecosystem engineers, though the organic matter provision. Belowground, they act as soil engineers through their root system and as carbon sinks. Different effects may however be experience, depending on the soil and weeds management regimes (tillage or no-tillage) and the type of pest management and nutrition technologies applied (organic *vs* conventional farming). A soil conservation engineering study carried out on hilly areas in South-East China showed that highly biodiverse soil covers, with native plant communities, improved the sustainability of *Citrus* sp. crops and their resilience to thermal stress in the hot-dry seasons, by increasing soil moisture and reducing erosion (Chen et al. 2004). When studying organic citrus orchards in Sicily (Italy), in comparison with conventional management, Canali et al. (2009) found that organic soils had a higher content of all the organic matter pools, increasing the carbon supply for soil metabolic processes. Similarly, organic citrus orchards in Brazil showed higher levels of microbial activity and mycorrizal biodiversity (Franca et al. 2007).

The above- and belowground biodiversities have to be considered as dynamic structures. Their status, including the interactive relationships among different organisms, may change in a short time, depending on many external factors, including those with a social or economic nature. Olive orchards provide a significant example of a traditional, environment-protective culture, given olive trees adaptability and productive capacities even in the most difficult environments, often integrated in a diversified crop environment. Olive, through the income and other social benefits provided, sustains social and cultural traditions in many areas along the whole Mediterranean region. The recent outbreak of a severe pathogenic bacterium, *Xylella fastidiosa* subsp. *pauca* in Apulia (Italy) on olive and other plants such as oleander, cherry, myrtleleaf and rosemary, revealed its vulnerability in relation to external factors such as the globalization of commerce and mobilization of plant materials. Further *X. fastidiosa* subsp. *multiplex* and other strains have been detected also in Southern France in Corsica and the Baleary Islands, also in this case likely due to imported, asymptomatic ornamental plants (Legendre et al. 2014; Loconsole et al. 2016).

The introduction of new crop plants or ornamentals (i.e. *Polygala myrtifolia*) for commercial purposes in Apulia and other Mediterranean regions induced a shift in the environment biodiversity levels in the recipient space, due to the associated invasive species. The bacterium was undetected until disease field transmission to highly susceptible olive trees and/or other plants became evident. Transmission occurs by an indigenous endemic hemipteran, the polyphagous splittlebug *Philaenus spumarius*. This species had a secondary importance as herbivore, but became an important pest due to its newly acquired vector status (Saponari et al. 2014; Loconsole et al. 2016). The olive disease complex (Martelli et al. 2015) occurred in spite of the pathogen detection, and of the media and institutional follow-up, including GPS-based epidemics tracking and monitoring seehttp://webapps.sit.puglia.it/freewebapps/MonitoraggioXFSintesi/).

The change in the olive crops biodiversity in Apulia produced an outcome highly disruptive, and most demanding efforts concerned the slowdown of the northbound epidemic spread, since the pathogen decimated thousand olive trees, in some cases more than a century old and central to a traditional social system. The crops in this case could not show any resilient response, although genetic resistance has been recently identified and proposed as a possible remedy to an otherwise disastrous situation.

In a general perspective, biodiversity levels at global and local scales interact in a complex way, as different gradients are present for the above- and belowground species. When an evaluation about the effects of changes in habitats and ecosystems functioning is required, these gradients need to be identified also in relation to the habitats locations, the services produced and the density changes of the organisms involved. Plant species numbers and diversity have a strong impact on the soil communities and related biodiversity, as plants *i)* provide a fundamental functional link between the two habitats and *ii)* react to the many feedback mechanisms characterizing the interactions with, and within, other functional groups (i.e. herbivores, symbionts, decomposers). Plants thus are the main driver affecting nutrients cycles, resource use and availability. Although the effects of drivers such as energy flow and water availability on the above- and belowground diversities have not been completely elucidated, they are considered to act on the two habitats with different mechanisms and weights. The complexity of these relationships also depends on the network of indirect or feedback effects linking functional groups, acting on different space and time scales, including years. Some examples are the impact of aboveground herbivores on plants and, indirectly, on the species composition of other belowground herbivores, pathogens and decomposers, or *vice versa* the effects of mycorrhizae and other root symbionts on the composition of aboveground communities of plants and related feeders. Moreover, above- and belowground biodiversity levels differ by several orders of magnitude, as soil communities are more diverse and richer in terms of species and functional groups (De Deyn and Van der Putten 2005).

In this view, monocultures represent extreme and unbalanced habitats. The severity of the effects of monocultures on wildlife and related biodiversity losses have been identified in the oil palm producing countries in Southeast Asia. In particular, the growing world demand increased cultivated areas in Indonesia and Malesia, that cover more than 3/4 of the total world oil palm production. These countries are also home to high biodiversity rich tropical forests and environments, which are mostly lost due to conversion to oil palm crops. This situation is the opposite of the diversified crop environment considered as the less disturbing and most resilient agricultural approach (Lin 2011). Its negative effects promoted the identification of a number of mitigation tools and approaches, including a sustainability certification initiative, the stakeholder-based Roundtable for Sustainable Oil Palm and other mitigation and conservation initiatives (Yaap et al. 2010). In a meta-study on biodiversity losses, these authors found that almost 85% of natural forest species were lost in the conversion, concerning all vertebrate groups. Invertebrates showed a different response to conversion, with some groups, such as bees, that showed an increase in their species numbers. except the honeybees, with an impact on pollination. Other invertebrates such as the more specialized, niche inhabiting groups showed local extinctions in favour of more generalist, invasive species (Yaap et al. 2010).

20.3.2 Deforestation

Main factors responsible for deforestation are the burning for conversion of forest land to agriculture. Deforestation occurred in the last decades with an impressive progression, at the rate of an area equivalent to the surface of a country like Switzerland or Austria lost every year (Fig. 20.2). The first pressure underpinning these actions is often the loss or decline in productivity in the already cultivated tropical soils, due to their biological erosion and to water runoff, flanked by nutrients and soil particles losses. However, the search for new land, and the consequential destruction of tropical forests (and related aboveground biodiversity and services), are only part of a more complex problem, which has its roots also in further demographic and social pressures. These include the intensification and expansion of farms related to a growing global demand for food (including additional livestock for meat productions), arising from recent urbanization trends, and the forest clearing for the production of export-oriented, commodity tree crops (Sanchez et al. 2005).

Fig. 20.2 Satellite image showing alternated deforested areas in Rondonia (Brazil). *Scale bar* = 50 km (Image available on April 11th, 2017 at https://www.google.it/maps/@-10.450878,-63.882636,613895m/data=!3m1!1e3)

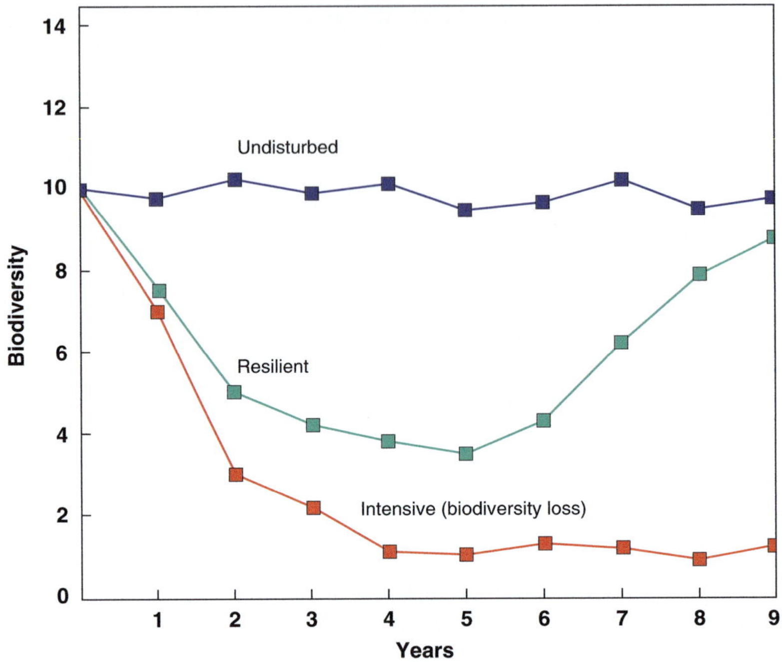

Fig. 20.3 A generic scheme showing possible local outcomes of forest clearing. These are: progressive loss of biodiversity in time, as observed in slash-and-burn clearing for intensive commodity-oriented tree crops (*red*); a resilient reaction, as in the shifting cultivation system (slash-and-burn clearing with a decadal fallow, *green*). In the last case the biodiversity levels are restored, approaching those of undisturbed areas (*blue*), at the end of fallow (Adapted from Sanchez et al. 2005)

The fallow period that characterized the conventional slash-and-burn practice of poor rural or indigenous communities in the Tropics was considered enough to ensure, on a decadal scale, the restoration of the original soil biodiversity levels, after initial induced fluctuations of soil bacteria (Adeniyi 2010). Of much more concern is the replacement of the forest environment with more intensive crops and a reduced fallow period, or the more intensive, trade-oriented monocultures with no or minimal fallow, which determine severe biodiversity losses (Fig. 20.3).

Although the latter situation represents an extreme and mostly irreversible evolution of the forest ecosystem, it is possible to identify alternative land conservation and exploitation systems for small farmers, allowing the conservation of soil fertility and biodiversity, by saving the farmers income. A number of policies aiming at ecosystem rehabilitation, reforestation, sustainable tree crops productions resulted in an increased soil productivity and in the reconstruction of habitats. Considering the wide range of agroecosystems, the several causal factor and processes leading to soil degradation and reduction in biodiversity (Lal 1997), it is not possible to have a general recipe for such interventions. However, it is generally assumed that social and policy-oriented initiatives needed to solve degradation issues should consider acceptable farmer incomes, secure land tenure and ownership, and forest rescue initiatives, possibly supported by interested urban communities and environment-friendly pressure groups (Sanchez et al. 2005).

The links between changes in soil biodiversity and agricultural intensification are not clearly delineated, and in many cases a severe disturbance and biodiversity loss may be also observed in poorly managed subsistence agriculture or in extensive cropping systems, requiring low intensity inputs (Giller et al. 1997; Lal 1997). However, there are examples showing that a direct relationship occurs, with impacts on plant productivity. A functional loss due to changes in belowground biodiversity has been documented in tropical soils after their conversion to pasture, in which the single survivor earthworm species was found associated to increased levels of soil compaction (Fragoso et al. 1997).

Similarly, the positive correlation between biodiversity levels and ecosystem functions may not be linear, as situations occur in which only a few species may support a number of functions, equal to that of a more biodiversity rich environment (Schwartz et al. 2000).

20.4 Conclusions

Knowledge-based soil and natural resource management are fundamental to any strategy aiming at restoration or recovery of lost biodiversity levels. Acting on multiple levels, including the use of multiple species and function restoration plans may be required to restore forest ecosystems services. The links between soil biodiversity and aboveground resilience, however, have been only partially studied and defined (Aerts and Honnay 2011). Social factors are often the most prominent cause of the belowground biodiversity changes encountered. Although a general rule for biodiversity management is not available, some basic concepts and the collection and use of scientific data may provide helpful clues for sustainable solutions. Given soil disturbance and biodiversity erosions are mainly human-driven, the links between the economic dimension of the problems and its effects on the success of the solutions proposed should be investigated in detail.

References

Addiscott, T. M., & Mirza, N. A. (1998). New paradigms for modelling mass transfers in soils. *Soil and Tillage Research, 47*, 105–109.

Adeniyi, A. S. (2010). Effects of slash and burning on soil microbial diversity and abundance in the tropical rainforest ecosystem, Ondo State, Nigeria. *African Journal of Plant Science, 4*, 322–329.

Aerts, R., & Honnay, O. (2011). Forest restoration, biodiversity and ecosystem functioning. *BMC Ecology, 11*, 29.

Akkermans A., Hahn D., & Zoon F. (1989). Interactions between root symbionts, root pathogens and actinorhizal plants. *Annals of Forest Science, 46*(Suppl), 765s–771s.

Antle, J. M., Stoorvogel, J. J., & Valdivia, R. O. (2006). Multiple equilibria, soil conservation investments and the resilience of agricultural systems. *Environment and Development Economics, 11*, 477–492.

Bardgett, R. D. (2002). Causes and consequences of biological diversity in soil. *Zoology, 105*, 367–374.

Barnes, A. D., et al. (2014). Consequences of tropical land use for multitrophic biodiversity and ecosystem functioning. *Nature Communications, 5*, 5351.

Behie, S. W., Zelisko, P. M., & Bidochka, M. J. (2012). Endophytic insect-parasitic fungi translocate nitrogen directly from insects to plants. *Science, 336*, 1576–1577.

Bender, S. F., Wagg, C., & van der Heijden, M. G. A. (2016). An underground revolution: Biodiversity and soil ecological engineering for agricultural sustainability. *Trends in Ecology and Evolution, 31*, 440–452.

Bennett, J. A., et al. (2017). Plant-soil feedbacks and mycorrhizal type influence temperate forest population dynamics. *Science, 355*, 181–184.

Bignell D. E., et al. (2005). Below-ground biodiversity assessment: Developing a key functional group approach in best-bet alternatives to blash and burn. In C. A. Palm, S. A. Vosti, P. A. Sanchez, & P. J. Ericksen (Eds), *Slash-and-burn agriculture: The search for alternatives* (pp. 119–142). New York: Columbia University Press.

Bongers, T. (1990). The maturity index, an ecological measure of environmental disturbance based on nematode species composition. *Oecologia, 83*, 14–19.

Bongers, T., & Ferris, H. (1999). Nematode community structure as a bioindicator in environmental monitoring. *Trends in Ecology and Evolution, 14*, 224–228.

Bonkowski, M., & Roy, J. (2005). Soil microbial diversity and soil functioning affect competition among grasses in experimental microcosms. *Oecologia, 143*, 232–240.

Botero, C., Dor, R., McCain, C. M., & Safran, R. J. (2013). Environmental harshness is positively correlated with intraspecific divergence in mammals and birds. *Molecular Ecology, 23*, 259–268.

Brandon, K., Gorenflo, L. J., Rodrigues, A. S. L., & Waller, R. W. (2005). Reconciling biodiversity conservation, people, protected areas, and agricultural suitability in Mexico. *World Development, 33*, 1403–1418.

Brown, G. G., et al. (1999). Effects of earthworms on plana production in the tropics. In L. Brussaard & P. Hendrix (Eds.), *Earthworm management in tropical agroecosystems. Lavelle P* (pp. 87–147). Wallingford: CAB International Edition.

Brunner, I., Herzog, C., Dawes, M. A., Arend, M., & Sperisen, C. (2015). How tree roots respond to drought. *Frontiers in Plant Science, 6*, 547.

Brussaard, L., Pulleman, M. M., Ouédraogo, E., Abdoulaye Mando, A., & Sixe, J. (2007). Soil fauna and soil function in the fabric of the food web. *Pedobiologia, 50*, 447–462.

Canali, S., et al. (2009). Effect of different management strategies on soil quality of citrus orchards in Southern Italy. *Soil Use and Management, 25*, 34–42.

Carter, M. R. (2004). Researching structural complexity in agricultural soils. *Soil and Tillage Research, 79*, 1–6.

Chen, X., Yang, Y., & Tang, J. (2004). Species-diversified plant cover enhances orchard ecosystem resistance to climatic stress and soil erosion in subtropical hillside. *Journal of the Zhejiang University Science, 5*, 1191–1198.

Daniels, R. (2003). Impact of tea cultivation on anurans in the Western Ghats. *Current Science, 85*, 1415–1422.

De Deyn, G. B., & Van der Putten, W. H. (2005). Linking aboveground and belowground diversity. *TRENDS in Ecology and Evolution, 20*, 625–633.

Dhandapani, S. (2015). *Biodiversity loss asoociated with oil palm plantations in Malaysia: Serving the need versus saving the nature*. British Ecological Society, Agriculture Ecology Annual Conference, Queens University, Belfast, UK.

Eayre, C. G., Jaffee, B. A., & Zehr, E. I. (1987). Suppression of *Criconemella xenoplax* by the nematophagous fungus *Hirsutella rhossiliensis*. *Plant Disease, 71*, 832–834.

Elliott L.F., & Lynch J. M. (1994). Biodiversity and soil resilience. In: D. J. Greenland, & I. Szabolcs (Eds.), *Soil resilience and sustainable land use* (pp. 353–364). Wallingford: CAB International.

Ferris, H., & Bongers, T. (2006). Nematode indicators of organic enrichment. *Journal of Nematology, 38*, 3–12.

Fragoso, C., et al. (1997). Agricultural intensification, soil biodiversity and agroecosystem function in the tropics: The role of earthworms. *Applied Soil Ecology, 6*, 17–35.

Franca, S. C., Gomes-da-Costa, S. M., & Silveira, A. P. D. (2007). Microbial activity and arbuscular mycorrhizal fungal diversity in conventional and organic citrus orchards. *Biological Agriculture and Horticulture, 25*, 91–102.

Giller, P. S. (1996). The diversity of soil communities the 'poor man's tropical rain forest'. *Biodiversity Conservation, 5*, 135–168.

Giller, K. E., Beare, M. H., Lavelle, P., Izac, A. M. N., & Swift, M. J. (1997). Agricultural intensification, soil biodiversity and agroecosystem function. *Applied Soil Ecology, 6*, 3–16.

Gobbi, J. A. (2000). Is biodiversity-friendly coffee financially viable? An analysis of five different coffee production systems in western El Salvador. *Ecological Economics, 33*, 267–281.

Gordon, C., Manson, R., Sundberg, J., & Cruz-Angón, A. (2007). Biodiversity, profitability, and vegetation structure in a Mexican coffee agroecosystem. *Agriculture, Ecosystems & Environment, 118*, 256–266.

Gurr, G. M., Wratten, S. D., & Luna, J. M. (2003). Multi-function agricultural biodiversity: Pest management and other benefits. *Basic Applied Ecology, 4*, 107–116.

Hacquard, S., et al. (2013). Laser microdissection and microarray analysis of *Tuber melanosporum* ectomycorrhizas reveal functional heterogeneity between mantle and Hartig net compartments. *Environmental Microbiology, 15*, 1853–1869.

Hågvar, S. (1998). The relevance of the Rio-Convention on biodiversity to conserving the biodiversity of soils. *Applied Soil Ecology, 9*, 1–7.

Havlicek E., Mitchell E.A.D. 2014. Soils supporting biodiversity. In J. Dighton & J. A. Krumins (Eds.), *Interactions in soil: Promoting plant* growth (Biodiversity, Community and Ecosystems, Vol. 1, pp. 27–58). Dordrecht: Springer.

Hawksworth, D. L. (1991). The fungal dimension of diversity: Magnitude significance and conservation. *Mycological Research, 95*, 641–655.

Hawksworth, D. L. (2001). The magnitude of fungal diversity: The 1.5 million species estimate revisted. *Mycological Research, 105*, 1422–1431.

Head, I. M., Saunders, J. R., & Pickup, R. W. (1998). Microbial evolution, diversity, and ecology: A decade of ribosomal RNA analysis of uncultivated microorganisms. *Microbial Ecology, 35*, 1–21.

Hohberg, K. (2003). Soil nematode fauna of afforested mine sites: Genera distribution, trophic structure and functional guilds. *Applied Soil Ecology, 22*, 113–126.

Itoh, H., et al. (2014). Bacterial population succession and adaptation affected by insecticide application and soil spraying history. *Frontiers in Microbiology, 5*, 457.

Joschko, M., et al. (2006). Spatial analysis of earthworm biodiversity at the regional scale. *Agriculture Ecosystems & Environment, 112*, 367–380.

Kampichler, C. (1999). Fractal concepts in studies of soil fauna. *Geoderma, 88*, 283–300.

Kellenberger, E. (2001). Exploring the unknown: The silent revolution of microbiology. *EMBO Reports, 2*, 5–7.

Kennedy, A. C., & Smith, K. L. (1995). Soil microbial diversity and the sustainability of agricultural soils. *Plant and Soil, 170*, 75–86.

Lal, R. (1997). Degradation and resilience of soils. *Philosophical Transactions of the Royal Society of London, B, 352*, 997–1010.

Lavelle P., & Spain A.V. (2006). *Soil ecology* (2nd ed., 651p). Amsterdam: Kluwer Scientific Publications.

Lavelle, P., et al. (2004). Plant parasite control and soil fauna diversity. *CR Academic Science Biology, 327*, 629–638.
Lavelle, P., et al. (2006). Soil invertebrates and ecosystem services. *European Journal of Soil Biology, 42*, S3–S15.
Lavelle, P., et al. (2016). Ecosystem engineers in a self-organized soil: A review of concepts and future research questions. *Soil Science, 181*, 91–109.
Legendre, B., et al. (2014). Identification and characterisation of *Xylella fastidiosa* isolated from coffee plants in France. *Journal of Plant Pathology, 96*, S4.100.
Liang, W., Beattie, G. A. C., Meats, A., & Spooner-Hart, R. (2007). Impact on soil-dwelling arthropods in citrus orchards of spraying horticultural mineral oil, carbaryl or methidathion. *Australian Journal of Entomology, 46*, 79–85.
Lin, B. B. (2011). Resilience in agriculture through crop diversification: Adaptive management for environmental change. *BioScience, 61*, 183–193.
Loconsole, G., et al. (2016). Intercepted isolates of *Xylella fastidiosa* in Europe reveal novel genetic diversity. *European Journal of Plant Pathology, 146*, 85–94.
Martelli, G. P., Boscia, D., Porcelli, F., & Saponari, M. (2015). The olive quick decline syndrome in south-east Italy: A threatening phytosanitary emergency. *European Journal of Plant Pathology, 144*, 235–243.
Morris, E. K., et al. (2014). Choosing and using diversity indices: insights for ecological applications from the German Biodiversity Exploratories. *Ecology and Evolution, 4*, 3514–3524.
Obando, V. (2002). *Biodiversidad en Costa Rica: Estado del Conocimiento y Gestión* (76 pp). Santo Domingo, Heredia, Editorial INBio: MINAE-SINAC & INBio.
Oldeman, L. R., Hakkeling, R. T. A., & Sombroek, W. G. (1990). World map of the status of human induced soil degradation: Explanatory note. In *Wageningen, Netherlands: International Soil Reference and Information Centre (ISIRC) and Nairobi*. Kenya: United Nations Environment Programme (UNEP).
Peeters, L. Y. K., Soto-Pinto, L., Perales, H., Montoya, G., & Ishiki, M. (2003). Coffee production, timber, and firewood in traditional and *Inga*-shaded plantations in Southern Mexico. *Agriculture, Ecosystems and Environment, 95*, 481–493.
Perfecto, I., Vandermeer, J., Mas, A., & Soto, P. L. (2005). Biodiversity, yield, and shade coffee certification. *Ecological Economics, 54*, 435–446.
Rasmann, S., et al. (2005). Recruitment of entomopathogenic nematodes by insect-damaged maize roots. *Nature, 434*, 732–737.
Roesch, L. F. W., et al. (2007). Pyrosequencing enumerates and contrasts soil microbial diversity. *ISME Journal, 1*, 283–290.
Rosello-Mora, R., et al. (2001). The species concept for prokaryotes. *FEMS Microbiology Reviews, 25*, 39–67.
Sanchez P.A., Palm C.A., Vosti S.A., Tomich T.P., Kasyoki J. 2005. Alternatives to slash and burn: Challenge and approaches of an International Consortium. In: C.A. Palm, S. A. Vosti, P. A. Sanchez, & P. J. Ericksen (Eds), *Slash-and-burn agriculture: The search for alternatives* (pp. 3–40). New York: Columbia University Press.
Saponari, M., et al. (2014). Infectivity and transmission of *Xylella fastidiosa* Salento strain by *Philaenus spumarius* L. (Hemiptera: Aphrophoridae) in Apulia, Italy. *Journal of Economic Entomology, 107*, 1316–1319.
Savilaakso, S., et al. (2014). Systematic review of effects on biodiversity from oil palm production. *Environmental Evidence, 3*, 4.
Schulz, B. (2006). Mutualistic interactions with fungal root endophytes. In B. Schulz, C. Boyle, & T. N. Sieber (Eds.), *Microbial root endophytes* (pp. 261–279). Berlin: Springer.
Schulz, B., & Boyle, C. (2005). The endophytic continuum. *Mycological Research, 109*, 661–686.
Schwartz, M. W., Bringham, C. A., Hoeksema, J. D., Lyons, K. G., Mills, M. H., & van Mantgem, P. J. (2000). Linking biodiversity to ecosystem function: implications for conservation ecology. *Oecologia, 122*, 297–305.

Soto-Pinto, L., Perfecto, I., & Caballero-Nieto, J. (2002). Shade over coffee: Its effects on berry borer, leaf rust and spontaneous herbs in Chiapas, Mexico. *Agroforestry Systems, 55*, 37–45.
Stirling, G. R., & Mankau, R. (1978). *Dactylella oviparasitica*, a new fungal parasite of *Meloidogyne* eggs. *Mycologia, 70*, 774–783.
Torsvik, V., Goksøyr, J., & Daae, F. L. (1990). High diversity in DNA of soil bacteria. *Applied Environmental Microbiology, 56*, 782–787.
Vijay, V., Pimm, S. L., Jenkins, C. N., & Smith, S. J. (2016). The impacts of oil palm on recent deforestation and biodiversity loss. *PLOS ONE*. doi:10.1371/journal.pone.0159668.
Wall, D. H., Bardgett, R. D., & Kelly, E. F. (2010). Biodiversity in the dark. *Nature Geoscience, 3*, 297–298.
Wall, D. H., Nielsen, U. N., & Six, J. (2015). Soil biodiversity and human health. *Nature, 528*, 69–76.
Wardle, D. A., & Jonsson, M. (2014). Long-term resilience of above- and belowground ecosystem components among contrasting ecosystems. *Ecology, 95*, 1836–1849.
Wilson, E. O. (2000). A global biodiversity map. *Science, 289*, 2279.
Yaap, B., Struebig, M. J., Paoli, G., & Pin Koh, L. (2010). Mitigating the biodiversity impacts of oil palm development. *CAB Reviews: Perspectives in Agriculture, Veterinary Science, Nutrition and Natural Resources, 5*, No. 019.

Index

M. Lukac et al. (eds.), *Soil Biological Communities and Ecosystem Resilience*, Sustainability in Plant and Crop Protection, DOI 10.1007/978-3-319-63336-7